移动开发经典丛书

Android 4 编程入门经典
——开发智能手机与平板电脑应用

[美] Wei-Meng Lee 著
何晨光 李洪刚 译

清华大学出版社
北京

Wei-Meng Lee

Beginning Android 4 Application Development

EISBN：978-1-118-19954-1

Copyright © 2012 by Wiley Publishing, Inc.

All Rights Reserved. This translation published under license.

本书中文简体字版由Wiley Publishing, Inc. 授权清华大学出版社出版。未经出版者书面许可，不得以任何方式复制或抄袭本书内容。

北京市版权局著作权合同登记号 图字：01-2012-3836

本书封面贴有Wiley公司防伪标签，无标签者不得销售。

版权所有，侵权必究。侵权举报电话：010-62782989 13701121933

图书在版编目(CIP)数据

Android 4编程入门经典——开发智能手机与平板电脑应用/(美)李伟梦 著；何晨光，李洪刚 译.
—北京：清华大学出版社，2012.11（2015.10重印）

（移动开发经典丛书）

书名原文：Beginning Android 4 Application Development

ISBN 978-7-302-30151-6

Ⅰ.①A… Ⅱ.①李… ②何… ③李… Ⅲ.①移动终端—应用程序—程序设计 Ⅳ.①TN929.53

中国版本图书馆CIP数据核字(2012)第222884号

责任编辑：王　军　杨信明
装帧设计：牛艳敏
责任校对：成凤进
责任印制：沈　露

出版发行：清华大学出版社
网　　址：http://www.tup.com.cn, http://www.wqbook.com
地　　址：北京清华大学学研大厦A座　　　　邮　编：100084
社 总 机：010-62770175　　　　　　　　　　邮　购：010-62786544
投稿与读者服务：010-62776969, c-service@tup.tsinghua.edu.cn
质 量 反 馈：010-62772015, zhiliang@tup.tsinghua.edu.cn

印 装 者：北京密云胶印厂
经　　销：全国新华书店
开　　本：185mm×260mm　　　　印　张：32.25　　　　字　数：785千字
版　　次：2012年11月第1版　　　　　　　　印　次：2015年10月第12次印刷
印　　数：24001~26000
定　　价：68.00元

产品编号：046719-01

作者简介

Wei-Meng Lee 是 Developer Learning Solutions 公司(www.learn2develop.net)的创始人和技术专家,这家技术公司专门从事最新移动技术的培训。Wei-Meng Lee 具有多年的培训经验,他的培训课程特别强调实践学习法。这种动手学习编程的方法比通过阅读书籍、教程和文档来理解主题要容易得多。

Wei-Meng Lee 还是 *Beginning iOS 5 Application Development*(Wrox, 2010)和 *Beginning Android Application Development*(Wrox, 2011)的作者。读者可以通过 weimenglee@learn2develop.net 与他联系。

技术编辑简介

Chaim Krause 是 US Army's Command and General Staff College 学院的模拟专家(Simulation Specialist),他为该学院开发了运行在多种平台(从 iOS 和 Android 设备到 Windows 桌面操作系统和 Linux 服务器)的各种各样的软件产品,并且还担负其他一些工作。Python 是他最喜欢的语言,但是他本人擅长使用多种语言,比如使用 Java 和 JavaScript/HTML5/CSS 等编写代码。很幸运,他的软件开发职业生涯是在 Borland 开始的,当时他是 Delphi 语言的高级开发支持工程师。除了计算机相关的工作,Chaim 喜欢 techno 音乐和 dubstep 音乐,以及和自己的两条雪橇犬 Dasher 和 Minnie 玩踏板车。

致　　谢

　　写这本书的过程就像是在坐过山车。使用发布不久的软件在任何时候都是一种巨大的挑战。当我刚开始创作这本书时，Android 4 SDK 刚刚发布，浏览文档获取信息的过程就像是大海捞针。更加糟糕的是，Android 平板电脑模拟器很慢，很不稳定，这让开发过程变得十分辛苦。

　　现在书终于写完了。希望您的旅途能够一帆风顺。我所做的就像任何好的向导一样，领您进入 Android 平板电脑开发的大门，享受一个丰富多彩而又收获颇丰的学习过程。您手里拿着的这本书是很多人共同工作的成果，我要借此机会感谢他们的付出。

　　首先，特别感谢 Wrox 的执行编辑 Bob Elliott。Bob 是一个耐心的倾听者，随时准备为我提供帮助。和 Bob 合作的过程十分愉快，在我一起合作过的人当中，他是响应最积极的人之一。Bob，谢谢你的帮助和指导！

　　当然，不应该忘记我的朋友，编辑 Ami Sullivan，和她一起工作的时候我总是感到很愉快。合作完成了 4 本书后，我们已经非常了解对方，甚至不必打开收到的电子邮件就知道里面在说什么。谢谢你，Ami！

　　还不能忘记的是那些幕后的英雄：文字编辑 Luann Rouff 和技术编辑 Chaim Krause。他们在编辑本书时所具有的锐利眼光，保证了每一句话的准确性——无论是在语法方面还是在技术方面。Luann 和 Chaim，谢谢你们！

　　最后，同样重要的是，我要感谢我的父母以及妻子 SzeWa 所给予我的全力支持。在我创作本书的那段时间里，他们无私地调整自己的日程安排来迁就我。SzeWa 在我为满足期限要求拼命工作的无数个夜晚总是一直陪我熬夜。因此，我要对我的妻子和我的父母说一声："我爱你们！"最后要感谢我们可爱的小狗 Ookii，谢谢你陪伴着我们。

译 者 序

Android 的发展呈现逐渐加速的趋势。从 2008 年 Google 在 I/O 开发者大会上提出的 Android 1.0 开始，经过 Google 进军"甜品业"的标志：Android 1.5，直到目前刚刚发布的最新版 Android 4.1，短短 4 个年头不到，小机器人已经经历了十几个大小版本的成长历程。正如本书内容所展现的那样，年初关于 Android 3 系列的入门开发经典刚刚付梓，现在这本介绍 Android 4 的升级版本又和大家见面了，这无疑也是 Android 社区和应用日益活跃和发展的集中体现，众人拾柴火焰高。尽管和相对封闭的 iOS 相比，在版本"碎片"控制上由于其开放性无法做到完全杜绝，但也正因为如此，持续的新鲜血液和新型应用才能不断得到实践的机会和用户的最终检验。Android 的生态系统也许在一时的盈利水平上还无法全面和苹果抗衡，但凭借 Google 的技术、管理优势和以全世界 34 家服务运营商、设备提供商为肇始的"开放手机联盟"的广泛覆盖能力，Android 将会把越来越多的江湖大哥变为传说。摩托罗拉、诺基亚、高通、三星这些传统的业界大佬自不必提，仅就中国境内的移动和联通两大运营商先后加入联盟，并频繁推出 Android 设备平台上的各种增值应用，就可想象在世界最大的移动电话用户国家 Android 的前景。一组数字说明了这一点：截止 2011 年 9 月份，Android 系统的应用数目已经达到了 48 万，而在智能手机市场，Android 系统的占有率已经达到了 43%，继续排在移动操作系统首位。因此，对 Android 的开发者来说，现在首要的工作与其说是技术的革新和应用，不如说是寻找新的盈利模式，把技术的优势转化为盈利的优势是 Android 生态系统富有挑战性的课题。希望本书的读者在 Android 的世界里循序渐进，在紧盯屏幕之余多思考一下现实世界，因为那里才是所有开发工作的源头和归宿。

本书是一本有关基于 Android 4 手机操作系统进行应用程序开发的入门读物，作者 Wei-Meng Lee 在移动操作系统平台的项目开发和培训上具有丰富的实践经验。他采用图文并茂、上手性极强的步步引导的方式将门外汉领入 Android 的大千世界的同时，又为他们展示了较为广阔的视野，避免了初学者常常具有的只见树木不见森林的缺憾，堪称一大特色。本书从 Android 的发展沿革讲起，通过对其中关键概念深入浅出的介绍，用大量的实例概括了 Android 应用程序的构成、表现形式以及运行原理，为读者构建了较为完整的 Android 开发蓝图。在此基础上，引入了一些高级组件和功能的介绍，为读者进一步的实践和开发高价值的应用指明了方向，再辅以每章课后的练习，除了使读者巩固所学知识之外，通篇口语化的表达方式也拉近了本书与读者的距离。

和早先的版本相比，本书所使用的 Android 4.0 平台系统拥有全新的系统解锁界面，小插件也进行了重新设计，最特别的就是系统的任务管理器可以显示出程序的缩略图，便于用户准确快速地关闭无用的程序。总的来看，新版本具有统一的 UI 框架、沟通与共享能力和全新的连接类型，全新的输入方式及文本服务，并增强了媒体处理能力，尤其是在应用程序及内容安全性方面做了不少功课，一定程度上补齐了开源平台常见的短板，将带给用户全新的体验，这对于开发者来说无疑是颠覆应用设计新的契机。

  当然，由于是面向初学者，Android 本身的体系结构和原理并未过多介绍，对于想对此有更深入了解的读者可以访问 Android 社区以及海量的互联网相关资源。限于译者水平，译文定有很多不当之处，敬请读者批评指正。译者的博客是 http://blog.csdn.net/charlieking，欢迎大家一起交流。

译者于深圳

2012 年 6 月 22 日

前 言

我最开始玩 Android SDK 是在其正式版本 1.0 发布以前。那时,工具还不完善,SDK 中的 API 不稳定,文档也很缺乏。经过三年半时间的快速发展,现在的 Android 已经成为一个和 iPhone 相比毫不逊色的强大的移动操作系统。由于经历过 Android 成长的所有痛苦,我想现在是开始学习 Android 编程的最好时机——API 已经稳定,工具也有了改善。但是仍然存在一个挑战:对许多人来说,入门仍是一个可望而不可及的目标。这一挑战在我脑海里徘徊许久,也成为我写本书的动力,它也许可以给 Android 初级程序员带来益处,并使他们能够逐步编写更复杂的应用程序。但是,对很多人来说学习 Android 仍然不太容易。而且,Google 最近发布了 Android SDK 的最新版本——4.0,这是同时可用于智能手机和平板电脑的一个统一的移动操作系统。Android 4.0 SDK 包含原来平板电脑开发人员可用的一些新功能,初学者理解这些新功能需要付出一些努力。

正是考虑到了初学者面临的这种挑战,我决定创作本书,让 Android 编程初学者能够逐步掌握开发复杂应用程序的方法。

由于本书是写给 Android 初级开发人员的,为的是使他们能够快速上手,因此我以线性方式涵盖了必要的主题,这样可以使您建立起自己的知识体系而不会被细节淹没。我采取的哲学观点是:最好的学习方法是实践——因此,每一章的"试一试"部分将首先教您如何构建一些东西,然后解释其工作原理。我利用创作本书的机会对本书的上一版进行了修订和更新,加入了读者的反馈和对 Android 初学者很重要的一些主题。

尽管 Android 编程是一个宏大的主题,但本书要实现三重目标:帮助读者从最基本的原理入手、使读者理解 SDK 的底层架构以及领会事情要按特定方式完成的原因。任何一本书都不能面面俱到地介绍有关 Android 编程的知识,但我确信当您阅读完此书(并做了练习)之后,将有充分的准备来应对下一个 Android 编程的挑战。

本书读者对象

本书针对的是打算使用 Google 的 Android SDK 来开发应用程序的 Android 初级开发人员。为了从本书中真正获益,您应该在编程方面具有一些背景知识,并且至少熟悉面向对象编程的概念。如果对 Java(Android 开发所用的语言)一无所知,那么您也许应该首先学习一门 Java 编程课程,或者阅读有关 Java 编程方面的优秀书籍。以我的经验,如果您已经了解 C#或 VB.NET,学习 Java 就比较轻松;只要按照"试一试"的步骤就可以使您的学习过程顺利进行。

对于那些对所有编程概念都一无所知的人来说,我知道开发移动应用程序并赚到钱是很有诱惑力的。然而，在尝试本书的示例之前，我想首先学习一些基本的编程知识才是更好的着手点。

注意：本书中讨论的所有示例均使用 Android SDK 4.0 版本编写和测试。尽管我们已经努力保证本书中所有用到的工具都是最新的，但当您阅读本书时，还是很可能有更新版本的工具可用。如果是这样，某些指示和/或屏幕截图会有少许不同。不过，任何改变都应是可控的。

本书主要内容

本书涵盖了使用 Android SDK 进行 Android 编程的基本概念，共分为 12 章和 3 个附录。

"第 1 章：Android 编程入门"介绍了 Android 操作系统的基本概念和当前发展状况。您可以了解 Android 设备的各种功能以及市场上一些比较流行的设备。还可以学习如何下载和安装所有必需的工具来开发 Android 应用程序并在 Android 模拟器上进行测试。

"第 2 章：活动、碎片和意图"使您熟悉 Android 编程中的这三个最重要的概念。活动和碎片是 Android 应用程序的构建块。您将学习如何使用意图将活动链接起来形成一个完整的 Android 应用程序。这是 Android 操作系统的独特特征之一。

"第 3 章：Android 用户界面"介绍了 Android 应用程序的用户界面的不同组成部分。您将学习到用来构建应用程序的用户界面的不同布局，以及当用户和应用程序交互时与用户界面相关联的多种事件。

"第 4 章：使用视图设计用户界面"介绍了可用于构建 Android 用户界面的各种基本视图。该章将学习 3 组主要的视图：基本视图、选取器视图和列表视图，还将学习 Android 3.0 和 Android 4.0 中可用的特殊碎片。

"第 5 章：使用视图显示图片和菜单"继续研究视图。您将了解到如何使用不同的图像视图来显示图像，以及在应用程序中显示选项和上下文菜单。该章最后将额外介绍一些很酷的视图，可以用它们来为您的应用程序锦上添花。

"第 6 章：数据持久化"教您如何在 Android 应用程序中保存或存储数据。除了学习使用不同的技术来存储用户数据外，您将学习到文件操作以及如何把文件保存到内部或外部存储器(SD 卡)上。此外，还将学习到如何在 Android 应用程序中创建和使用 SQLite 数据库。

"第 7 章：内容提供者"讨论了在 Android 设备的不同应用程序间如何共享数据。您将学习如何使用内容提供者并自己创建一个。

"第 8 章：消息传递"研究了移动编程中最有趣的两个主题——发送 SMS 消息和电子邮件。您将学习如何以编程方式发送和接收 SMS 消息和电子邮件，以及如何拦截传入的 SMS 消息，使内置的 Messaging 应用程序不能收到任何消息。

"第 9 章：基于位置的服务"描述了如何使用 Google Maps 来构建基于位置的服务应用程序。您还将学习到如何获取地理位置数据并在地图上显示该位置。

"第 10 章：联网"研究了如何连接 Web 服务器来下载数据。您将看到如何在 Android 应用程序中使用 XML 和 JSON Web 服务。本章还将介绍套接字编程，以及如何在 Android 中构建一个聊天客户端。

"第 11 章：开发 Android 服务"将向您展示如何使用服务来编写应用程序。服务是运行于后台且没有用户界面的应用程序。您将了解如何在一个单独的线程中以异步方式运行您的服务，以及活动与之通信的方法。

"第 12 章：发布 Android 应用程序"讨论了您在准备好发布 Android 应用程序时可以采用的不同方法。您还将了解到在 Android Market 上发布并出售应用程序的必要步骤。

"附录 A：使用 Eclipse 进行 Android 开发"简要概述了 Eclipse 中的许多功能。

"附录 B：使用 Android 模拟器"提供了有关使用 Android 模拟器进行应用程序测试方面的一些提示和技巧。

"附录 C：练习答案"包含了每章最后的练习的答案。

本书结构安排

本书将学习 Android 编程的任务分解为若干个更小的环节，使您能够在钻研更高级的内容之前消化每一个主题。

如果您对于 Android 编程完全是个新手，那就首先从第 1 章开始。一旦熟悉基本概念，就可以转到附录去阅读更多有关 Eclipse 和 Android 模拟器的知识。当完成这些之后，可以再从第 2 章继续，并按部就班地学习更高级的主题。

本书一大特色就是每章的所有示例代码都独立于先前章节所讨论的内容。这样，您可以灵活地转入到所感兴趣的主题并按照"试一试"的项目内容开始练习。

使用本书的前提条件

本书中的所有示例都在 Android 模拟器(作为 Android SDK 的一部分)中运行。当然，为了从本书中得到更多收获，拥有一个真实的 Android 设备还是很有益的(尽管这不是绝对必要的)。

源代码

在读者学习本书中的示例时，可以手动输入所有代码，也可以使用本书附带的源代码文件。本书使用的所有源代码都可以从本书合作站点 http://www.wrox.com/ 或 http://www.tupwk.com.cn/ downpage 上下载。登录到站点 http://www.wrox.com/，使用 Search

工具或使用书名列表就可以找到本书。接着单击 Download Code 链接，就可以获得所有的源代码。

 注意： 因为很多书都有类似的书名，通过书号可以很容易找到本书，本书的 EISBN 为 978-1-118-19954-1。

在下载代码后，只需要用解压缩软件对它进行解压缩即可。另外，也可以进入 http://www.wrox.com/dynamic/books/download.aspx 上的 Wrox 代码下载主页，查看本书和其他 Wrox 图书的所有代码。

勘误表

尽管我们已经尽了各种努力来保证文章或代码中不出现错误，但是错误总是难免的，如果您在本书中找到了错误，例如拼写错误或代码错误，请告诉我们，我们将非常感激。通过勘误表，可以让其他读者避免受挫，当然，这还有助于提供更高质量的信息。

要在网站上找到本书英文版的勘误表，可以登录 http://www.wrox.com，通过 Search 工具或书名列表查找本书，然后在本书的细目页面上，单击 Book Errata 链接。在这个页面上可以查看到 Wrox 编辑已提交和粘贴的所有勘误项。完整的图书列表还包括每本书的勘误表，网址是 www.wrox.com/misc-pages/booklist.shtml。

如果您发现的错误在我们的勘误表里还没有出现的话，请登录 www.wrox.com/contact/techsupport.shtml 并完成那里的表格，把您发现的错误发送给我们。我们会检查您的反馈信息，如果正确，我们将在本书的勘误表页面张贴该错误消息，并在本书的后续版本加以修订。

p2p. wrox.com

要与作者和同行讨论，请加入 p2p.wrox.com 上的 P2P 论坛。这个论坛是一个基于 Web 的系统，便于您张贴与 Wrox 图书相关的消息和相关技术，与其他读者和技术用户交流心得。该论坛提供了订阅功能，当论坛上有新的消息时，它可以给您传送感兴趣的论题。Wrox 作者、编辑和其他业界专家以及读者都会到这个论坛上探讨问题。

在 p2p.wrox.com 上有许多不同的论坛，它们不仅有助于阅读本书，还有助于开发自己的应用程序。要加入论坛，可以遵循下面的步骤。

(1) 进入 p2p.wrox.com，单击 Register 链接。
(2) 阅读使用协议，并单击 Agree 按钮。
(3) 填写加入该论坛所需要的信息和自己希望提供的其他可选信息，单击 Submit 按钮。
(4) 您会收到一封电子邮件，其中的信息描述了如何验证账户，完成加入过程。

注意：不加入 P2P 也可以阅读论坛上的消息，但要张贴自己的消息，就必须先加入该论坛。

加入论坛后，就可以张贴新消息，响应其他用户张贴的消息。可以随时在 Web 上阅读消息。如果要让该网站给自己发送特定论坛中的消息，可以单击论坛列表中该论坛名旁边的 Subscribe to this Forum 图标。

要想了解更多的有关论坛软件的工作情况，以及 P2P 和 Wrox 图书的许多常见问题的解答，就一定要阅读 FAQ，只需要在任意 P2P 页面上单击 FAQ 链接即可。

目　　录

第 1 章　Android 编程入门 ················ 1
1.1　Android 简介 ································ 2
　　1.1.1　Android 版本 ······················ 2
　　1.1.2　Android 功能 ······················ 3
　　1.1.3　Android 架构 ······················ 3
　　1.1.4　市场上的 Android 设备 ······ 4
　　1.1.5　Android Market ··················· 7
　　1.1.6　Android 开发社区 ·············· 7
1.2　获得所需工具 ······························ 8
　　1.2.1　Android SDK ······················ 8
　　1.2.2　安装 Android SDK 工具 ····· 9
　　1.2.3　配置 Android SDK
　　　　　Manager ··························· 10
　　1.2.4　Eclipse ································ 11
　　1.2.5　Android 开发工具 ············ 12
　　1.2.6　创建 Android 虚拟
　　　　　设备(AVD) ······················· 14
1.3　创建第一个 Android 应用
　　程序 ··· 17
1.4　Android 应用程序剖析 ············· 24
1.5　本章小结 ··································· 28

第 2 章　活动、碎片和意图 ··············· 31
2.1　理解活动 ··································· 31
　　2.1.1　如何对活动应用样式
　　　　　和主题 ····························· 36
　　2.1.2　隐藏活动标题 ··················· 37
　　2.1.3　显示对话框窗口 ··············· 38
　　2.1.4　显示进度对话框 ··············· 43
　　2.1.5　显示更复杂的进度对话框 ···· 46
2.2　使用意图链接活动 ··················· 50
　　2.2.1　解决意图筛选器的冲突 ········ 54
　　2.2.2　从意图返回结果 ··············· 56
　　2.2.3　使用意图对象传递数据 ········ 59
2.3　碎片 ··· 65
　　2.3.1　动态添加碎片 ··················· 70
　　2.3.2　碎片的生命周期 ··············· 72
　　2.3.3　碎片之间进行交互 ··········· 76
2.4　使用意图调用内置应用程序 ···· 80
　　2.4.1　理解意图对象 ··················· 85
　　2.4.2　使用意图筛选器 ··············· 86
　　2.4.3　添加类别 ··························· 91
2.5　显示通知 ··································· 93
2.6　本章小结 ··································· 98

第 3 章　Android 用户界面 ··············· 101
3.1　了解屏幕的构成 ······················ 101
　　3.1.1　视图和视图组 ················· 102
　　3.1.2　LinearLayout ···················· 103
　　3.1.3　AbsoluteLayout ················ 109
　　3.1.4　TableLayout ····················· 110
　　3.1.5　RelativeLayout ················· 111
　　3.1.6　FrameLayout ···················· 113
　　3.1.7　ScrollView ······················· 115
3.2　适应显示方向 ························· 118
　　3.2.1　锚定视图 ························· 119
　　3.2.2　调整大小和重新定位 ········ 121
3.3　管理屏幕方向的变化 ·············· 124
　　3.3.1　配置改变时保持状态
　　　　　信息 ································· 127
　　3.3.2　检测方向改变 ················· 128
　　3.3.3　控制活动的方向 ············· 129
3.4　使用 Action Bar ······················· 130
　　3.4.1　向 Action Bar 添加动作项 ·· 132

3.4.2　定制动作项和应用
　　　　　　程序图标 ················· 138
3.5　以编程方式创建用户界面 ······ 139
3.6　侦听用户界面通知 ··············· 142
　　　3.6.1　重写活动中定义的方法 ····· 142
　　　3.6.2　为视图注册事件 ············ 146
3.7　本章小结 ···························· 149

第4章　使用视图设计用户界面········ 151
4.1　基本视图 ···························· 151
　　　4.1.1　TextView 视图 ············ 152
　　　4.1.2　Button、ImageButton、EditText、
　　　　　　CheckBox、ToggleButton、
　　　　　　RadioButton 和 RadioGroup
　　　　　　视图 ························· 152
　　　4.1.3　ProgressBar 视图 ········· 163
　　　4.1.4　AutoCompleteTextView
　　　　　　视图 ························· 169
4.2　选取器视图 ························· 171
　　　4.2.1　TimePicker 视图 ··········· 171
　　　4.2.2　DatePicker 视图 ··········· 176
4.3　使用列表视图显示长列表 ······ 183
　　　4.3.1　ListView 视图 ············· 183
　　　4.3.2　使用 Spinner 视图 ········· 191
4.4　了解特殊碎片 ······················ 194
　　　4.4.1　使用 ListFragment ········· 194
　　　4.4.2　使用 DialogFragment ····· 199
　　　4.4.3　使用 PreferenceFragment ···· 202
4.5　本章小结 ···························· 206

第5章　使用视图显示图片和菜单 ······ 209
5.1　使用图像视图显示图片 ········· 209
　　　5.1.1　Gallery 和 ImageView
　　　　　　视图 ························· 209
　　　5.1.2　ImageSwitcher ············· 217
　　　5.1.3　GridView ··················· 222
5.2　将菜单和视图一起使用 ········· 225

　　　5.2.1　创建辅助方法 ··············· 226
　　　5.2.2　选项菜单 ····················· 228
　　　5.2.3　上下文菜单 ·················· 230
5.3　其他一些视图 ······················ 233
　　　5.3.1　AnalogClock 和 DigitalClock
　　　　　　视图 ························· 233
　　　5.3.2　WebView ···················· 234
5.4　本章小结 ···························· 240

第6章　数据持久化 ······················ 243
6.1　保存和加载用户首选项 ········· 243
　　　6.1.1　使用活动访问首选项 ······· 244
　　　6.1.2　通过编程检索和修改首
　　　　　　选项值 ······················ 250
　　　6.1.3　修改首选项文件的默认
　　　　　　名称 ························· 252
6.2　将数据持久化到文件中 ········· 254
　　　6.2.1　保存到内部存储器 ········· 254
　　　6.2.2　保存到外部存储器
　　　　　　(SD 卡) ···················· 259
　　　6.2.3　选择最佳存储选项 ········· 262
　　　6.2.4　使用静态资源 ·············· 263
6.3　创建和使用数据库 ··············· 264
　　　6.3.1　创建 DBAdapter 辅助类 ··· 265
　　　6.3.2　以编程方式使用数据库 ··· 270
　　　6.3.3　预创建数据库 ·············· 277
6.4　本章小结 ···························· 281

第7章　内容提供者 ······················ 283
7.1　在 Android 中共享数据 ········· 283
7.2　使用内容提供者 ··················· 284
　　　7.2.1　预定义查询字符串常量 ··· 289
　　　7.2.2　投影 ·························· 292
　　　7.2.3　筛选 ·························· 293
　　　7.2.4　排序 ·························· 294
7.3　创建自己的内容提供者 ········· 295
7.4　使用内容提供者 ··················· 304

| 7.5 | 本章小结 | 309 |

第8章 消息传递 311
- 8.1 SMS 消息传递 311
  - 8.1.1 以编程方式发送 SMS 消息 312
  - 8.1.2 发送消息后获取反馈 315
  - 8.1.3 使用意图发送 SMS 消息 318
  - 8.1.4 接收 SMS 消息 319
  - 8.1.5 说明和警告 334
- 8.2 发送电子邮件 335
- 8.3 本章小结 338

第9章 基于位置的服务 339
- 9.1 显示地图 339
  - 9.1.1 创建项目 340
  - 9.1.2 获取 Maps API 密钥 340
  - 9.1.3 显示地图 343
  - 9.1.4 显示缩放控件 346
  - 9.1.5 改变视图 349
  - 9.1.6 导航到特定位置 350
  - 9.1.7 添加标记 353
  - 9.1.8 获取触摸的位置 356
  - 9.1.9 地理编码和反向地理编码 358
- 9.2 获取位置数据 362
- 9.3 监控一个位置 371
- 9.4 项目——创建一个位置跟踪应用程序 372
- 9.5 本章小结 378

第10章 联网 381
- 10.1 通过 HTTP 使用 Web 服务 381
  - 10.1.1 下载二进制数据 384
  - 10.1.2 下载文本内容 390
  - 10.1.3 通过 GET 方法访问 Web 服务 392
- 10.2 使用 JSON 服务 397
- 10.3 套接字编程 405
- 10.4 本章小结 414

第11章 开发 Android 服务 417
- 11.1 创建自己的服务 417
  - 11.1.1 在服务中执行长时间运行的任务 421
  - 11.1.2 在服务中执行重复的任务 426
  - 11.1.3 使用 IntentService 在单独的线程上执行异步任务 430
- 11.2 在服务和活动之间通信 433
- 11.3 将活动绑定到服务 437
- 11.4 理解线程 442
- 11.5 本章小结 448

第12章 发布 Android 应用程序 451
- 12.1 为发布做准备 451
  - 12.1.1 版本化 451
  - 12.1.2 对 Android 应用程序进行数字签名 454
- 12.2 部署 APK 文件 459
  - 12.2.1 使用 adb.exe 工具 459
  - 12.2.2 使用 Web 服务器 461
  - 12.2.3 在 Android Market 上发布 462
- 12.3 本章小结 467

附录 A 使用 Eclipse 进行 Android 开发 469

附录 B 使用 Android 模拟器 483

附录 C 练习答案 595

第 1 章

Android 编程入门

本章将介绍以下内容：
- Android 简介
- Android 版本及其功能集
- Android 架构
- 市场上的各种 Android 设备
- Android Market 应用程序商店
- 如何获得开发 Android 应用程序的工具和 SDK(软件开发工具包)
- 如何开发您的第一个 Android 应用程序

欢迎阅读本书！当我撰写自己的第一本关于 Android 的图书时，曾提到 Android 取代了 Apple 的 iPhone，在美国智能手机市场中排名第二，仅次于 Research In Motion(RIM)的 BlackBerry。那本书付印后不久，comScore(数字世界评估的全球领先者，是数字世界的首选信息息源)发布的报告称 Android 超过了 BlackBerry，成为美国最受欢迎的智能手机平台。

几个月后，Google 发布了 Android 3.0，代号为 Honeycomb(蜂巢)。在这个版本中，Google 将重点放到了新的软件开发套件上，引入了几个专为宽屏设备(特别是平板电脑)设计的新功能。如果是为 Android 智能手机开发应用程序，Android 3.0 的用处并不大，因为智能手机不支持它提供的新功能。在 Android 3.0 发布的同时，Google 开始开发下一个版本的 Android，致力于让它在智能手机和平板电脑上都可使用。2011 年 10 月，Google 发布了 Android 4.0，代号为 Ice Cream Sandwich(冰激凌三明治)，本书将重点介绍这个版本。

本章将介绍 Android 到底是什么，以及是什么让开发人员和设备制造商都有如此大的兴趣。您也将开始开发您的第一个 Android 应用程序，并学会如何获得必要的工具并对其设置，以便可以在 Android 4.0 模拟器上测试应用程序。在本章结尾，您将具备进一步探索更尖端的技术和技巧以开发您的下一个杀手级的 Android 应用程序所需的基础知识。

1.1 Android 简介

Android 是一款基于 Linux 修订版本的移动操作系统。它最初是由同名的 Android 有限公司作为进入移动市场的战略的一部分于 2005 年开发的。Google 收购了 Android 公司，并接管了它的开发工作(包括整个开发团队)。

Google 要求 Android 系统是开放和免费的。因此，大部分 Android 代码在 Apache License 开源协议下都公开了，这意味着任何想使用 Android 的人都可以下载 Android 的全部源代码。此外，供应商(特别是硬件制造商)可以添加他们自己专有的 Android 扩展，通过定制 Android 以区别于其他厂商的产品。这一简单的开发模型使 Android 非常有吸引力，并因此引起了许多供应商的兴趣。Apple 公司 iPhone 产品的巨大成功彻底改变了智能手机产业，这深深影响到了诸如摩托罗拉和索爱这一类多年只开发自己的移动操作系统的公司。当 iPhone 发布时，这些大部分厂商不得不争相寻找振兴自己产品的新出路。他们将 Android 视为一种解决方案——继续设计自己的硬件，同时将 Android 用作支持硬件的操作系统。

使用 Android 的主要优势是它提供了统一的应用程序开发方法。开发人员只需要为 Android 进行开发，开发出的应用程序可以运行在许多不同的设备上，只要这些设备用的是 Android 系统。在智能手机界，应用程序是成功链中的最重要一环。因此，为了应对已经占据大量应用程序市场的 iPhone 带来的巨大冲击，设备制造商对 Android 寄予了厚望。

1.1.1 Android 版本

自首次发布以来，Android 已历经了相当多数量的更新版本。表 1-1 列出了 Android 的不同版本及其相应代号。

表 1-1 Android 版本简史

Android 版本	发 布 日 期	代　　号
1.1	2009 年 2 月 9 日	
1.5	2009 年 4 月 30 日	Cupcake(纸杯蛋糕)
1.6	2009 年 9 月 15 日	Donut(炸面圈)
2.0/2.1	2009 年 10 月 26 日	Eclair(长松饼)
2.2	2010 年 5 月 20 日	Froyo(冻酸奶)
2.3	2010 年 12 月 6 日	Gingerbread(姜饼)
3.0/3.1/3.2	2011 年 2 月 22 日	Honeycomb(蜂巢)
4.0	2011 年 10 月 19 日	Ice Cream Sandwich(冰激凌三明治)

2011 年 2 月，Google 发布了 Android 3.0，它支持宽屏设备，是一种只针对平板电脑的版本。Android 3.0 的关键变化包括：

- 针对平板电脑进行优化的新用户界面
- 使用新的小组件的 3D 桌面

- 优化的多任务功能
- 新的 Web 浏览器功能，例如标签式浏览、表单自动填充、书签同步和隐私浏览
- 支持多核处理器

为 Android 3.0 之前的版本编写的应用程序在 Android 3.0 设备上可以直接运行，无须修改。但是，使用了 Android 3.0 的新功能编写的 Android 3.0 平板电脑应用程序是不能在较早的设备上运行的。为了确保 Android 3.0 平板电脑应用程序可以在各种版本的设备上运行，必须从编程方面入手确保只使用 Android 的特定版本支持的功能。

在 2011 年 11 月，Google 发布了 Android 4.0，让智能手机也具有了 Android 3.0 中引入的所有功能，并且还提供了一些新功能，包括面部识别解锁功能、数据使用监控、近距离通信(Near Field Communication，NFC)等。

1.1.2 Android 功能

鉴于 Android 的开源以及制造商可对其自由定制的特点，因此没有固定的软硬件配置。然而，Android 本身支持如下功能：

- 存储——使用 SQLite(轻量级的关系数据库)进行数据存储，第 6 章将对数据存储进行详细讨论。
- 连接性——支持 GSM/EDGE、IDEN、CDMA、EV-DO、UMTS、Bluetooth(包括 A2DP 和 AVRCP)、WiFi、LTE 和 WiMAX。第 8 章将详细讨论联网。
- 消息传递——支持 SMS 和 MMS，也在第 8 章进行详细探讨。
- Web 浏览器——基于开源的 WebKit，并集成 Chrome 的 V8 JavaScript 引擎。
- 媒体支持——支持以下媒体：H.263、H.264(在 3GP 或 MP4 容器中)、MPEG-4 SP、AMR、AMR-WB(在 3GP 容器中)、AAC、HE-AAC(在 MP4 或 3GP 容器中)、MP3、MIDI、OggVorbis、WAV、JPEG、PNG、GIF 和 BMP。
- 硬件支持——加速度传感器、摄像头、数字式罗盘、接近传感器和全球定位系统(GPS)。
- 多点触摸——支持多点触摸屏幕。
- 多任务——支持多任务应用。
- Flash 支持——Android 2.3 支持 Flash 10.1。
- tethering——支持作为有线/无线热点实现 Internet 连接共享。

1.1.3 Android 架构

为了理解 Android 的工作方式，可以参看图 1-1，该图描述了构成 Android 操作系统(OS)的各个层。

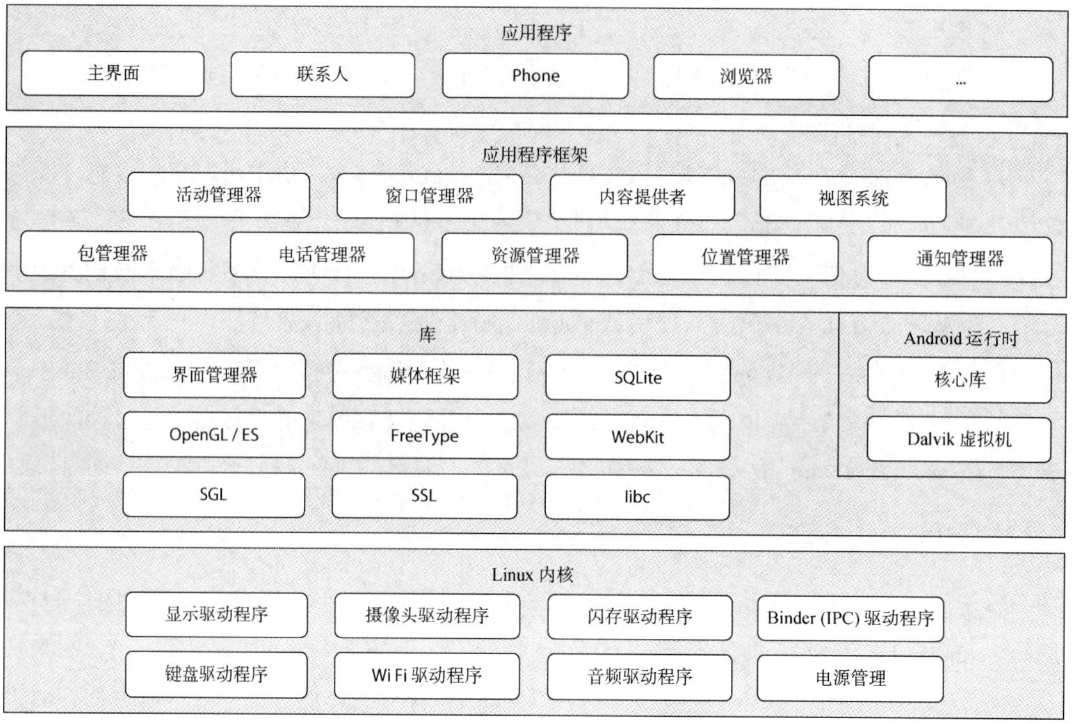

```
                          应用程序
   ┌──────┐  ┌──────┐  ┌──────┐  ┌──────┐  ┌──────┐
   │ 主界面│  │ 联系人│  │ Phone│  │浏览器 │  │  …   │
   └──────┘  └──────┘  └──────┘  └──────┘  └──────┘

                        应用程序框架
   ┌────────┐ ┌────────┐ ┌────────┐ ┌────────┐
   │活动管理器│ │窗口管理器│ │内容提供者│ │视图系统 │
   └────────┘ └────────┘ └────────┘ └────────┘
   ┌────────┐ ┌────────┐ ┌────────┐ ┌────────┐ ┌────────┐
   │包管理器 │ │电话管理器│ │资源管理器│ │位置管理器│ │通知管理器│
   └────────┘ └────────┘ └────────┘ └────────┘ └────────┘

              库                              Android 运行时
   ┌────────┐ ┌────────┐ ┌────────┐          ┌────────┐
   │界面管理器│ │媒体框架 │ │ SQLite │          │ 核心库  │
   └────────┘ └────────┘ └────────┘          └────────┘
   ┌────────┐ ┌────────┐ ┌────────┐          ┌────────┐
   │OpenGL/ES│ │FreeType│ │ WebKit │          │Dalvik虚拟机│
   └────────┘ └────────┘ └────────┘          └────────┘
   ┌────────┐ ┌────────┐ ┌────────┐
   │  SGL   │ │  SSL   │ │  libc  │
   └────────┘ └────────┘ └────────┘

                         Linux 内核
   ┌────────┐ ┌──────────┐ ┌──────────┐ ┌──────────────┐
   │显示驱动 │ │摄像头驱动 │ │闪存驱动程序│ │Binder(IPC)驱动│
   │ 程序   │ │ 程序     │ │          │ │   程序       │
   └────────┘ └──────────┘ └──────────┘ └──────────────┘
   ┌────────┐ ┌──────────┐ ┌──────────┐ ┌──────────┐
   │键盘驱动 │ │WiFi驱动  │ │音频驱动   │ │电源管理   │
   │ 程序   │ │ 程序     │ │ 程序     │ │          │
   └────────┘ └──────────┘ └──────────┘ └──────────┘
```

图 1-1

Android 操作系统大致可以在 4 个主要层面上分为以下 5 个部分：

- Linux 内核——这是 Android 所基于的核心。这一层包括了一个 Android 设备的各种硬件组件的所有低层设备驱动程序。
- 库——包括了提供 Android 操作系统的主要功能的全部代码。例如，SQLite 库提供了支持应用程序进行数据存储的数据库。WebKit 库为浏览 Web 提供了众多功能。
- Android 运行时——它与库同处一层，提供了一组核心库，可以使开发人员使用 Java 编程语言来写 Android 应用程序。Android 运行时还包括 Dalvik 虚拟机，这使得每个 Android 应用程序都在它自己的进程中运行，都拥有一个自己的 Dalvik 虚拟机实例(Android 应用程序被编译成 Dalvik 可执行文件)。Dalvik 是特别为 Android 设计，并为内存和 CPU 受限的电池供电的移动设备进行过优化的专门的虚拟机。
- 应用程序框架——对应用程序开发人员公开了 Android 操作系统的各种功能，使他们可以在应用程序中使用这些功能。
- 应用程序——在这个最顶层中，可以找到 Android 设备自带的应用程序(例如电话、联系人、浏览器等)，以及可以从 Android Market 应用程序商店下载和安装的应用程序。您所写的任何应用程序都处于这一层。

1.1.4 市场上的 Android 设备

Android 设备有各种样式和大小。截至 2011 年 11 月底，Android 操作系统可以支持如下类型的设备：

- 智能手机
- 平板电脑
- 电子阅读器
- 上网本
- MP4 播放器
- 互联网电视

而您目前很可能已经至少拥有其中一种设备。图 1-2 (从左到右)展示了 Samsung Galaxy S II、Motorola Atrix 4G 和 HTC EVO 4G 智能手机。

图 1-2

制造商都趋之若鹜的另一类流行的设备是平板电脑。平板电脑通常有两种尺寸： 7 英寸和 10 英寸(指对角线长度)。图 1-3 展示了 Samsung Galaxy Tab 10.1(左)和 Asus Eee Pad Transformer TF101(右)(二者都是 10.1 英寸的平板电脑)。Samsung Galaxy 10.1 和 Asus Eee Pad Transfer TF101 都使用 Android 3 作为操作系统。

图 1-3

除了智能手机和平板电脑外，Android 也开始出现在专用设备中，如电子书阅读器。

图 1-4 展示了两款运行 Android 操作系统的彩色电子书阅读器产品——Barnes & Noble 公司的 NOOK Color(左)和 Amazon 的 Kindle Fire(右)。

图 1-4

除了这些流行的移动设备，Android 也正慢慢进入到您的客厅。瑞典公司 People of Lava 开发了一款基于 Android 的电视机，名为 Scandinavia，如图 1-5 所示。

Google 还涉足了基于 Android 的专有的智能电视平台，并和诸如英特尔、索尼、罗技等公司进行共同开发。图 1-6 展示了索尼公司的 Google 电视。

图 1-5   图 1-6

在撰写本书时，Samsung Galaxy Nexus(见图 1-7)是唯一一款运行 Android 4.0 的设备。但是，Google 承诺可以使现有的设备(例如 Nexus S)升级到 Android 4.0。在您读到本书时，可能已经有了很多运行 Android 4.0 的设备。

图 1-7

1.1.5 Android Market

如前所述,决定一个智能手机平台成功的主要因素之一是支持它的应用程序。从 iPhone 的成功可以清楚地看出,应用程序在决定一个新的平台是成功还是失败方面扮演了一个非常关键的角色。此外,使这些应用程序能为广大用户访问也是极为重要的。

因此,在 2008 年 8 月,Google 宣布将在同年 10 月份为用户提供一个适用于 Android 设备的在线应用程序商店:Android Market。使用预装于 Android 设备上的 Market 应用程序,用户可以很方便地把第三方应用程序直接下载到他们的设备上。付费和免费的应用程序在 Android Market 上都是受支持的,不过付费的应用程序由于法律问题只提供给某些国家的用户。

同样,在一些国家,用户可以从 Android Market 购买付费的应用程序,但开发人员不能在该国销售。例如,在写作本书时,印度的用户可以从 Android Market 购买应用程序,但印度的开发人员却不能在 Android Market 上出售应用程序。相反的情况也可能是存在的。例如,韩国的用户不能购买应用程序,但韩国的开发人员可以在 Android Market 上出售应用程序。

 注意:第 12 章讨论了更多有关 Android Market 的内容,以及如何在上面出售自己的应用程序。

1.1.6 Android 开发社区

Android 迎来了第 4 个版本,它已经在世界范围内形成了庞大的开发社区。现在更容易寻求问题的解决办法,并找到有类似想法的开发人员来分享关于应用程序的想法和经验。

如果在进行 Android 开发的过程中遇到问题,可以到下面列出的开发社区/网站来寻找帮助:

- Stack Overflow(www.stackoverflow.com)——Stack Overflow 是一个协作编辑性的问答网站，可以解决开发人员的各种问题。您遇到的关于 Android 的问题，很可能已经在 Stack Overflow 中被讨论过并已经有了解决方案了。最好的地方是，其他开发人员可以投票选出最佳答案，这样您就知道哪些答案是可以相信的。
- Google Android Training(http://developer.android.com/training/index.html)——Google 创建了 Android Training 网站，按主题分类包含了许多实用的课程。在写作本书时，这些课程大多包含了实用的代码片段，当 Android 开发人员掌握了基础后，会发现这些代码十分有用。学习完本书介绍的基础知识后，强烈建议您去看看这些课程。
- Android Discuss(http://groups.google.com/group/android-discuss)——Android Discuss 是 Google 通过 Google Groups 服务管理的一个讨论组。在这里可以讨论关于 Android 编程的各种话题。Google 的 Android 团队密切监管着这个组，所以这是解决自己的疑问和学习新技巧的一个好地方。

1.2 获得所需工具

既然已了解了 Android 的概念和其功能集，您也许渴望亲自动手试一试，并开始写些应用程序。然而，在您写第一个应用程序之前，需要下载所需的工具和 SDK。

对于 Android 开发，可以使用 Mac、Windows PC 或 Linux 机器。所有必需的工具都可以通过网络免费下载。除了少数需要访问硬件的例子以外，本书提供的大多数例子都可以在 Android 模拟器中运行得很好。本书中，我将使用运行 Windows 7 操作系统的计算机来演示所有的代码示例。如果您用的是 Mac 或 Linux 计算机，除了存在一些细微的差别，屏幕截图应该是很相似的，您应该可以毫无困难地按照本书的指导来练习。

那么，让我们开始有趣的学习之旅吧！

> **Java JDK**
>
> Android SDK 使用 Java SE 开发工具包(JDK)。因此。如果您的计算机上没有安装 JDK，那么应该通过 www.oracle.com/technetwork/java/javase/downloads/index.html 地址下载并在阅读下一小节前进行安装。

1.2.1 Android SDK

需要下载的第一个、也是最重要的软件自然是 Android SDK。Android SDK 包含了一个调试器、库、一个模拟器、文档、示例代码和教程。

可以从 http://developer.android.com/sdk/index.html 下载 Android SDK，如图 1-8 所示。

Android SDK 打包在一个 zip 文件中。下载完成后，将其内容(android-sdk-windows 文件夹)解压到一个文件夹中，例如 C:\Android 4.0\。对于 Windows 用户，Google 建议下载 installer_r15-windows.exe 文件，它可以自动设置所需的工具。下面的步骤将介绍使用这种方法进行安装的过程。

第 1 章　Android 编程入门

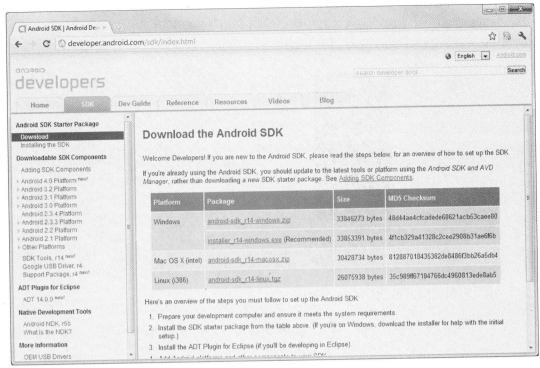

图 1-8

1.2.2　安装 Android SDK 工具

下载了 installer_r15-windows.exe 文件后，双击它开始安装 Android 工具。在 Setup Wizard 的欢迎界面，单击 Next 按钮继续操作。

如果计算机中没有安装 Java，会看到如图 1-9 所示的出错对话框。但是，即使安装了 Java，仍可能看到此错误。此时，单击 Report error 按钮，然后单击 Next 按钮。

此时需要指定一个目标文件夹来安装 Android SDK 工具。输入目标路径(如图 1-10 所示)，然后单击 Next 按钮。

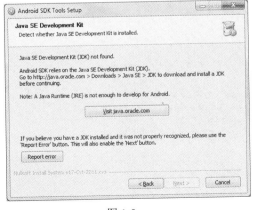

图 1-9

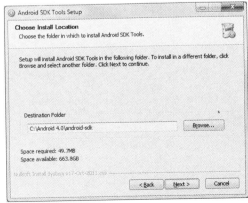

图 1-10

当向导要求选择一个 Start Menu 文件夹来创建程序的快捷方式时，选择默认的 Android

SDK Tools，然后单击 Install。安装完成后，单击 Start SDK Manager(to download system images, etc.)选项，然后单击 Finish 按钮(如图 1-11 所示)。这将启动 SDK Manager。

图 1-11

1.2.3 配置 Android SDK Manager

Android SDK Manager 管理计算机上目前安装的各种版本的 Android SDK。启动 Android SDK Manager 后，会看到一个项目列表，以及当前计算机中是否安装了它们，如图 1-12 所示。

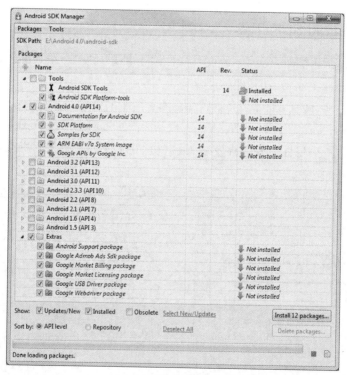

图 1-12

选中您的项目需要的工具、文档和平台。然后，单击 Install 按钮下载它们。因为从 Google 的服务器下载需要一些时间，只下载迫切需要的东西，而等到有更多时间时再下载其他东西是一个好主意。现在，可以只选中图中显示的项。

 注意：一开始，至少应该选中最新的 Android 4.0 SDK 平台和 Extras。在创作本书时，最新的 SDK 平台是 SDK Platform Android 4.0，API 14。

每个版本的 Android OS 都通过一个 API 级别号标识。例如，Android 2.3.3 是级别 10(API 10)，而 Android 3.0 是级别 11(API 11)等。对于每个级别，有两个平台可用，例如，级别 14 提供了以下两个平台：

- SDK Platform
- Google 公司的 Google APIs

两者的关键区别是，Google APIs 平台包含 Google 提供的附加 API(如 Google Maps 库)。因此，如果所编写的应用程序需要使用 Google Maps，就需要使用 Google APIs 平台创建一个 AVD(第 9 章将详细介绍这方面的知识)。

需要选择要安装的包(如图 1-13 所示)。选中 Accept All 选项，然后单击 Install 按钮。

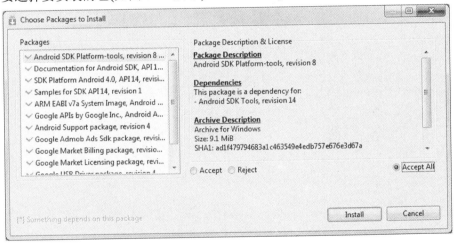

图 1-13

SDK Manager 将继续下载选中的包。安装过程需要一些时间，所以要保持耐心。所有的包安装完成后，会要求重启 ADB(Android Debug Bridge)。单击 Yes 按钮。

1.2.4 Eclipse

下一步是获得 Android 应用程序开发的集成开发环境(IDE)。就 Android 来说，推荐使用 Eclipse。它是一个多语言的软件开发环境，有一个可扩展的插件系统。通过它可以用 Java、Ada、C、C++、COBOL、Python 等语言开发各种类型的应用程序。

对于 Android 的开发，要下载 Eclipse IDE for Java EE Developers (www.eclipse.org/downloads/)。目前有 6 个版本可用：Windows(32 位和 64 位)、Mac OS X (Cocoa 32 和 64) 以及 Linux (32 位和 64 位)。只要选择与您的操作系统相对应的那个版本进行安装即可。本书

所有示例均使用 Windows 平台下的 32 位版本的 Eclipse 进行过测试。

下载 Eclipse IDE 后，把其内容(eclipse 文件夹)解压到一个文件夹下，比如 C:\Android 4.0\。图 1-14 显示了 eclipse 文件夹的内容。

要启动 Eclipse，可以双击 eclipse.exe 文件。这时系统会要求指定工作区。Eclipse 中的工作区就是存储所有项目的一个文件夹。接受建议的默认工作区(也可以指定自己的文件夹作为工作区)，然后单击 OK 按钮。

1.2.5 Android 开发工具

启动 Eclipse 后，选择 Help | Install New Software 菜单项(如图 1-15 所示)来安装 Eclipse 的 Android Development Tools(ADT)插件。

图 1-14

图 1-15

ADT 是对 Eclipse IDE 的扩展，用以支持 Android 应用程序的创建和调试。使用 ADT，可以在 Eclipse 中做如下工作：

- 创建新的 Android 应用程序项目
- 访问 Android 模拟器和设备的存取工具
- 编译和调试 Android 应用程序
- 将 Android 应用程序导出到 Android 包(APK)
- 创建数字证书来对 APK 进行代码签名

在出现的 Install 窗口中的文本框内输入 https://dl-ssl.google.com/android/eclipse/，然后按 Enter 键。稍后，您将看到在窗口的正中央显示出 Developer Tools 项(如图 1-16 所示)。展开后，显示出以下内容：Android DDMS、Android Development Tools、Android Hierarchy

Viewer 和 Android Traceview。全选并单击 Next 按钮两次。

图 1-16

注意：如果在下载 ADT 过程中遇到任何问题，可以在 http://developer.android.com/sdk/eclipse-adt.html#installing 上查找 Google 的帮助。

您会被要求查看并接受工具的许可证。选中 I accept the terms of the license agreements 选项并单击 Finish 按钮。ADT 安装完毕后，将会提示您重启 Eclipse，重启它即可。

重启后，将提示您配置 Android SDK，如图 1-17 所示。因此前一节已经下载了 Android SDK，所以选中 Use existing SDKs 选项，并指定 Android SDK 的安装路径。单击 Next 按钮。

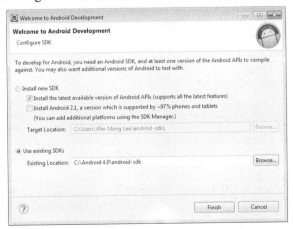

图 1-17

完成这一步后，将询问是否把使用数据发送给 Google。做出选择后，单击 Finish 按钮。

 注意：每个新版本的 SDK 发布时，安装步骤都稍有变化。如果您发现安装步骤与这里的介绍不同，也不用担心，只要按照屏幕上的提示进行操作即可。

1.2.6 创建 Android 虚拟设备(AVD)

下一步是创建用于测试 Android 应用程序的 AVD。AVD 表示 Android 虚拟设备(Android Virtual Device)。AVD 是一个模拟器实例，可以用来模拟一个真实的设备。每一个 AVD 包含一个硬件配置文件、一个到系统映像的映射，以及模拟存储器(例如安全数字(SD)卡)。

您打算测试多少个不同配置的应用程序，就可以创建多少个 AVD。这种测试对于确定应用程序在有着不同功能的不同设备上运行时的行为是很重要的。

 注意：附录 B 将讨论 Android 模拟器的部分功能。

为了创建 AVD，选择 Window | AVD Manager，如图 1-18 所示。

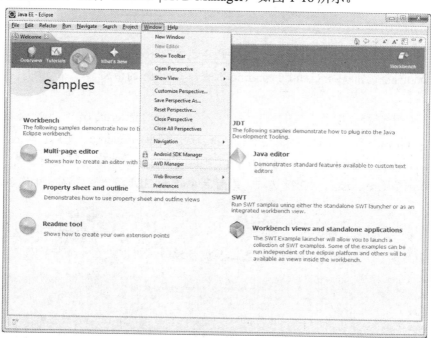

图 1-18

在 Android Virtual Device Manager 对话框(如图 1-19 所示)中，单击 New…按钮来创建一个新的 AVD。

在 Create new Android Virtual Device(AVD)窗口中，输入如图 1-20 所示的各项内容。完成后单击 Create AVD 按钮。

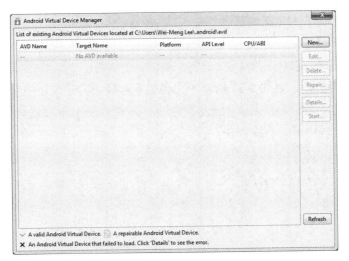

图 1-19

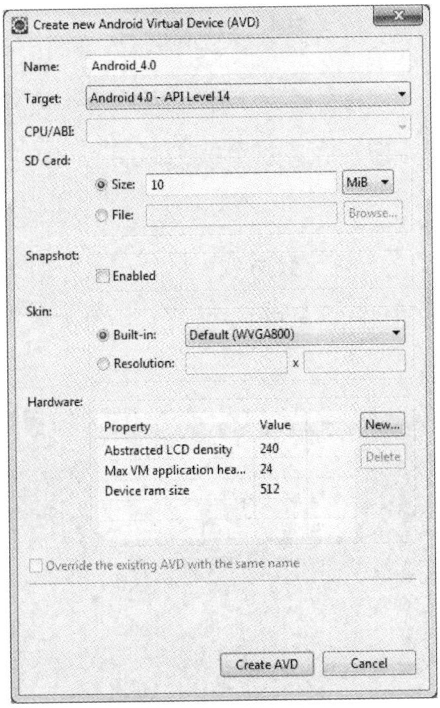

图 1-20

在这里,您已经创建了一个 AVD(简言之,一个 Android 模拟器),可以用来模拟运行 4.0 版本操作系统且内置了 10-MB SD 卡的 Android 设备。除了所创建的 AVD 之外,还可以选择模拟具有不同屏幕像素密度和分辨率的设备。

 注意:附录 B 介绍了如何模拟不同类型的 Android 设备。

创建一些具有不同 API 级别和硬件配置的 AVD 会更好些，这样您的应用程序可以在不同版本的 Android OS 上得到测试。

创建了 ADV 以后，就应该测试它。选择想要测试的 AVD，然后单击 Start…按钮。Launch Options 对话框将会显示，如图 1-21 所示。如果显示器较小，建议选中 Scale display to real size 选项，这样可以将模拟器设为一个较小的尺寸。单击 Launch 按钮启动模拟器。

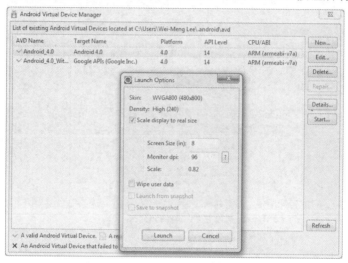

图 1-21

Android 模拟器将会启动，等待一会后就可以使用了，如图 1-22 所示。现在可以试试模拟器的用法。它就像实际的 Android 设备一样。下一节将学习如何编写你的第一个 Android 应用程序。

图 1-22

1.3 创建第一个 Android 应用程序

在所有的工具和 SDK 都下载和安装好以后，现在是开动马达的时候了。和所有的编程书籍一样，第一个示例是用无所不在的 Hello World 应用程序。这将有助于您详细了解构成一个 Android 项目的不同组件。

试一试 创建第一个 Android 应用程序

HelloWorld.zip 代码文件可以在 Wrox.com 上下载

(1) 启动 Eclipse，选择菜单 File | New | Project...创建一个新项目(如图 1-23 所示)。

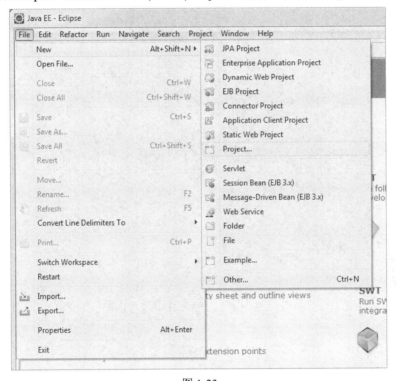

图 1-23

 注意：在创建了您的第一个 Android 应用程序后，以后的 Android 项目可以通过依次选择菜单项 File | New | Android Project 来创建。

(2) 展开 Android 文件夹，选择 Android Project(如图 1-24 所示)，单击 Next 按钮。
(3) 按图 1-25 所示将 Android 项目命名为 HelloWorld，然后单击 Next 按钮。

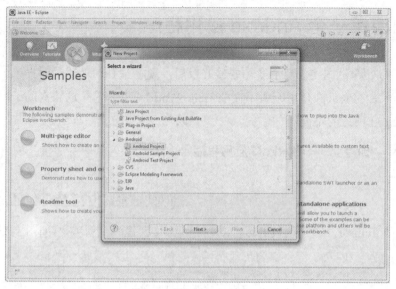

图 1-24

图 1-25

(4) 选择 Android 4.0 目标，然后单击 Next 按钮。

(5) 按照图 1-26 所示填写 Application Info，单击 Finish 按钮。

 注意：您要在包的名称中至少包含一个句点(.)。包名称的惯例是使用反向域名，项目名称紧随其后。例如，本书作者公司的域名是 learn2develop.net，因此作者的包的名称应该是 net.learn2develop.HelloWorld。

图 1-26

(6) 此时，Eclipse IDE 应该如图 1-27 所示。

(7) 在 Package Explorer 窗口(位于 Eclipse IDE 的左边)中，单击项目中每个项左侧显示的各种箭头，展开 HelloWorld 项目，如图 1-28 所示。在 res/layout 文件夹中，双击 main.xml 文件。

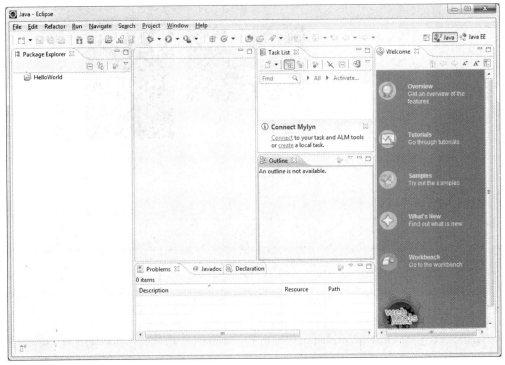

图 1-27

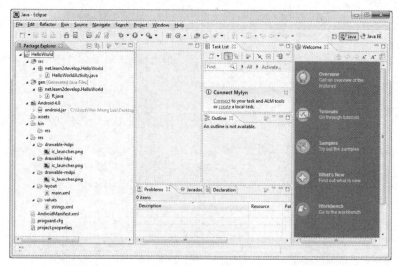

图 1-28

(8) main.xml 文件定义了应用程序的用户界面(UI)。默认视图是 Layout 视图,以图形化的方式显示了活动。要修改该用户界面,可单击位于底部的 main.xml 选项卡(如图 1-29 所示)。

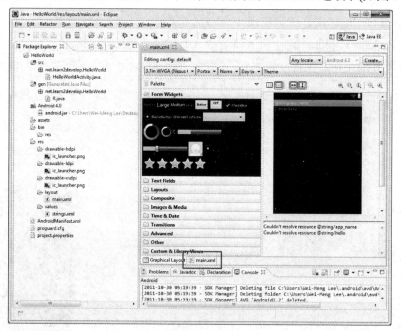

图 1-29

(9) 把下列粗体显示的代码添加到 main.xml 文件中:

```
<?xml version="1.0" encoding="utf-8"?>
<LinearLayout xmlns:android="http://schemas.android.com/apk/res/
android"
    android:layout_width="fill_parent"
    android:layout_height="fill_parent"
    android:orientation="vertical" >
```

```xml
<TextView
    android:layout_width="fill_parent"
    android:layout_height="wrap_content"
    android:text="@string/hello" />

<TextView
    android:layout_width="fill_parent"
    android:layout_height="wrap_content"
    android:text="This is my first Android Application!" />

<Button
    android:layout_width="fill_parent"
    android:layout_height="wrap_content"
    android:text="And this is a clickable button!" />

</LinearLayout>
```

(10) 按 Ctrl+S 组合键保存对项目的修改。

(11) 现在可以着手准备在 Android 模拟器上测试应用程序了。在 Eclipse 中右击项目名称，并选择 Run As | Android Application，如图 1-30 所示。

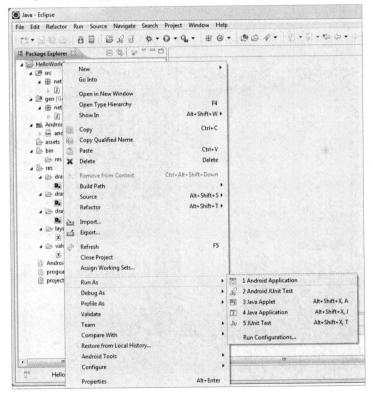

图 1-30

(12) 如果项目中没有错误，现在就应该看到应用程序已被安装并在 Android 模拟器中运行，如图 1-31 所示。

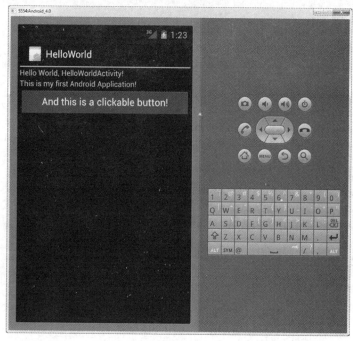

图 1-31

(13) 单击 Home 按钮(位于键盘上左下角的房子图标),将显示主屏内容(如图 1-32 所示)。

图 1-32

(14) 单击应用程序的启动器图标来显示已安装到设备上的应用程序列表。注意,HelloWorld 现在已安装在应用程序启动器中(如图 1-33 所示)。

第 1 章　Android 编程入门

图 1-33

将用哪一个 AVD 测试您的应用程序

　　回忆一下先前使用 AVD Manager 创建的几个 AVD。那么，当运行一个 Android 应用程序时，Eclipse 将启动哪一个呢？Eclipse 会检查您(当创建一个新项目时)指定的目标，将其与已经创建好的 AVD 列表对照，然后启动第一个匹配的 AVD 来运行您的应用程序。

　　如果在调试应用程序前有多个合适的 AVD 正在运行，Eclipse 将显示 Android Device Chooser 窗口，使您可以从中选择想要使用的模拟器或设备来调试应用程序(如图 1-34 所示)。

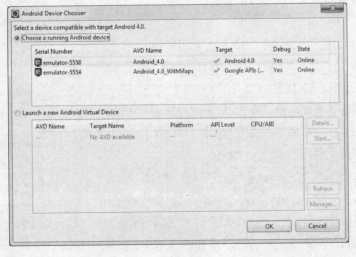

图 1-34

示例说明

为了使用 Eclipse 创建一个 Android 项目，需要提供如表 1-2 所示的信息。

表 1-2 默认方式创建的项目文件

属　　性	描　　述
Project name	项目的名称
Application name	用户友好的应用程序名称
Package name	包的名称，必须使用反向域名
Create Activity	应用程序中第一个活动的名称
Min SDK Version	项目所需的最低版本的 SDK

在 Android 中，Activity(活动)是一个包含应用程序的用户界面的窗口。一个应用程序可以有零个或多个活动。本例中，应用程序包含一个活动：HelloWorldActivity。这个 HelloWorldActivity 是应用程序的入口点，在应用程序启动时显示。第 2 章将详细讨论活动。

在这个简单的示例中，修改 main.xml 文件以显示字符串 "This is my first Android Application!" 和一个按钮。main.xml 文件包含了这个活动的用户界面，此界面在 HelloWorldActivity 加载时显示。

当在 Android 模拟器上调试应用程序时，应用程序会自动在模拟器上进行安装——没错，您已经开发了您的第一个 Android 应用程序！

下一节将揭示您的 Android 项目中所有这些不同的文件是如何一起工作来使应用程序正常运行的。

1.4 Android 应用程序剖析

既然已经创建了您的第一个 Hello World Android 应用程序，那就该分析一下 Android 项目的内部结构，检查一下使之工作的所有部件。

首先，请注意在 Eclipse 的 Package Explorer 中所显示的构成 Android 项目的不同文件(如图 1-35 所示)。

这些不同的文件夹及其文件如下：

- src——包含项目的.java 源文件。在本例中，有一个文件：HelloWorldActivity.java。HelloWorld Activity.java 文件是活动的源文件，您将在这个文件中编写应用程序的代码。这个 Java 文件在项目的包名下列出，在本例中这个包名为 net.learn2develop.HelloWorld。
- gen——包含了由编译器生成的 R.java 文件，它

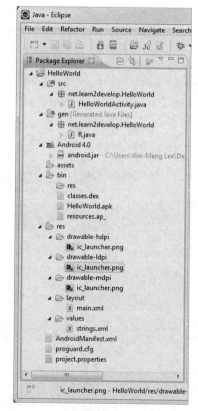

图 1-35

引用在项目中能找到的全部资源。不要修改此文件。项目中的所有资源会自动编译到这个类中，所以可以使用这个类引用它们。

- Android 4.0 库——这一项中有一个 android.jar 文件，包含了一个 Android 应用程序所需的所有类库。
- assets——这个文件夹包含了应用程序所用到的所有资产，例如 HTML、文本文件、数据库等。
- bin——这个文件夹包含了生成过程中 ADT 生成的文件。特别是，它会生成.apk 文件(Android 包)。.apk 是 Android 应用程序的二进制文件，包含运行 Android 应用程序所需的一切。
- res——这个文件夹包含了应用程序中使用的所有资源。它还包含了几个子文件夹：drawable-<resolution>、layout 和 values。第 3 章将进一步讨论如何支持具有不同屏幕分辨率和像素密度的设备。
- AndroidManifest.xml——这是 Android 应用程序的清单文件。在这一文件中，可以指定应用程序所需的权限，还可以指定其他特性(如意图筛选器、接收者等)。第 2 章将详细讨论 AndroidManifest.xml 文件的使用。

main.xml 文件定义了活动的用户界面。注意观察以下代码的粗体字部分：

```
<TextView
    android:layout_width="fill_parent"
    android:layout_height="wrap_content"
    android:text="@string/hello" />
```

这里，@string 指的是位于 res/values 文件夹下的 strings.xml 文件。因此，@string/hello 指的是在 strings.xml 文件中定义的 hello 字符串，即 "Hello World, HelloWorldActivity!"：

```
<?xml version="1.0" encoding="utf-8"?>
<resources>

    <string name="hello">Hello World, HelloWorldActivity!</string>
    <string name="app_name">HelloWorld</string>

</resources>
```

建议您将应用程序中所有的字符串常量存储于这个 strings.xml 文件中，并用@string 标识符引用这些字符串。这样，如果需要将您的应用程序本地化为另一种语言，则只需要备份整个 values 文件夹，并用目标语言替换 strings.xml 文件中存储的字符串。在图 1-36 中，使用了另外一个名为 values-fr 的文件夹，其 strings.xml 文件中包含用法语表示的相同的 hello 字符串。

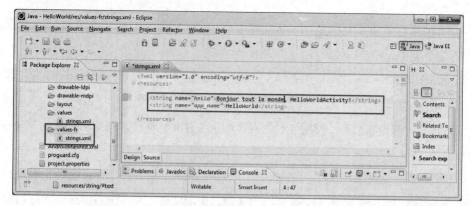

图 1-36

如果用户在一台被配置为默认显示法语的手机上加载这个应用程序，应用程序将自动显示法语表示的 hello 字符串。

在 Android 项目中，下一个重要的文件是清单文件。观察一下 AndroidManifest.xml 文件的内容：

```xml
<?xml version="1.0" encoding="utf-8"?>
<manifest xmlns:android="http://schemas.android.com/apk/res/android"
    package="net.learn2develop.HelloWorld"
    android:versionCode="1"
    android:versionName="1.0" >

    <uses-sdk android:minSdkVersion="14" />

    <application
        android:icon="@drawable/ic_launcher"
        android:label="@string/app_name" >
        <activity
            android:label="@string/app_name"
            android:name=".HelloWorldActivity" >
            <intent-filter >
                <action android:name="android.intent.action.MAIN" />

                <category android:name="android.intent.category.LAUNCHER" />
            </intent-filter>
        </activity>
    </application>

</manifest>
```

文件 AndroidManifest.xml 包含了关于应用程序的详细信息。
- 它定义了应用程序的包名：net.learn2develop.HelloWorld。
- 应用程序的版本代码为 1(通过 android:versionCode 属性设置)。这个值是用来标识您的应用程序的版本号。它可用于以编程方式确定应用程序是否需要升级。

- 应用程序的版本名称是 1.0(通过 android:versionName 属性设置)。此字符串值主要用来显示给用户。这个值应该采用以下格式：*<major>*.*<minor>*.*<point>*。
- <uses-sdk>元素的 android:minSdkVersion 属性指定了应用程序运行所需的操作系统的最低版本。
- 应用程序使用位于 drawable 文件夹下的图像 ic_launcher.png。
- 应用程序的名称是在 strings.xml 文件中定义的名为 app_name 的字符串。
- HelloWorldActivity.java 文件代表了应用程序中的一项活动。代表这项活动的标签名称与应用程序的名称相同。
- 在这项活动的定义中，有一个名为<intent-filter>的元素：
  - 意图筛选器的动作名称为 android.intent.action.MAIN，表明了这项活动是应用程序的入口点。
  - 意图筛选器的类别名称为 android.intent.category.LAUNCHER，表明了应用程序可从设备的启动器图标启动。第 2 章将详细讨论意图。

在向项目中加入更多的文件和文件夹后，Eclipse 将自动生成 R.java 的内容，而目前包含以下内容：

```
/* AUTO-GENERATED FILE.  DO NOT MODIFY.
 *
 * This class was automatically generated by the
 * aapt tool from the resource data it found.  It
 * should not be modified by hand.
 */

package net.learn2develop.HelloWorld;

public final class R {
    public static final class attr {
    }
    public static final class drawable {
        public static final int ic_launcher=0x7f020000;
    }
    public static final class layout {
        public static final int main=0x7f030000;
    }
    public static final class string {
        public static final int app_name=0x7f040001;
        public static final int hello=0x7f040000;
    }
}
```

建议您不要修改 R.java 文件的内容。当您修改项目时，Eclipse 会自动为您生成相应内容。

注意：如果手动删除了 R.java 文件，Eclipse 会为您立即再重新生成一个。注意，为了使 Eclipse 可以生成 R.java 文件，您的项目不能包含任何错误。如果在删除 R.java 后发觉 Eclipse 没有重新生成这个文件，那么您需要再检查一遍项目。代码中可能包含语法错误或者 XML 文件(如 AndroidManifest.xml、main.xml 等)的格式不良好。

最后，把活动连接到用户界面(main.xml)的代码是位于 HelloWorldActivity.java 文件中的 setContentView()方法：

```java
package net.learn2develop.HelloWorld;

import android.app.Activity;
import android.os.Bundle;

public class HelloWorldActivity extends Activity {
    /** Called when the activity is first created. */
    @Override
    public void onCreate(Bundle savedInstanceState) {
        super.onCreate(savedInstanceState);
        setContentView(R.layout.main);
    }
}
```

这里，R.layout.main 指的是位于 res/layout 文件夹下的 main.xml 文件。在向 res/layout 文件夹添加额外的 XML 文件时，R.java 中将自动生成这个文件名。onCreate()方法是活动加载时被调用的多个方法之一。第 2 章将详细讨论活动的生命周期。

1.5 本章小结

本章介绍了 Android 的概况，并强调了它的一些功能。如果您已经按照本章前面所述下载了工具和 SDK，那么现在应该有了一个工作系统——一个能够开发其他比 Hello World 更有趣的 Android 应用程序的系统。在第 2 章，您将学习到有关活动和意图的概念以及这些概念在 Android 中所扮演的重要角色。

练 习

1. 什么是 AVD？
2. AndroidManifest.xml 文件中的 android:versionCode 和 android:versionName 属性有什么区别？
3. strings.xml 文件的作用是什么？

练习答案参见附录 C。

本章主要内容

主　　题	关　键　概　念
Android 操作系统	Android 是一个基于 Linux 的开源的手机操作系统。它可以供任何打算使之在其自己设备上运行的用户使用
Android 应用程序开发语言	使用 Java 编程语言开发 Android 应用程序。编写的应用程序被编译成可在 Dalvik 虚拟机之上运行的 Dalvik 可执行文件
Android Market	Android Market 包括了由第三方开发人员编写的各种 Android 应用程序
Android 应用程序开发工具	Eclipse IDE、Android SDK 和 ADT
活动	Android 应用程序中的一个屏幕代表一个活动。每一个应用程序可以有零个或多个活动
Android 清单文件	AndroidMainifest.xml 文件包含了应用程序的详细配置信息。随着应用程序变得更加复杂，您需要不断修改这个文件，同时在本书的学习过程中，您将看到可添加到这个文件的不同信息

第 2 章

活动、碎片和意图

本章将介绍以下内容：
- 活动的生命周期
- 如何使用碎片定制 UI
- 如何对活动应用样式和主题
- 如何将活动显示为对话框窗口
- 理解意图的概念
- 如何使用 Intent 对象链接活动
- 意图筛选器如何使您有选择地链接到其他活动
- 如何使用通知向用户显示警报

在第 1 章中，您已经知道了活动就是一个包含应用程序的用户界面的窗口。一个应用程序可以包含零个或多个活动。通常，应用程序具有一个或多个活动，活动的主要目的就是与用户交互。一个活动的生命周期是指从在屏幕上显示那一刻起一直到最后隐藏所经历的若干个阶段。理解活动的生命周期对确保应用程序正确地工作是极其关键的。除了活动，Android 4.0 还支持在 Android 3.0(用于平板电脑)中引入的一种功能：碎片。可以把碎片看做"微缩版"的活动，它们可以组合到一起形成活动。本章将学习活动和碎片是如何协同工作的。

Android 中的另一个独特的概念就是意图。一个意图从根本上来说就是能够将来自不同应用程序的不同活动无缝连接在一起工作的"胶水"，确保这些任务执行起来像是都属于一个单一的应用程序。在本章后面的部分，您将学习到有关这一重要概念的更多内容，并学会如何使用它来调用诸如 Browser、Phone、Maps 等内置应用程序。

2.1 理解活动

首先让我们看看是如何创建一个活动的。要创建一个活动，需要创建一个扩展 Activity

基类的 Java 类：

```java
package net.learn2develop.Activity101;

import android.app.Activity;
import android.os.Bundle;

public class Activity101Activity extends Activity {
    /** Called when the activity is first created. */
    @Override
    public void onCreate(Bundle savedInstanceState) {
        super.onCreate(savedInstanceState);
        setContentView(R.layout.main);
    }
}
```

随后，您自己的活动类将使用在 res/layout 文件夹下定义的 XML 文件加载此活动的用户界面(UI)组件。本例中，将使用 main.xml 文件来加载用户界面：

```java
setContentView(R.layout.main);
```

应用程序中的每一个活动必须在 AndroidManifest.xml 文件中声明，如下所示：

```xml
<?xml version="1.0" encoding="utf-8"?>
<manifest xmlns:android="http://schemas.android.com/apk/res/android"
    package="net.learn2develop.Activity101"
    android:versionCode="1"
    android:versionName="1.0" >

    <uses-sdk android:minSdkVersion="14" />

    <application
        android:icon="@drawable/ic_launcher"
        android:label="@string/app_name" >
        <activity
            android:label="@string/app_name"
            android:name=".Activity101Activity" >
            <intent-filter >
                <action android:name="android.intent.action.MAIN" />

                <category android:name="android.intent.category.LAUNCHER" />
            </intent-filter>
        </activity>
    </application>

</manifest>
```

Activity 基类定义了管理一个活动的生命周期的一系列事件。该类定义了如下事件：
- onCreate()——当活动首次被创建时调用。

- onStart()——当活动对用户可见时调用。
- onResume()——当活动与用户开始交互时调用。
- onPause()——在当前活动被暂停并恢复以前的活动时调用。
- onStop()——当活动不再对用户可见时调用。
- onDestroy()——在活动被系统销毁(手动或由系统执行以节省内存)前调用。
- onRestart()——在活动已停止并要再次启动时调用。

默认情况下，所创建的活动包含 OnCreate()事件。在这个事件处理程序中含有帮助显示屏幕的用户界面元素的代码。

图 2-1 展示了一个活动的生命周期及其所经历的各个阶段——从活动开始直到结束。

本图是依据 Creative Commons 2.5 的署名许可所描述的条款，复制了 Android 开源项目创建并共享的工作内容，详见 http://developer.android.com/reference/android/app/Activity.html。

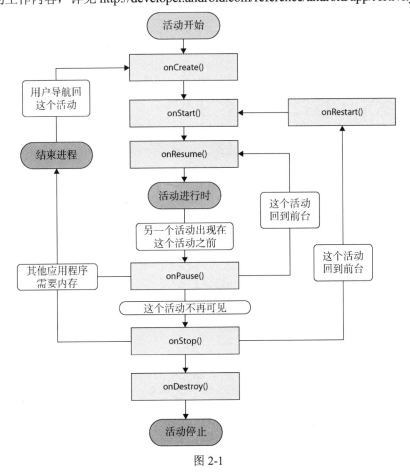

图 2-1

要了解一个活动所经历的各个阶段，最好的办法是创建一个新项目，实现各种事件，然后使活动经受各种用户交互的考验。

试一试　理解一个活动的生命周期

Activity101.zip 代码文件可以在 Wrox.com 上下载

(1) 启动 Eclipse，创建一个新的 Android 项目，命名为 Activity101。

(2) 在 Activity101Activity.java 文件中添加下列粗体显示的语句：

```java
package net.learn2develop.Activity101;

import android.app.Activity;
import android.os.Bundle;
import android.util.Log;

public class Activity101Activity extends Activity {
    String tag = "Lifecycle";

    /** Called when the activity is first created. */
    @Override
    public void onCreate(Bundle savedInstanceState) {
        super.onCreate(savedInstanceState);
        setContentView(R.layout.main);
        Log.d(tag, "In the onCreate() event");
    }

    public void onStart()
    {
        super.onStart();
        Log.d(tag,"In the onStart() event");
    }

    public void onRestart()
    {
        super.onRestart();
        Log.d(tag,"In the onRestart() event");
    }

    public void onResume()
    {
        super.onResume();
        Log.d(tag,"In the onResume() event");
    }

    public void onPause()
    {
        super.onPause();
        Log.d(tag,"In the onPause() event");
    }

    public void onStop()
```

```
    {
        super.onStop();
        Log.d(tag,"In the onStop() event");
    }

    public void onDestroy()
    {
        super.onDestroy();
        Log.d(tag, "In the onDestroy() event");
    }
}
```

(3) 按 F11 键在 Android 模拟器上调试应用程序。

(4) 当活动第一次被加载时,应该可以在 LogCat 窗口中看到与下面类似的内容(单击 Debug 透视图,参见图 2-2):

```
11-16 06:25:59.396: D/Lifecycle(559): In the onCreate() event
11-16 06:25:59.396: D/Lifecycle(559): In the onStart() event
11-16 06:25:59.396: D/Lifecycle(559): In the onResume() event
```

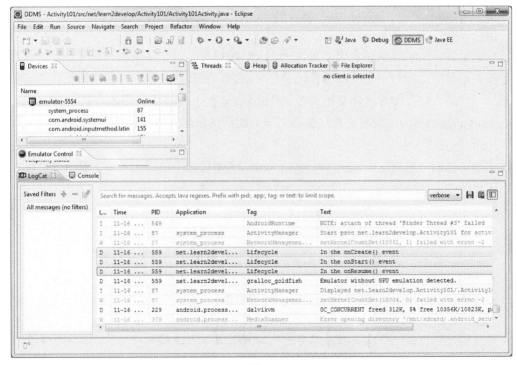

图 2-2

(5) 如果在 Android 模拟器上按 Back 按钮,可以观察到以下显示内容:

```
11-16 06:29:26.665: D/Lifecycle(559): In the onPause() event
11-16 06:29:28.465: D/Lifecycle(559): In the onStop() event
11-16 06:29:28.465: D/Lifecycle(559): In the onDestroy() event
```

(6) 按住 Home 按钮不放，同时单击 Activities 图标可以看到以下内容：

```
11-16 06:31:08.905: D/Lifecycle(559): In the onCreate() event
11-16 06:31:08.905: D/Lifecycle(559): In the onStart() event
11-16 06:31:08.925: D/Lifecycle(559): In the onResume() event
```

(7) 按下 Android 模拟器上的 Phone 按钮，当前活动就会被推到后台，观察 LogCat 窗口中的输出：

```
11-16 06:32:00.585: D/Lifecycle(559): In the onPause() event
11-16 06:32:05.015: D/Lifecycle(559): In the onStop() event
```

(8) 注意，onDestroy()事件并没有被调用，表明这个活动仍旧在内存中。按 Back 按钮退出电话拨号程序，活动又再次显示了。观察 LogCat 窗口中的输出：

```
11-16 06:32:50.515: D/Lifecycle(559): In the onRestart() event
11-16 06:32:50.515: D/Lifecycle(559): In the onStart() event
11-16 06:32:50.515: D/Lifecycle(559): In the onResume() event
```

onRestart()事件被激活，随后是 onStart()和 onResume()事件。

示例说明

从这个简单的示例可以看出，当按下 Back 按钮时，一个活动就被销毁了。知道这一点是至关重要的，因为无论活动当前处于什么状态，它都将丢失。因此，需要在您的活动中额外写一些代码以便在活动要销毁时保存其状态(第 3 章将告诉您怎么做)。在这里，注意 onPause()事件在两个情况下都将被调用——当活动被送入后台以及用户按了 Back 按钮而终止活动时。

当一个活动开始时，onStart()和 onResume()事件总是会被调用，而不管这个活动是从后台恢复的还是新创建的。当活动第一次创建时，会调用 onCreate()方法。

从这个示例中，可以获知以下指导原则：

- 使用 onCreate()方法创建和实例化将在应用程序中使用的对象。
- 使用 onResume()方法启动当活动位于前台时需要运行的任何服务或代码。
- 使用 onPause()方法停止当活动不在前台时不需要运行的任何服务或代码。
- 使用 onDestroy()方法在活动销毁前释放资源。

注意：即使一个应用程序只有一个活动并且这个活动被终止了，该应用程序仍旧会运行于内存中。

2.1.1 如何对活动应用样式和主题

默认情况下，一个活动占据整个屏幕。然而，也可以对活动应用一个对话框主题，使其显示为一个浮动对话框。例如，您打算定制一个活动，以弹出窗口的形式显示它，用来提醒用户将执行的一些操作。在这种情况下，以对话框形式显示活动以引起用户的注意是个不错的方法。

要对活动应用对话框主题，只要修改 AndroidManifest.xml 文件中的<Activity>元素，添加 android:theme 属性：

```xml
<?xml version="1.0" encoding="utf-8"?>
<manifest xmlns:android="http://schemas.android.com/apk/res/android"
    package="net.learn2develop.Activity101"
    android:versionCode="1"
    android:versionName="1.0" >

    <uses-sdk android:minSdkVersion="14" />

    <application
        android:icon="@drawable/ic_launcher"
        android:label="@string/app_name"
        android:theme="@android:style/Theme.Dialog">
        <activity
            android:label="@string/app_name"
            android:name=".Activity101Activity" >
            <intent-filter >
                <action android:name="android.intent.action.MAIN" />

                <category android:name="android.intent.category.LAUNCHER" />
            </intent-filter>
        </activity>
    </application>

</manifest>
```

这样就可以使活动显示为一个对话框，如图 2-3 所示。

2.1.2 隐藏活动标题

如果需要的话，还可以隐藏一个活动的标题(例如当您打算向用户显示状态更新时)。要做到这一点，可以使用 requestWindowFeature()方法，传递 Window.FEATURE_NO_TITLE 常量，如下所示：

```java
import android.app.Activity;
import android.os.Bundle;
import android.util.Log;
import android.view.Window;

public class Activity101Activity extends Activity {
    String tag = "Lifecycle";
```

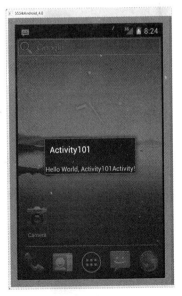

图 2-3

```
/** Called when the activity is first created. */
@Override
public void onCreate(Bundle savedInstanceState) {
    super.onCreate(savedInstanceState);
    //---hides the title bar---
    requestWindowFeature(Window.
    FEATURE_NO_TITLE);

    setContentView(R.layout.main);
    Log.d(tag, "In the onCreate()
    event");
}
}
```

这样,标题栏就被隐藏了,如图 2-4 所示。

2.1.3 显示对话框窗口

您经常会需要显示一个对话框窗口,以便从用户那里得到确认。这时,可以重写在 Activity 基类中定义的受保护的 onCreateDialog()方法来显示一个对话框窗口。下面的"试一试"会告诉您怎么做。

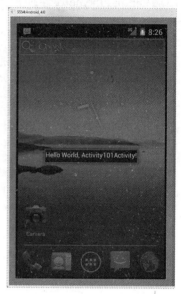

图 2-4

试一试　　使用活动显示一个对话框

Dialog.zip 代码文件可以在 Wrox.com 上下载

(1) 打开 Eclipse,创建一个新的 Android 项目,并将其命名为 Dialog。
(2) 将下列粗体显示的语句添加到 main.xml 文件中:

```
<?xml version="1.0" encoding="utf-8"?>
<LinearLayout xmlns:android="http://schemas.android.com/apk/res/android"
    android:layout_width="fill_parent"
    android:layout_height="fill_parent"
    android:orientation="vertical" >

<Button
    android:id="@+id/btn_dialog"
    android:layout_width="fill_parent"
    android:layout_height="wrap_content"
    android:text="Click to display a dialog"
    android:onClick="onClick" />

</LinearLayout>
```

(3) 将下列粗体显示的语句添加到 DialogActivity.java 文件中:

```
package net.learn2develop.Dialog;
```

```java
import android.app.Activity;
import android.app.AlertDialog;
import android.app.Dialog;
import android.content.DialogInterface;
import android.os.Bundle;
import android.view.View;
import android.widget.Toast;

public class DialogActivity extends Activity {
    CharSequence[] items = { "Google", "Apple", "Microsoft" };
    boolean[] itemsChecked = new boolean [items.length];

    /** Called when the activity is first created. */
    @Override
    public void onCreate(Bundle savedInstanceState) {
        super.onCreate(savedInstanceState);
        setContentView(R.layout.main);
    }

    public void onClick(View v) {
        showDialog(0);
    }

    @Override
    protected Dialog onCreateDialog(int id) {
        switch (id) {
        case 0:
            return new AlertDialog.Builder(this)
            .setIcon(R.drawable.ic_launcher)
            .setTitle("This is a dialog with some simple text...")
            .setPositiveButton("OK",
                new DialogInterface.OnClickListener() {
                    public void onClick(DialogInterface dialog, int
                    whichButton)
                    {
                        Toast.makeText(getBaseContext(),
                            "OK clicked!", Toast.LENGTH_SHORT).show();
                    }
                }
            )
            .setNegativeButton("Cancel",
                new DialogInterface.OnClickListener() {
                    public void onClick(DialogInterface dialog, int
                    whichButton)
                    {
                        Toast.makeText(getBaseContext(),
                            "Cancel clicked!", Toast.LENGTH_SHORT).
                                show();
                    }
```

```
                    }
                )
                .setMultiChoiceItems(items, itemsChecked,
                    new DialogInterface.OnMultiChoiceClickListener() {
                        public void onClick(DialogInterface dialog,
                            int which, boolean isChecked) {
                            Toast.makeText(getBaseContext(),
                                items[which] + (isChecked ? " checked!":"
                                unchecked!"),
                                Toast.LENGTH_SHORT).show();
                        }
                    }
                ).create();


    }
    return null;
}
```

(4) 按 F11 键在 Android 模拟器上调试应用程序。单击按钮显示对话框(如图 2-5 所示)。选中不同的复选框会使 Toast 类显示那些选中/未选中项的文本。要关闭对话框，可单击 OK 或 Cancel 按钮。

图 2-5

示例说明

要显示一个对话框，首先需要实现 Activity 类中的 onCreateDialog()方法：

```
@Override
protected Dialog onCreateDialog(int id) {
    //...
}
```

在调用 showDialog()方法时调用这个方法：

```
public void onClick(View v) {
    showDialog(0);
}
```

onCreateDialog()是一个用于创建由活动管理的对话框的回调方法。当调用 showDialog()方法时，将调用这个回调方法。showDialog()方法接受一个整型参数，用来标识要显示的特定对话框。本例中使用了一个 switch 语句来标识要创建的不同类型的对话框，不过这里只创建了一种类型的对话框。后面的"试一试"将扩展这个示例，创建不同类型的对话框。

要创建一个对话框，需要使用 AlertDialog 类的 Builder 构造函数来设置不同的属性，例如图标、标题、按钮以及复选框：

```java
@Override
protected Dialog onCreateDialog(int id) {
    switch (id) {
    case 0:
        return new AlertDialog.Builder(this)
            .setIcon(R.drawable.ic_launcher)
            .setTitle("This is a dialog with some simple text...")
            .setPositiveButton("OK",
                new DialogInterface.OnClickListener() {
                    public void onClick(DialogInterface dialog, int
                    whichButton)
                    {
                        Toast.makeText(getBaseContext(),
                            "OK clicked!", Toast.LENGTH_SHORT).show();
                    }
                }
            )
            .setNegativeButton("Cancel",
                new DialogInterface.OnClickListener() {
                    public void onClick(DialogInterface dialog, int
                    whichButton)
                    {
                        Toast.makeText(getBaseContext(),
                            "Cancel clicked!", Toast.LENGTH_SHORT).show();
                    }
                }
            )
            .setMultiChoiceItems(items, itemsChecked,
                new DialogInterface.OnMultiChoiceClickListener() {
                    public void onClick(DialogInterface dialog,
                    int which, boolean isChecked) {
                        Toast.makeText(getBaseContext(),
                            items[which] + (isChecked ? " checked!":"
                            unchecked!"),
                            Toast.LENGTH_SHORT).show();
                    }
                }
            ).create();

    }
    return null;
}
```

以上代码使用 setPositiveButton()和 setNegativeButton()方法分别设置了两个按钮：OK 和 Cancel。还可以通过 setMultiChoiceItems()方法设置一个复选框列表供用户选择。对于 setMultiChoiceItems()方法，需要传入两个数组：一个是要显示的项列表；另一个包含了表明每个项是否被选中的值。当选中一个项时，使用 Toast 类来显示一条信息，表明哪一项被选中。

前面创建对话框的代码看起来有些复杂，但是很容易把它们重写为如下所示的代码：

```java
package net.learn2develop.Dialog;

import android.app.Activity;
import android.app.AlertDialog;
import android.app.AlertDialog.Builder;
import android.app.Dialog;
import android.content.DialogInterface;
import android.os.Bundle;
import android.view.View;
import android.widget.Toast;

public class DialogActivity extends Activity {
    CharSequence[] items = { "Google", "Apple", "Microsoft" };
    boolean[] itemsChecked = new boolean [items.length];

    /** Called when the activity is first created. */
    @Override
    public void onCreate(Bundle savedInstanceState) {
        super.onCreate(savedInstanceState);
        setContentView(R.layout.main);
    }

    public void onClick(View v) {
        showDialog(0);
    }

    @Override
    protected Dialog onCreateDialog(int id) {
        switch (id) {
        case 0:
            Builder builder = new AlertDialog.Builder(this);
            builder.setIcon(R.drawable.ic_launcher);
            builder.setTitle("This is a dialog with some simple text...");
            builder.setPositiveButton("OK",
                new DialogInterface.OnClickListener() {
                    public void onClick(DialogInterface dialog,   int whichButton) {
                        Toast.makeText(getBaseContext(),
                            "OK clicked!",
                             Toast.LENGTH_SHORT).show();
                    }
                }
            );

            builder.setNegativeButton("Cancel",
                new DialogInterface.OnClickListener() {
                    public void onClick(DialogInterface dialog, int whichButton) {
                        Toast.makeText(getBaseContext(),
                            "Cancel clicked!", Toast.LENGTH_SHORT).
```

```
                    show();
                }
            }
        );

        builder.setMultiChoiceItems(items, itemsChecked,
            new DialogInterface.OnMultiChoiceClickListener() {
                public void onClick(DialogInterface dialog,
                    int which, boolean isChecked) {
                    Toast.makeText(getBaseContext(),
                        items[which] + (isChecked ? " checked!":"
                        unchecked!"),
                        Toast.LENGTH_SHORT).show();
                }
            }
        );
        return builder.create();
    }
    return null;
}
```

上下文对象

在 Android 中常常会遇到 Context 类和其实例。Context 类的实例常用来给应用程序提供引用。例如，在本示例中，Toast 类的第一个参数接受一个 Context 对象。

```
.setPositiveButton("OK",
    new DialogInterface.OnClickListener() {
        public void onClick(DialogInterface dialog, int whichButton)
        {
            Toast.makeText(getBaseContext(),
                "OK clicked!", Toast.LENGTH_SHORT).show();
        }
    }
```

但是，由于 Toast() 类并没有在活动中直接使用(而是在 AlertDialog 类中使用)，因此需要使用 getBaseContext() 方法返回一个 Context 类的实例。

另一个会遇到 Context 类的地方是当在一个活动中动态创建视图时。例如，您可能想通过代码动态创建一个 TextView 视图。因此，使用如下语句实例化 TextView 类：

```
TextView tv = new TextView(this);
```

TextView 类的构造函数接受一个 Context 对象，因为 Activity 类是 Context 类的子类，因此可以使用 this 关键字来代表这个 Context 对象。

2.1.4 显示进度对话框

Android 设备的另外一个常见的用户界面功能是在应用程序执行长时间运行的任务时

显示的 Please wait 对话框。例如，应用程序可能需要在登录到服务器以后才能让用户使用，或者需要在执行计算后才能显示结果给用户。在这类情况中，显示"进度对话框"很有帮助，这样用户可以知道操作正在进行中。

下面的"试一试"将告诉您如何显示一个进度对话框。

试一试　　显示一个进度(Please Wait)对话框

(1) 使用前一节创建的同一个项目，在 main.xml 文件中添加下列粗体显示的语句：

```xml
<?xml version="1.0" encoding="utf-8"?>
<LinearLayout
xmlns:android="http://schemas.android.com/apk/res/android"
    android:layout_width="fill_parent"
    android:layout_height="fill_parent"
    android:orientation="vertical" >

<Button
    android:id="@+id/btn_dialog"
    android:layout_width="fill_parent"
    android:layout_height="wrap_content"
    android:text="Click to display a dialog"
    android:onClick="onClick" />

<Button
    android:id="@+id/btn_dialog2"
    android:layout_width="fill_parent"
    android:layout_height="wrap_content"
    android:text="Click to display a progress dialog"
    android:onClick="onClick2" />

</LinearLayout>
```

(2) 在 DialogActivity.java 文件中添加下列粗体显示的语句：

```java
package net.learn2develop.Dialog;

import android.app.Activity;
import android.app.AlertDialog;
import android.app.AlertDialog.Builder;
import android.app.Dialog;
import android.app.ProgressDialog;
import android.content.DialogInterface;
import android.os.Bundle;
import android.view.View;
import android.widget.Toast;

public class DialogActivity extends Activity {
    CharSequence[] items = { "Google", "Apple", "Microsoft" };
    boolean[] itemsChecked = new boolean [items.length];
```

```
/** Called when the activity is first created. */
@Override
public void onCreate(Bundle savedInstanceState) {
    super.onCreate(savedInstanceState);
    setContentView(R.layout.main);
}

public void onClick(View v) {
    showDialog(0);
}

public void onClick2(View v) {
    //---show the dialog---
    final ProgressDialog dialog = ProgressDialog.show(
        this, "Doing something", "Please wait...", true);
    new Thread(new Runnable(){
        public void run(){
            try {
                //---simulate doing something lengthy---
                Thread.sleep(5000);
                //---dismiss the dialog---
                dialog.dismiss();
            } catch (InterruptedException e) {
                e.printStackTrace();
            }
        }
    }).start();
}

@Override
protected Dialog onCreateDialog(int id)
{ ... }

}
```

(3) 按 F11 键在 Android 模拟器上对应用程序进行调试。单击第二个按钮会显示进度对话框，如图 2-6 所示。5 秒钟后它会消失。

示例说明

为了创建一个进度对话框，首先要创建一个 ProgressDialog 类的实例并调用其 show()方法：

图 2-6

```
//---show the dialog---
final ProgressDialog dialog = ProgressDialog.show(
    this, "Doing something", "Please wait...", true);
```

这时会显示刚才看到的进度对话框。这是一个模态对话框，所以会阻塞用户界面，直到关闭它为止。为在后台执行长时间运行的任务，使用了一个 Runnable 代码块创建了一个

Thread 线程(第 11 章将详细介绍线程)。run()方法中的代码将在一个单独的线程中执行，本例中在 run()方法中使用 sleep()方法插入延迟，模拟了一个执行 5 秒钟的操作：

```
new Thread(new Runnable(){
    public void run(){
        try {
            //---simulate doing something lengthy---
            Thread.sleep(5000);
            //---dismiss the dialog---
            dialog.dismiss();
        } catch (InterruptedException e) {
            e.printStackTrace();
        }
    }
}).start();
```

5 秒钟过后，调用 dismiss()方法关闭进度对话框。

2.1.5　显示更复杂的进度对话框

除了前一节创建的一般性的 please wait 对话框以外，还可以创建一个显示操作进度(如下载状态)的对话框。

下面的"试一试"示范了如何显示一个有特定用途的进度对话框。

试一试　显示操作的进度

(1) 使用前一节创建的同一个项目，在 main.xml 文件中添加下面的粗体代码：

```
<?xml version="1.0" encoding="utf-8"?>
<LinearLayout xmlns:android="http://schemas.android.com/apk/res/android"
    android:layout_width="fill_parent"
    android:layout_height="fill_parent"
    android:orientation="vertical" >

<Button
    android:id="@+id/btn_dialog"
    android:layout_width="fill_parent"
    android:layout_height="wrap_content"
    android:text="Click to display a dialog"
    android:onClick="onClick" />

<Button
    android:id="@+id/btn_dialog2"
    android:layout_width="fill_parent"
    android:layout_height="wrap_content"
    android:text="Click to display a progress dialog"
    android:onClick="onClick2" />

<Button
    android:id="@+id/btn_dialog3"
    android:layout_width="fill_parent"
```

```xml
            android:layout_height="wrap_content"
            android:text="Click to display a detailed progress dialog"
            android:onClick="onClick3" />

</LinearLayout>
```

(2) 在 DialogActivity.java 文件中添加下面的粗体代码：

```java
package net.learn2develop.Dialog;

import android.app.Activity;
import android.app.AlertDialog;
import android.app.AlertDialog.Builder;
import android.app.Dialog;
import android.app.ProgressDialog;
import android.content.DialogInterface;
import android.os.Bundle;
import android.view.View;
import android.widget.Toast;

public class DialogActivity extends Activity {
    CharSequence[] items = { "Google", "Apple", "Microsoft" };
    boolean[] itemsChecked = new boolean [items.length];

    ProgressDialog progressDialog;

    /** Called when the activity is first created. */
    @Override
    public void onCreate(Bundle savedInstanceState) { ... }

    public void onClick(View v) { ... }

    public void onClick2(View v) { ... }

    public void onClick3(View v) {
        showDialog(1);
        progressDialog.setProgress(0);

        new Thread(new Runnable(){
            public void run(){
                for (int i=1; i<=15; i++) {
                    try {
                        //---simulate doing something lengthy---
                        Thread.sleep(1000);
                        //---update the dialog---
                        progressDialog.incrementProgressBy((int)(100/15));
                    } catch (InterruptedException e) {
                        e.printStackTrace();
                    }
                }
                progressDialog.dismiss();
            }
        }).start();
```

```
    }

    @Override
    protected Dialog onCreateDialog(int id) {
        switch (id) {
        case 0:
            return new AlertDialog.Builder(this)
                //...
            ).create();

        case 1:
            progressDialog = new ProgressDialog(this);
            progressDialog.setIcon(R.drawable.ic_launcher);
            progressDialog.setTitle("Downloading files...");
            progressDialog.setProgressStyle(ProgressDialog.STYLE_HORIZONTAL);
            progressDialog.setButton(DialogInterface.BUTTON_POSITIVE, "OK",
                new DialogInterface.OnClickListener() {
                public void onClick(DialogInterface dialog,
                int whichButton)
                {
                    Toast.makeText(getBaseContext(),
                        "OK clicked!", Toast.LENGTH_SHORT).show();
                }
            });
            progressDialog.setButton(DialogInterface.BUTTON_NEGATIVE,
            "Cancel",
                new DialogInterface.OnClickListener() {
                    public void onClick(DialogInterface dialog,
                    int whichButton)
                    {
                        Toast.makeText(getBaseContext(),
                                "Cancel clicked!", Toast.LENGTH_SHORT).
                                show();
                    }
            });
            return progressDialog;
        }

        return null;
    }
}
```

(3) 按 F11 键在 Android 模拟器上调试应用程序。单击第三个按钮显示进度对话框, 如图 2-7 所示。注意进度条将会逐渐增加到 100%。

示例说明

为创建一个对话框来显示操作进度, 首先创建 ProgressDialog 类的一个实例, 并设置其各个属性, 如图

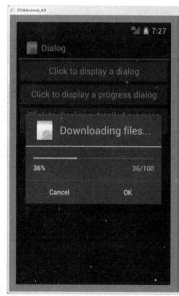

图 2-7

标、标题和样式:

```
progressDialog = new ProgressDialog(this);
progressDialog.setIcon(R.drawable.ic_launcher);
progressDialog.setTitle("Downloading files...");
progressDialog.setProgressStyle(ProgressDialog.STYLE_
    HORIZONTAL);
```

然后设置在进度对话框中显示的两个按钮。

```
progressDialog.setButton(DialogInterface.BUTTON_POSITIVE, "OK",
    new DialogInterface.
    OnClickListener() {
        public void onClick
        (DialogInterface dialog,
        int whichButton)
        {
            Toast.makeText(getBaseContext(),
                "OK clicked!", Toast.LENGTH_SHORT).show();
        }
});
progressDialog.setButton(DialogInterface.BUTTON_NEGATIVE, "Cancel",
    new DialogInterface.OnClickListener() {
        public void onClick(DialogInterface dialog,
            int whichButton)
        {
            Toast.makeText(getBaseContext(),
                "Cancel clicked!", Toast.LENGTH_SHORT).show();
        }
});
return progressDialog;
```

前面的代码段会显示一个进度对话框,如图 2-8 所示。

要在进度对话框中显示进度状态,可以使用 Thread 对象来运行一个 Runnable 代码块:

图 2-8

```
progressDialog.setProgress(0);

new Thread(new Runnable(){
    public void run(){
        for (int i=1; i<=15; i++) {
            try {
                //---simulate doing something lengthy---
                Thread.sleep(1000);
                //---update the dialog---
                progressDialog.incrementProgressBy((int)(100/15));
            } catch (InterruptedException e) {
                e.printStackTrace();
            }
        }
```

```
            progressDialog.dismiss();
        }
    }).start();
```

这段代码中让进度从 1 增加到 15，每隔一秒钟增加到下一个数字。incrementProgressBy()
方法增加进度对话框中的计数。当进度对话框到达 100%后，就关闭它。

2.2 使用意图链接活动

一个 Android 应用程序可以包含零或多个活动。当应用程序具有多个活动时，您可能
需要从一个活动导航到另一个活动。在 Android 中，活动之间的导航通过意图来完成。

要理解 Android 中这个非常重要又有些抽象的概念，最好的办法就是亲自去体验一下，
看看它能够实现什么。下面的"试一试"将告诉您如何在一个已有的项目中添加另一个活
动，并在这两个活动之间实现导航。

试一试　　使用意图链接活动

UsingIntent.zip 代码文件可以在 Wrox.com 上下载

(1) 使用 Eclipse 创建一个新的 Android 项目，命名为 UsingIntent。
(2) 右击 src 文件夹下的包名，依次选择 New | Class 命令(如图 2-9 所示)。

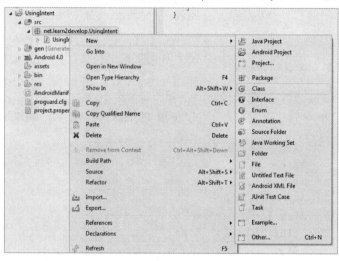

图 2-9

(3) 将新的类文件命名为 SecondActivity，单击 Finish 按钮。
(4) 将下列粗体显示的语句添加到 AndroidManifest.xml 文件中：

```
<?xml version="1.0" encoding="utf-8"?>
<manifest xmlns:android="http://schemas.android.com/apk/res/android"
    package="net.learn2develop.UsingIntent"
```

```xml
    android:versionCode="1"
    android:versionName="1.0" >

    <uses-sdk android:minSdkVersion="14" />

    <application
        android:icon="@drawable/ic_launcher"
        android:label="@string/app_name" >
        <activity
            android:label="@string/app_name"
            android:name=".UsingIntentActivity" >
            <intent-filter >
                <action android:name="android.intent.action.MAIN" />
                <category android:name="android.intent.category.
                    LAUNCHER" />
            </intent-filter>
        </activity>
        <activity
            android:label="Second Activity"
            android:name=".SecondActivity" >
            <intent-filter >
                <action android:name="net.learn2develop.SecondActivity" />
                <category android:name="android.intent.category.DEFAULT" />
            </intent-filter>
        </activity>
    </application>

</manifest>
```

(5) 右击 res/layout 文件夹中的 main.xml 文件并选择 Copy 命令创建一个副本。然后右击 res/layout 文件夹并选择 Paste。将副本文件命名为 secondactivity.xml。现在，res/layout 文件夹下就包含了 secondactivity.xml 文件(如图 2-10 所示)。

(6) 按如下所示修改 secondactivity.xml 文件：

图 2-10

```xml
<?xml version="1.0" encoding="utf-8"?>
<LinearLayout xmlns:android="http://schemas.android.com/apk/res/android"
    android:layout_width="fill_parent"
    android:layout_height="fill_parent"
    android:orientation="vertical" >

    <TextView
        android:layout_width="fill_parent"
        android:layout_height="wrap_content"
        android:text="This is the Second Activity!" />

</LinearLayout>
```

(7) 将下列粗体显示的语句添加到 SecondActivity.java 文件中：

```
package net.learn2develop.UsingIntent;
import android.app.Activity;
import android.os.Bundle;

public class SecondActivity extends Activity{
    @Override
    public void onCreate(Bundle savedInstanceState) {
        super.onCreate(savedInstanceState);
        setContentView(R.layout.secondactivity);
    }
}
```

(8) 将下列粗体显示的语句添加到 main.xml 文件中：

```
<?xml version="1.0" encoding="utf-8"?>
<LinearLayout xmlns:android="http://schemas.android.com/apk/res/android"
    android:layout_width="fill_parent"
    android:layout_height="fill_parent"
    android:orientation="vertical" >

<Button
    android:layout_width="fill_parent"
    android:layout_height="wrap_content"
    android:text="Display second activity"
    android:onClick="onClick"/>

</LinearLayout>
```

(9) 按下列粗体字内容修改 UsingIntentActivity.java 文件：

```
package net.learn2develop.UsingIntent;

import android.app.Activity;
import android.content.Intent;
import android.os.Bundle;
import android.view.View;

public class UsingIntentActivity extends Activity {
    /** Called when the activity is first created. */
    @Override
    public void onCreate(Bundle savedInstanceState) {
        super.onCreate(savedInstanceState);
        setContentView(R.layout.main);
    }

    public void onClick(View view) {
        startActivity(new Intent("net.learn2develop.SecondActivity"));
    }
}
```

(10) 按 F11 键在 Android 模拟器上调试应用程序。当第一个活动被加载时，单击按钮，第二个活动将开始加载，如图 2-11 所示。

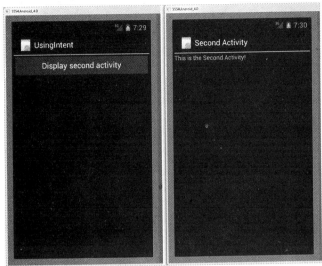

图 2-11

示例说明

正如已经学习过的，一个活动是由一个用户界面组件(例如，main.xml)和一个类组件(例如，UsingIntentActivity.java)组成的。因此，如果想向项目中添加另外的活动，需要创建这两个组件。

在 AndroidManifest.xml 文件中，专门添加了以下内容：

```
<activity
    android:label=" Second Activity"
    android:name=".SecondActivity" >
    <intent-filter >
        <action android:name="net.learn2develop.SecondActivity" />
        <category android:name="android.intent.category.DEFAULT" />
    </intent-filter>
</activity>
```

到这里，您已经给应用程序添加了一个新的活动。注意以下事项：

- 添加的新活动的名称(类)是 SecondActivity。
- 新活动显示的标签名称为 Second Activity。
- 新活动的意图筛选器的名称是 net.learn2develop.SecondActivity。其他活动将通过这个名称来调用这个活动。理想的情况下，您应该使用您公司的反向域名作为意图筛选器的名称，以减少同另一个应用程序具有相同意图筛选器名称的可能性(下一节将讨论当两个或更多个活动具有相同的意图筛选器时会发生什么)。
- 意图筛选器的类别是 android.intent.category.DEFAULT。您需要将类别添加给意图筛选器，使其他活动可以通过使用 startActivity()方法启动这个活动(稍候将作进一步介绍)。

单击按钮时,使用 startActivity()方法来显示 SecondActivity,可以通过创建一个 Intent 类的实例并将 SecondActivity 的意图筛选器名称(即 net.learn2develop.SecondActivity)传递给这个实例来完成:

```java
public void onClick(View view) {
    startActivity(new Intent("net.learn2develop.SecondActivity"));
}
```

Android 中的活动可以被设备上运行的任意应用程序调用。例如,可以创建一个新的 Android 项目,然后使用 SecondActivity 的意图筛选器 net.learn2develop.SecondActivity 来显示 SecondActivity。使一个应用程序容易地调用其他应用程序是 Android 中的基本概念之一。

如果要调用的活动是定义在同一个项目之中的,可以像下面这样重写先前的语句:

```java
startActivity(new Intent(this, SecondActivity.class));
```

不过,只有当要显示的活动与当前活动在同一个项目中时,这种方法才是适用的。

2.2.1 解决意图筛选器的冲突

在前一节中,我们知道<intent-filter>元素可以定义一个活动被另一个活动调用的方法。其他活动(在相同或一个单独的应用程序中)如果具有相同的筛选器名称会发生什么呢?例如,假设应用程序有另外一个名为 Activity3 的活动,在 AndroidManifest.xml 文件中具有以下入口:

```xml
<?xml version="1.0" encoding="utf-8"?>
<manifest xmlns:android="http://schemas.android.com/apk/res/android"
    package="net.learn2develop.UsingIntent"
    android:versionCode="1"
    android:versionName="1.0" >

    <uses-sdk android:minSdkVersion="14" />

    <application
        android:icon="@drawable/ic_launcher"
        android:label="@string/app_name" >
        <activity
            android:label="@string/app_name"
            android:name=".UsingIntentActivity" >
            <intent-filter >
                <action android:name="android.intent.action.MAIN" />
                <category android:name="android.intent.category.LAUNCHER" />
            </intent-filter>
        </activity>
        <activity
            android:label="Second Activity"
            android:name=".SecondActivity" >
            <intent-filter >
                <action android:name="net.learn2develop.SecondActivity" />
```

```xml
            <category android:name="android.intent.category.DEFAULT" />
        </intent-filter>
    </activity>

    <activity
        android:label="Third Activity"
        android:name=".ThirdActivity" >
        <intent-filter >
            <action android:name="net.learn2develop.SecondActivity" />
            <category android:name="android.intent.category.DEFAULT" />
        </intent-filter>
    </activity>

</application>

</manifest>
```

如果调用带有下列意图的 startActivity()方法，Android 操作系统会显示一个选择，如图 2-12 所示：

```
startActivity(new Intent("net.learn2develop.SecondActivity"));
```

如果您选中了 Use by default for this action 项来选择一个活动，那么下一回将再一次调用意图 net.learn2develop.SecondActivity，它将总是启动先前您选择过的活动。

为了清除这个默认值，转到 Android 中的 Settings 应用程序，依次选择 Apps | Manage applications 命令，选择应用程序名称(如图 2-13 所示)。当应用程序的详细信息显示出来时，将屏幕滚动到底部并单击 Clear defaults 按钮。

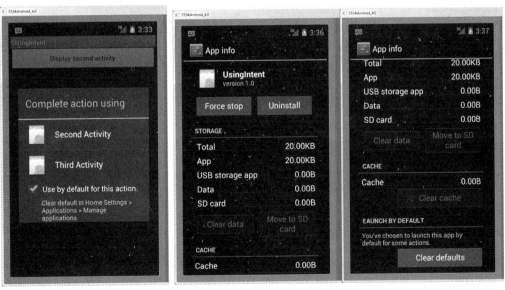

图 2-12   图 2-13

2.2.2 从意图返回结果

startActivity()方法调用另一个活动,但并没有返回结果给当前活动。例如,您可能有一个提示用户输入用户名和密码的活动。用户在这个活动中输入的信息需要回传给调用它的活动来作进一步的处理。如果需要从一个活动中回传数据,应该使用 startActivityForResult()方法。下面的"试一试"演示了这一过程。

试一试　　**从一个活动获得结果**

(1) 使用前一节创建的同一个项目,在 secondactivity.xml 文件中添加下列粗体显示的语句:

```xml
<?xml version="1.0" encoding="utf-8"?>
<LinearLayout xmlns:android="http://schemas.android.com/apk/res/android"
    android:layout_width="fill_parent"
    android:layout_height="fill_parent"
    android:orientation="vertical" >

<TextView
    android:layout_width="fill_parent"
    android:layout_height="wrap_content"
    android:text="This is the Second Activity!" />

<TextView
    android:layout_width="fill_parent"
    android:layout_height="wrap_content"
    android:text="Please enter your name" />

<EditText
    android:id="@+id/txt_username"
    android:layout_width="fill_parent"
    android:layout_height="wrap_content" />

<Button
    android:id="@+id/btn_OK"
    android:layout_width="fill_parent"
    android:layout_height="wrap_content"
    android:text="OK"
    android:onClick="onClick"/>

</LinearLayout>
```

(2) 将下列粗体显示的语句添加到 SecondActivity.java 文件中:

```java
package net.learn2develop.UsingIntent;

import android.app.Activity;
import android.content.Intent;
import android.net.Uri;
import android.os.Bundle;
```

```
import android.view.View;
import android.widget.EditText;

public class SecondActivity extends Activity{
    @Override
    public void onCreate(Bundle savedInstanceState) {
        super.onCreate(savedInstanceState);
        setContentView(R.layout.secondactivity);
    }

    public void onClick(View view) {
        Intent data = new Intent();

        //---get the EditText view---
        EditText txt_username =
            (EditText) findViewById(R.id.txt_username);

        //---set the data to pass back---
        data.setData(Uri.parse(
            txt_username.getText().toString()));
        setResult(RESULT_OK, data);

        //---closes the activity---
        finish();
    }
}
```

(3) 将下列粗体显示的语句添加到 UsingIntentActivity.java 文件中：

```
package net.learn2develop.UsingIntent;

import android.app.Activity;
import android.content.Intent;
import android.os.Bundle;
import android.view.View;
import android.widget.Toast;

public class UsingIntentActivity extends Activity {
    int request_Code = 1;

    /** Called when the activity is first created. */
    @Override
    public void onCreate(Bundle savedInstanceState) {
        super.onCreate(savedInstanceState);
        setContentView(R.layout.main);
    }

    public void onClick(View view) {
        //startActivity(new Intent("net.learn2develop.SecondActivity"));
        //or
        //startActivity(new Intent(this, SecondActivity.class));
```

```
        startActivityForResult(new Intent(
            "net.learn2develop.SecondActivity"),
            request_Code);
    }

    public void onActivityResult(int requestCode, int resultCode, Intent
    data)
    {
        if (requestCode == request_Code) {
            if (resultCode == RESULT_OK) {
                Toast.makeText(this,data.getData().toString(),
                    Toast.LENGTH_SHORT).show();
            }
        }
    }
```

(4) 按 F11 键在 Android 模拟器上调试应用程序。当加载第一个活动时，单击按钮，SecondActivity 将被加载。输入您的姓名(如图 2-14 所示)并单击 OK 按钮。这时，第一个活动会显示出您用 Toast 类输入的名字。

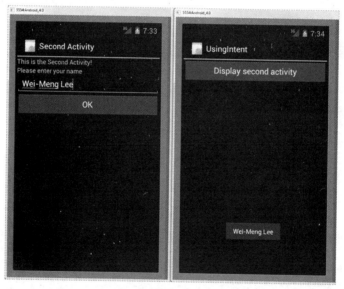

图 2-14

示例说明

调用一个活动并等待从此活动返回结果，需要使用 startActivityForResult()方法，如下所示：

```
startActivityForResult(new Intent(
    "net.learn2develop.SecondActivity"),
    request_Code);
```

除了传入一个 Intent 对象，还需要传入请求码。请求码仅仅是一个整数值，用来标识正在调用的活动。这是必须的，因为当一个活动返回一个值时，必须有办法将它标识出来。例如，您可能同时在调用多个活动，而一些活动可能没有立即返回(如正在等待服务器的应答)。当一个活动返回时，需要这个请求码来确定实际返回的是哪一个活动。

 注意：如果请求码设为-1，则使用 startActivityForResult()方法来调用活动与使用 startActivity()方法来调用是等同的。也就是说，没有结果返回。

为了使被调活动可以返回一个值给调用它的活动，可以通过 setData()方法使用一个 Intent 对象来回传数据：

```
Intent data = new Intent();

//---get the EditText view---
EditText txt_username =
    (EditText) findViewById(R.id.txt_username);

//---set the data to pass back---
data.setData(Uri.parse(
    txt_username.getText().toString()));
setResult(RESULT_OK, data);

//---closes the activity---
finish();
```

setResult()方法设置了一个结果码(RESULT_OK 或是 RESULT_CANCELLED)和回传给调用活动的数据(一个 Intent 对象)。finish()方法关闭活动并将控制返回给调用者活动。

在调用者活动中，需要实现 onActivityResult()方法，一个活动无论何时返回都要调用这个方法：

```
public void onActivityResult(int requestCode, int resultCode,
Intent data)
{
    if (requestCode == request_Code) {
        if(resultCode == RESULT_OK) {
            Toast.makeText(this,data.getData().toString(),
                Toast.LENGTH_SHORT).show();
        }
    }
}
```

这里，检验请求和结果码的正确性，并显示返回的结果。返回的结果通过 data 参数传入，并且通过 getData()方法来获得数据的细节。

2.2.3 使用意图对象传递数据

除了从活动返回数据外，也经常需要传递数据给活动。例如，在前面的示例中，您可能想在活动显示之前在 EditText 视图中设置一些默认文本。对此，可以使用 Intent 对象将这些数据传递给目标活动。

下面的"试一试"将展示在活动之间传递数据的各种方法：

试一试　　**将数据传递给目标活动**

(1) 使用 Eclipse 创建一个新的 Android 项目，命名为 PassingData。

(2) 将下列粗体显示的代码添加到 main.xml 文件中：

```xml
<?xml version="1.0" encoding="utf-8"?>
<LinearLayout xmlns:android="http://schemas.android.com/apk/res/android"
    android:layout_width="fill_parent"
    android:layout_height="fill_parent"
    android:orientation="vertical" >

<Button
    android:id="@+id/btn_SecondActivity"
    android:layout_width="fill_parent"
    android:layout_height="wrap_content"
    android:text="Click to go to Second Activity"
    android:onClick="onClick"/>

</LinearLayout>
```

(3) 在 res/layout 文件夹中添加一个新的 XML 文件，命名为 secondactivity.xml，并在文件中添加下面的代码：

```xml
<?xml version="1.0" encoding="utf-8"?>
<LinearLayout xmlns:android="http://schemas.android.com/apk/res/android"
    android:layout_width="fill_parent"
    android:layout_height="fill_parent"
    android:orientation="vertical" >

<TextView
    android:layout_width="fill_parent"
    android:layout_height="wrap_content"
    android:text="Welcome to Second Activity" />

<Button
    android:id="@+id/btn_MainActivity"
    android:layout_width="fill_parent"
    android:layout_height="wrap_content"
    android:text="Click to return to main activity"
    android:onClick="onClick"/>

</LinearLayout>
```

(4) 在包中添加一个新的 Class 文件，命名为 SecondActivity。在 SecondActivity.java 文件中添加下面的代码：

```java
package net.learn2develop.PassingData;

import android.app.Activity;
import android.content.Intent;
import android.net.Uri;
import android.os.Bundle;
import android.view.View;
import android.widget.Toast;

public class SecondActivity extends Activity {
    @Override
    public void onCreate(Bundle savedInstanceState) {
        super.onCreate(savedInstanceState);
        setContentView(R.layout.secondactivity);

        //---get the data passed in using getStringExtra()---
        Toast.makeText(this,getIntent().getStringExtra("str1"),
            Toast.LENGTH_SHORT).show();

        //---get the data passed in using getIntExtra()---
        Toast.makeText(this,Integer.toString(
            getIntent().getIntExtra("age1", 0)),
            Toast.LENGTH_SHORT).show();

        //---get the Bundle object passed in---
        Bundle bundle = getIntent().getExtras();

        //---get the data using the getString()---
        Toast.makeText(this, bundle.getString("str2"),
            Toast.LENGTH_SHORT).show();

        //---get the data using the getInt() method---
        Toast.makeText(this,Integer.toString(bundle.getInt("age2")),
            Toast.LENGTH_SHORT).show();
    }

    public void onClick(View view) {
        //---use an Intent object to return data---
        Intent i = new Intent();

        //---use the putExtra() method to return some
        // value---
        i.putExtra("age3", 45);

        //---use the setData() method to return some value---
        i.setData(Uri.parse(
            "Something passed back to main activity"));

        //---set the result with OK and the Intent object---
        setResult(RESULT_OK, i);
```

```
            //---destroy the current activity---
            finish();
        }
    }
}
```

(5) 在 AndroidManifest.xml 文件中添加下列粗体显示的代码:

```xml
<?xml version="1.0" encoding="utf-8"?>
<manifest xmlns:android="http://schemas.android.com/apk/res/android"
    package="net.learn2develop.PassingData"
    android:versionCode="1"
    android:versionName="1.0" >

    <uses-sdk android:minSdkVersion="14" />

    <application
        android:icon="@drawable/ic_launcher"
        android:label="@string/app_name" >
        <activity
            android:label="@string/app_name"
            android:name=".PassingDataActivity" >
            <intent-filter >
                <action android:name="android.intent.action.MAIN" />
                <category android:name="android.intent.category.LAUNCHER" />
            </intent-filter>
        </activity>
        <activity
            android:label="Second Activity"
            android:name=".SecondActivity" >
            <intent-filter >
                <action android:name="net.learn2develop.PassingData
                 SecondActivity" />
                <category android:name="android.intent.category.DEFAULT" />
            </intent-filter>
        </activity>
    </application>

</manifest>
```

(6) 在 PassingDataActivity.java 文件中添加下列粗体显示的代码:

```java
package net.learn2develop.PassingData;

import android.app.Activity;
import android.content.Intent;
import android.os.Bundle;
import android.view.View;
import android.widget.Toast;
```

```java
public class PassingDataActivity extends Activity {
    /** Called when the activity is first created. */
    @Override
    public void onCreate(Bundle savedInstanceState) {
        super.onCreate(savedInstanceState);
        setContentView(R.layout.main);
    }

    public void onClick(View view) {
        Intent i = new
                Intent("net.learn2develop.PassingDataSecondActivity");
        //---use putExtra() to add new name/value pairs---
        i.putExtra("str1", "This is a string");
        i.putExtra("age1", 25);

        //---use a Bundle object to add new name/values
        // pairs---
        Bundle extras = new Bundle();
        extras.putString("str2", "This is another string");
        extras.putInt("age2", 35);

        //---attach the Bundle object to the Intent object---
        i.putExtras(extras);

        //---start the activity to get a result back---
        startActivityForResult(i, 1);
    }

    public void onActivityResult(int requestCode,
    int resultCode, Intent data)
    {
        //---check if the request code is 1---
        if (requestCode == 1) {

            //---if the result is OK---
            if (resultCode == RESULT_OK) {

                //---get the result using getIntExtra()---
                Toast.makeText(this, Integer.toString(
                    data.getIntExtra("age3", 0)),
                    Toast.LENGTH_SHORT).show();

                //---get the result using getData()---
                Toast.makeText(this, data.getData().toString(),
                    Toast.LENGTH_SHORT).show();
            }
        }
    }

}
```

(7) 按 F11 键在 Android 模拟器上调试应用程序。单击每个活动的按钮，观察显示的值。

示例说明

虽然这种应用程序并不华丽，但是确实可以演示在活动之间传递数据的一些重要的方法。首先，可以使用 Intent 对象的 putExtra()方法添加一个键/值对：

```
//---use putExtra() to add new name/value pairs---
i.putExtra("str1", "This is a string");
i.putExtra("age1", 25);
```

前面的语句向 Intent 对象添加了两个键/值对：一个是 String 类型的；另一个是 integer 类型的。

除了使用 putExtra()方法，还可以创建一个 Bundle 对象，并使用 putExtras()方法将 Bundle 对象添加给 Intent 对象。可以把 Bundle 对象看做一个包含一组键/值对的字典对象。下面的语句创建了一个 Bundle 对象，然后向其添加了两个键/值对。然后把 Bundle 对象添加给 Intent 对象：

```
//---use a Bundle object to add new name/values pairs---
Bundle extras = new Bundle();
extras.putString("str2", "This is another string");
extras.putInt("age2", 35);

//---attach the Bundle object to the Intent object---
i.putExtras(extras);
```

在第二个活动中，为了获得使用 Intent 对象发送的数据，首先使用 getIntent()方法来获取该 Intent 对象，然后调用该对象的 getStringExtra()方法来获得使用 putExtra()方法设置的字符串值：

```
//---get the data passed in using getStringExtra()---
Toast.makeText(this,getIntent().getStringExtra("str1"),
    Toast.LENGTH_SHORT).show();
```

本例中，需要根据所设置数据的类型，调用合适的方法来提取键/值对。对于整数值，使用 getIntExtra()方法(如果在指定的名称中没有存储值，第二个参数会使用默认值)：

```
//---get the data passed in using getIntExtra()---
Toast.makeText(this,Integer.toString(
    getIntent().getIntExtra("age1", 0)),
    Toast.LENGTH_SHORT).show();
```

为了获取 Bundle 对象，需要使用 getExtras()方法：

```
//---get the Bundle object passed in---
Bundle bundle = getIntent().getExtras();
```

为了获得单独的键/值对，需要使用合适的方法。对于字符串值，使用 getString()方法：

```
//---get the data using the getString()---
   Toast.makeText(this, bundle.getString("str2"),
           Toast.LENGTH_SHORT).show();
```

类似地，使用 getInt()方法可以获取整数值：

```
//---get the data using the getInt() method---
   Toast.makeText(this,Integer.toString(bundle.getInt("age2")),
           Toast.LENGTH_SHORT).show();
```

另外一种给活动传递数据的方法是使用 setData()方法(前一节使用了这种方法)，如下所示：

```
//---use the setData() method to return some value---
   i.setData(Uri.parse(
           "Something passed back to main activity"));
```

通常，使用 setData()方法来设置 Intent 对象将会操作的数据(例如传递一个 URL 给 Intent 对象，使其能够调用 Web 浏览器来查看网页；更多示例请参看本章后面的"使用意图调用内置应用程序"一节)。

为获取使用 setData()方法设置的数据，需要使用 getData()方法(在本例中 data 是一个 Intent 对象)：

```
//---get the result using getData()---
   Toast.makeText(this, data.getData().toString(),
       Toast.LENGTH_SHORT).show();
```

2.3 碎片

前一节学习了什么是活动和如何使用活动。在小屏幕设备(例如智能手机)上，活动通常会填满整个屏幕，显示构成应用程序用户界面的各个视图。活动本质上是视图的一个容器。但是，在大屏幕设备(例如平板电脑)上显示活动时，就有些不太合适了。因为屏幕增大了，所以必须排列活动中的所有视图，以便充分利用增加的空间，这导致需要对视图层次做复杂的变动。更好的方法是使用"微活动"，让每个微活动包含自己的一组视图。在运行时，根据持有设备的屏幕方向，一个活动可以包含一个或者多个这样的微活动。在 Android 3.0 及更高版本中，这种微活动被称为"碎片"。

可以把碎片看做另外一种形式的活动。就像活动一样，您创建碎片来包含视图。碎片总是嵌入在活动中。例如，图 2-15 显示了两个碎片。碎片 1 可能包含一个 ListView，显示一个书名列表。碎片 2 可能包含某些 TextView 和 ImageView，显示一些文本和图片。

现在，假设应用程序运行在 Android 平板电脑(或 Android 智能手机)中并处于纵向模式。此时，碎片 1 可能嵌入在一个活动中，而碎片 2 嵌入在另一个活动中，如图 2-16 所示。当用户在碎片 1 的列表中选择一项时，碎片 2 将会启动。

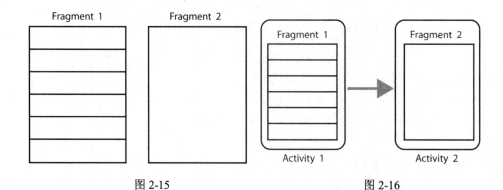

图 2-15   图 2-16

如果应用程序在平板电脑中改为以横向模式显示，两个碎片可以嵌入在一个活动中，如图 2-17 所示。

从上面的讨论中可以看到，碎片为 Android 应用程序用户界面的创建提供了一种灵活的方式。它们可以作为用户界面的基本单元，在活动中动态地添加或移除，从而为目标设备创建最佳的用户体验。

下面的"试一试"演示了碎片的基本用法。

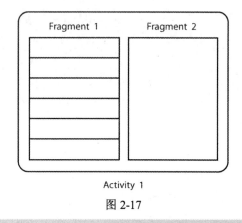

图 2-17

试一试   使用碎片

Fragments.zip 代码文件可以在 Wrox.com 上下载

(1) 使用 Eclipse 创建一个新的 Android 项目，命名为 Fragments。

(2) 在 res/layout 文件夹中，创建一个新文件，命名为 fragment1.xml，并在该文件中添加下面的代码：

```xml
<?xml version="1.0" encoding="utf-8"?>
<LinearLayout
    xmlns:android="http://schemas.android.com/apk/res/android"
    android:orientation="vertical"
    android:layout_width="fill_parent"
    android:layout_height="fill_parent"
    android:background="#00FF00"
    >
<TextView
    android:layout_width="fill_parent"
    android:layout_height="wrap_content"
    android:text="This is fragment #1"
    android:textColor="#000000"
    android:textSize="25sp" />
</LinearLayout>
```

(3) 仍然在 res/layout 文件夹中，创建另外一个新文件，命名为 fragment2.xml，并在该文件中添加下面的代码：

```xml
<?xml version="1.0" encoding="utf-8"?>
<LinearLayout
    xmlns:android="http://schemas.android.com/apk/res/android"
    android:orientation="vertical"
    android:layout_width="fill_parent"
    android:layout_height="fill_parent"
    android:background="#FFFE00"
    >
<TextView
    android:layout_width="fill_parent"
    android:layout_height="wrap_content"
    android:text="This is fragment #2"
    android:textColor="#000000"
    android:textSize="25sp" />
</LinearLayout>
```

(4) 在 main.xml 中添加下面的粗体代码：

```xml
<?xml version="1.0" encoding="utf-8"?>
<LinearLayout xmlns:android="http://schemas.android.com/apk/res/android"
    android:layout_width="fill_parent"
    android:layout_height="fill_parent"
    android:orientation="horizontal" >

    <fragment
        android:name="net.learn2develop.Fragments.Fragment1"
        android:id="@+id/fragment1"
        android:layout_weight="1"
        android:layout_width="0px"
        android:layout_height="match_parent" />
    <fragment
        android:name="net.learn2develop.Fragments.Fragment2"
        android:id="@+id/fragment2"
        android:layout_weight="1"
        android:layout_width="0px"
        android:layout_height="match_parent" />

</LinearLayout>
```

(5) 在 net.learn2develop.Fragments 包名下，添加两个 Java 类文件，分别命名为 Fragment1.java 和 Fragment2.java，如图 2-18 所示。

(6) 在 Fragment1.java 类文件中添加下面的代码：

```java
package net.learn2develop.Fragments;

import android.app.Fragment;
import android.os.Bundle;
```

图 2-18

```
import android.view.LayoutInflater;
import android.view.View;
import android.view.ViewGroup;

public class Fragment1 extends Fragment {
    @Override
    public View onCreateView(LayoutInflater inflater,
    ViewGroup container, Bundle savedInstanceState) {
        //---Inflate the layout for this fragment---
        return inflater.inflate(
            R.layout.fragment1, container, false);
    }
}
```

(7) 在 Fragment2.java 类文件中添加下面的代码:

```
package net.learn2develop.Fragments;

import android.app.Fragment;
import android.os.Bundle;
import android.view.LayoutInflater;
import android.view.View;
import android.view.ViewGroup;

public class Fragment2 extends Fragment {
    @Override
    public View onCreateView(LayoutInflater inflater,
    ViewGroup container, Bundle savedInstanceState) {
        //---Inflate the layout for this fragment---
        return inflater.inflate(
            R.layout.fragment2, container, false);
    }
}
```

(8) 按 F11 键在 Android 模拟器上调试应用程序。图 2-19 显示了活动中包含的两个碎片。

图 2-19

示例说明

碎片的行为与活动十分相似：它有一个 Java 类，并从一个 XML 文件加载 UI。这个 XML 文件中包含了在活动中常见的 UI 元素：TextView、EditText、Button 等。碎片的 Java 类需要继承 Fragment 基类：

```
public class Fragment1 extends Fragment {
}
```

注意：除了 Fragment 基类，碎片还可以继承 Fragment 类的几个子类，例如 Dialogfragment、ListFragment 和 PreferenceFragment。第 4 章将详细讨论这些类型的碎片。

为了绘制碎片的 UI，重写了 onCreateView()方法。该方法需要返回一个 View 对象，如下所示：

```
public View onCreateView(LayoutInflater inflater,
ViewGroup container, Bundle savedInstanceState) {
    //---Inflate the layout for this fragment---
    return inflater.inflate(
    R.layout.fragment1, container, false);
}
```

这里使用一个 LayoutInflater 对象来增大指定 XML 文件(本例中为 R.layout.fragment1)中的 UI。container 参数引用父 ViewGroup，即准备用于嵌入碎片的活动。savedInstanceState 参数允许将碎片还原到前一次保存的状态。

为向活动添加一个碎片，使用了<fragment>元素：

```
<?xml version="1.0" encoding="utf-8"?>
<LinearLayout xmlns:android="http://schemas.android.com/apk/res/android"
    android:layout_width="fill_parent"
    android:layout_height="fill_parent"
    android:orientation="horizontal" >

    <fragment
        android:name="net.learn2develop.Fragments.Fragment1"
        android:id="@+id/fragment1"
        android:layout_weight="1"
        android:layout_width="0px"
        android:layout_height="match_parent" />
    <fragment
        android:name="net.learn2develop.Fragments.Fragment2"
        android:id="@+id/fragment2"
        android:layout_weight="1"
        android:layout_width="0px"
        android:layout_height="match_parent" />
```

```
</LinearLayout>
```

注意每个碎片都需要一个唯一标识符,这可以通过 android:id 或 android:tag 属性进行设置。

2.3.1 动态添加碎片

将 UI 分割为多个可配置的部分是碎片的优势之一,但是其真正的强大之处在于可在运行时动态地把它们添加到活动。前一节看到了在设计时通过修改 XML 文件可以向活动添加碎片。在实际应用中,如果能够创建碎片并在运行时把它们添加到活动会更加有用。这样就可以为应用程序创建可以自定义的用户界面。例如,如果应用程序运行在智能手机上,可能会在一个活动中填充一个碎片;如果应用程序运行中平板电脑上,可能在一个活动中填充两个或更多个碎片,因为平板电脑的屏幕空间比智能手机大得多。

下面的"试一试"演示了如何以编程方式在运行时向活动添加碎片。

试一试　在运行时添加碎片

(1) 使用前一节创建的同一个项目,在 main.xml 文件中注释掉两个<fragment>元素:

```xml
<?xml version="1.0" encoding="utf-8"?>
<LinearLayout xmlns:android="http://schemas.android.com/apk/res/android"
    android:layout_width="fill_parent"
    android:layout_height="fill_parent"
    android:orientation="horizontal" >

    <!--
    <fragment
        android:name="net.learn2develop.Fragments.Fragment1"
        android:id="@+id/fragment1"
        android:layout_weight="1"
        android:layout_width="0px"
        android:layout_height="match_parent" />
    <fragment
        android:name="net.learn2develop.Fragments.Fragment2"
        android:id="@+id/fragment2"
        android:layout_weight="1"
        android:layout_width="0px"
        android:layout_height="match_parent" />
    -->
</LinearLayout>
```

(2) 在 FragmentsActivity.java 文件中添加下面的粗体代码:

```java
package net.learn2develop.Fragments;

import android.app.Activity;
import android.app.FragmentManager;
import android.app.FragmentTransaction;
import android.os.Bundle;
```

```java
import android.view.Display;
import android.view.WindowManager;

public class FragmentsActivity extends Activity {
    /** Called when the activity is first created. */
    @Override
    public void onCreate(Bundle savedInstanceState) {
        super.onCreate(savedInstanceState);

        FragmentManager fragmentManager = getFragmentManager();
        FragmentTransaction fragmentTransaction =
                fragmentManager.beginTransaction();

        //---get the current display info---
        WindowManager wm = getWindowManager();
        Display d = wm.getDefaultDisplay();
        if (d.getWidth() > d.getHeight())
        {
            //---landscape mode---
            Fragment1 fragment1 = new Fragment1();
            // android.R.id.content refers to the content
            // view of the activity
            fragmentTransaction.replace(
                    android.R.id.content, fragment1);
        }
        else
        {
            //---portrait mode---
            Fragment2 fragment2 = new Fragment2();
            fragmentTransaction.replace(
                    android.R.id.content, fragment2);
        }
        fragmentTransaction.commit();
    }
}
```

(3) 按 F11 键在 Android 模拟器上运行应用程序。注意观察当模拟器处于纵向模式时，会显示碎片 2(黄色)，如图 2-20 所示。如果按 Ctrl+F11 组合键将模拟器的方向改为横向，则会显示碎片 1(绿色)，如图 2-21 所示。

示例说明

为向活动添加碎片，使用了 FragmentManager 类。首先获得该类的一个实例：

```
FragmentManager fragmentManager = getFragmentManager();
```

还需要在活动中使用 FragmentTransaction 类来执行碎片操作(例如添加、删除或替换)：

```
FragmentTransaction fragmentTransaction =
        fragmentManager.beginTransaction();
```

图 2-20

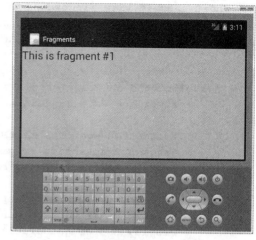
图 2-21

在本例中,使用 WindowManager 来确定设备当前处于纵向模式还是横向模式。一旦确定以后,向活动添加合适的碎片。方法是创建碎片,然后调用 FragmentTransaction 对象的 replace()方法将碎片添加到指定的视图容器(在本例中,android.R.id.content 引用活动的内容视图):

```
//---landscape mode---
Fragment1 fragment1 = new Fragment1();
// android.R.id.content refers to the content
// view of the activity
fragmentTransaction.replace(
        android.R.id.content, fragment1);
```

使用 replace()方法本质上相当于调用 FragmentTransaction 对象的 remove()方法,然后调用其 add()方法。为了确保更改生效,需要调用 commit()方法:

```
fragmentTransaction.commit();
```

2.3.2 碎片的生命周期

与活动类似,碎片具有自己的生命周期。理解了碎片的生命周期后,您可以在碎片被销毁时正确地保存其实例,在碎片被重新创建时将其还原到前一个状态。

下面的"试一试"展示了碎片的各个状态。

试一试　　理解碎片的生命周期

Fragments.zip 代码文件可以在 Wrox.com 上下载

(1) 使用前一节创建的同一个项目,在 Fragment1.java 文件中添加下面的粗体代码:

```
package net.learn2develop.Fragments;
```

```java
import android.app.Activity;
import android.app.Fragment;
import android.os.Bundle;
import android.util.Log;
import android.view.LayoutInflater;
import android.view.View;
import android.view.ViewGroup;

public class Fragment1 extends Fragment {
    @Override
    public View onCreateView(LayoutInflater inflater,
    ViewGroup container, Bundle savedInstanceState) {

        Log.d("Fragment 1", "onCreateView");

        //---Inflate the layout for this fragment---
        return inflater.inflate(
            R.layout.fragment1, container, false);
    }

    @Override
    public void onAttach(Activity activity) {
        super.onAttach(activity);
        Log.d("Fragment 1", "onAttach");
    }

    @Override
    public void onCreate(Bundle savedInstanceState) {
        super.onCreate(savedInstanceState);
        Log.d("Fragment 1", "onCreate");
    }

    @Override
    public void onActivityCreated(Bundle savedInstanceState) {
        super.onActivityCreated(savedInstanceState);
        Log.d("Fragment 1", "onActivityCreated");
    }

    @Override
    public void onStart() {
        super.onStart();
        Log.d("Fragment 1", "onStart");
    }

    @Override
    public void onResume() {
        super.onResume();
        Log.d("Fragment 1", "onResume");
    }
```

```java
    @Override
    public void onPause() {
        super.onPause();
        Log.d("Fragment 1", "onPause");
    }

    @Override
    public void onStop() {
        super.onStop();
        Log.d("Fragment 1", "onStop");
    }

    @Override
    public void onDestroyView() {
        super.onDestroyView();
        Log.d("Fragment 1", "onDestroyView");
    }

    @Override
    public void onDestroy() {
        super.onDestroy();
        Log.d("Fragment 1", "onDestroy");
    }

    @Override
    public void onDetach() {
        super.onDetach();
        Log.d("Fragment 1", "onDetach");
    }
}
```

(2) 按 Ctrl+F11 组合键将 Android 模拟器切换到横向模式。

(3) 在 Eclipse 中按 F11 键，在 Android 模拟器上调试应用程序。

(4) 当应用程序加载到模拟器中后，LogCat 窗口(Windows | Show View | LogCat)会显示以下内容：

```
12-09 04:17:43.436: D/Fragment 1(2995): onAttach
12-09 04:17:43.466: D/Fragment 1(2995): onCreate
12-09 04:17:43.476: D/Fragment 1(2995): onCreateView
12-09 04:17:43.506: D/Fragment 1(2995): onActivityCreated
12-09 04:17:43.506: D/Fragment 1(2995): onStart
12-09 04:17:43.537: D/Fragment 1(2995): onResume
```

(5) 单击模拟器上的 Hombe 按钮。LogCat 窗口中会显示以下输出：

```
12-09 04:18:47.696: D/Fragment 1(2995): onPause
12-09 04:18:50.346: D/Fragment 1(2995): onStop
```

(6) 在模拟器上，单击并按住 Home 按钮。再次启动应用程序。这一次会显示以下输出：

```
12-09 04:20:08.726: D/Fragment 1(2995): onStart
12-09 04:20:08.766: D/Fragment 1(2995): onResume
```

(7) 最后,单击模拟器上的 **Back** 按钮。现在可以看到下面的输出:

```
12-09 04:21:01.426: D/Fragment 1(2995): onPause
12-09 04:21:02.346: D/Fragment 1(2995): onStop
12-09 04:21:02.346: D/Fragment 1(2995): onDestroyView
12-09 04:21:02.346: D/Fragment 1(2995): onDestroy
12-09 04:21:02.346: D/Fragment 1(2995): onDetach
```

示例说明

和活动一样,Android 中的碎片也有自己的生命周期。如您所见,当碎片被创建时,会经历以下状态:

- onAttach()
- onCreate()
- onCreateView()
- onActivityCreated()

当碎片变得可见时,会经历以下状态:

- onStart()
- onResume()

当碎片进入后台模式时,会经历以下状态:

- onPause()
- onStop()

当碎片被销毁(它当前所在的活动被销毁)时,会经历以下状态:

- onPause()
- onStop()
- onDestroyView()
- onDestroy()
- onDetach()

与活动一样,可以使用 Bundle 对象在以下状态中还原碎片的实例:

- onCreate()
- onCreateView()
- onActivityCreated()

注意:可以在 onSaveInstanceState()方法中保存碎片的状态。第 3 章将详细介绍这一主题。

碎片经历的状态大多数与活动类似。但是,有一些新状态是碎片独有的:

- onAttached()——当碎片与活动建立关联时调用。

- onCreateView()——用于创建碎片的视图。
- onActivityCreated()——当活动的 onCreate()方法被返回时调用。
- onDestroyView()——当碎片的视图被移除时调用。
- onDetach()——当碎片与活动的关联被移除时调用。

注意，活动和碎片之间存在一个主要的区别：当活动进入后台时，会被放到 back stack 中。这样当用户按 Back 按钮时，活动可以恢复。但是，碎片在进入后台时不会被自动放到 back stack 中。要实现这一目的，需要在碎片处理期间显式调用 addToBackStack()方法，如下所示：

```
//---get the current display info---
WindowManager wm = getWindowManager();
Display d = wm.getDefaultDisplay();
if (d.getWidth() > d.getHeight())
{
    //---landscape mode---
    Fragment1 fragment1 = new Fragment1();
    // android.R.id.content refers to the content
    // view of the activity
    fragmentTransaction.replace(
            android.R.id.content, fragment1);
}
else
{
    //---portrait mode---
    Fragment2 fragment2 = new Fragment2();
    fragmentTransaction.replace(
            android.R.id.content, fragment2);
}

//---add to the back stack---
fragmentTransaction.addToBackStack(null);
fragmentTransaction.commit();
```

这段代码确保了当把碎片添加到活动中后，用户可以单击 Back 按钮移除它。

2.3.3 碎片之间进行交互

很多时候，一个活动中会包含一个或者多个碎片，它们彼此协作，向用户展示一个一致的 UI。在这种情况中，碎片之间能够进行通信并交换数据十分重要。例如，一个碎片可能包含一个项目列表(例如某个 RSS 源的文章)，当用户点击该碎片中的某个项时，另外一个碎片中会显示选中项的细节。

下面的"试一试"演示了一个碎片如何访问另一个碎片中包含的视图。

试一试　碎片之间的通信

(1) 使用前一节创建的同一个项目，在 Fragment1.xml 文件中添加下面的粗体代码：

```
<?xml version="1.0" encoding="utf-8"?>
```

```
<LinearLayout
    xmlns:android="http://schemas.android.com/apk/res/android"
    android:orientation="vertical"
    android:layout_width="fill_parent"
    android:layout_height="fill_parent"
    android:background="#00FF00" >
<TextView
    android:id="@+id/lblFragment1"
    android:layout_width="fill_parent"
    android:layout_height="wrap_content"
    android:text="This is fragment #1"
    android:textColor="#000000"
    android:textSize="25sp" />
</LinearLayout>
```

(2) 在 fragment2.xml 中添加下面的粗体代码:

```
<?xml version="1.0" encoding="utf-8"?>
<LinearLayout
    xmlns:android="http://schemas.android.com/apk/res/android"
    android:orientation="vertical"
    android:layout_width="fill_parent"
    android:layout_height="fill_parent"
    android:background="#FFFE00" >
<TextView
    android:layout_width="fill_parent"
    android:layout_height="wrap_content"
    android:text="This is fragment #2"
    android:textColor="#000000"
    android:textSize="25sp" />

<Button
    android:id="@+id/btnGetText"
    android:layout_width="wrap_content"
    android:layout_height="wrap_content"
    android:text="Get text in Fragment #1"
    android:textColor="#000000"
    android:onClick="onClick" />

</LinearLayout>
```

(3) 将两个碎片重新添加到 main.xml 中:

```
<?xml version="1.0" encoding="utf-8"?>
<LinearLayout xmlns:android="http://schemas.android.com/apk/res/android"
    android:layout_width="fill_parent"
    android:layout_height="fill_parent"
    android:orientation="horizontal" >

    <fragment
        android:name="net.learn2develop.Fragments.Fragment1"
```

```xml
        android:id="@+id/fragment1"
        android:layout_weight="1"
        android:layout_width="0px"
        android:layout_height="match_parent" />
    <fragment
        android:name="net.learn2develop.Fragments.Fragment2"
        android:id="@+id/fragment2"
        android:layout_weight="1"
        android:layout_width="0px"
        android:layout_height="match_parent" />

</LinearLayout>
```

(4) 在 FragmentsActivity.java 文件中，注释掉前几节添加的代码。修改之后，该文件应该如下所示：

```java
public class FragmentsActivity extends Activity {
    /** Called when the activity is first created. */
    @Override
    public void onCreate(Bundle savedInstanceState) {
        super.onCreate(savedInstanceState);
        setContentView(R.layout.main);
        /*
        FragmentManager fragmentManager = getFragmentManager();
        FragmentTransaction fragmentTransaction =
                fragmentManager.beginTransaction();

        //---get the current display info---
        WindowManager wm = getWindowManager();
        Display d = wm.getDefaultDisplay();
        if (d.getWidth() > d.getHeight())
        {
            //---landscape mode---
            Fragment1 fragment1 = new Fragment1();
            // android.R.id.content refers to the content
            // view of the activity
            fragmentTransaction.replace(
                    android.R.id.content, fragment1);
        }
        else
        {
            //---portrait mode---
            Fragment2 fragment2 = new Fragment2();
            fragmentTransaction.replace(
                    android.R.id.content, fragment2);
        }
        //---add to the back stack---
        fragmentTransaction.addToBackStack(null);
        fragmentTransaction.commit();
        */
```

(5) 在 Fragment2.java 文件中添加下面的粗体代码：

```java
package net.learn2develop.Fragments;

import android.app.Fragment;
import android.os.Bundle;
import android.view.LayoutInflater;
import android.view.View;
import android.view.ViewGroup;
import android.widget.Button;
import android.widget.TextView;
import android.widget.Toast;

public class Fragment2 extends Fragment {
    @Override
    public View onCreateView(LayoutInflater inflater,
    ViewGroup container, Bundle savedInstanceState) {
        //---Inflate the layout for this fragment---
        return inflater.inflate(
            R.layout.fragment2, container, false);
    }

    @Override
    public void onStart() {
        super.onStart();
        //---Button view---
        Button btnGetText = (Button)
            getActivity().findViewById(R.id.btnGetText);
        btnGetText.setOnClickListener(new View.OnClickListener() {
            public void onClick(View v) {
                TextView lbl = (TextView)
                    getActivity().findViewById(R.id.lblFragment1);
                Toast.makeText(getActivity(), lbl.getText(),
                    Toast.LENGTH_SHORT).show();
            }
        });
    }
}
```

(6) 按 F11 键在 Android 模拟器上调试应用程序。在右侧的第二个碎片中单击按钮。可以看到 Toast 类显示的文本 This is fragment #1，如图 2-22 所示。

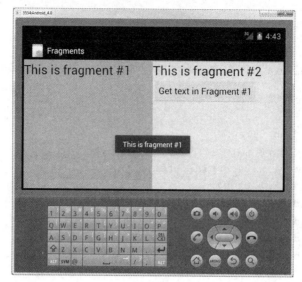

图 2-22

示例说明

因为碎片是嵌入在活动中的,所以可以通过先使用 getActivity()方法获得当前嵌入了该碎片的活动,然后使用 findViewById()方法定位该碎片中包含的视图:

```
TextView lbl = (TextView)
    getActivity().findViewById(R.id.lblFragment1);
Toast.makeText(getActivity(), lbl.getText(),
    Toast.LENGTH_SHORT).show();
```

getActivity()方法返回与当前碎片关联的活动。

另一种方法是,在 FragmentsActivity.java 文件中添加下面的方法:

```
public void onClick(View v) {
    TextView lbl = (TextView)
        findViewById(R.id.lblFragment1);
    Toast.makeText(this, lbl.getText(),
        Toast.LENGTH_SHORT).show();
}
```

2.4 使用意图调用内置应用程序

到目前为止,您已经了解了如何在自己的应用程序中调用活动。Android 编程的关键方面之一就是使用意图从其他应用程序中调用活动。特别是,您的应用程序可以调用 Android 设备内置的许多应用程序。例如,如果您的应用程序需要加载一个 Web 页面,就可以使用 Intent 对象调用内置的 Web 浏览器来显示该页面,而不是为此构建自己的 Web 浏览器。

下面的"试一试"将演示如何调用 Android 设备中一些常见的内置应用程序。

试一试   使用意图调用内置应用程序

Intents.zip 代码文件可以在 Wrox.com 上下载

(1) 打开 Eclipse，创建一个新的 Android 项目并命名为 Intents。

(2) 在 main.xml 文件中添加下列粗体显示的语句：

```xml
<?xml version="1.0" encoding="utf-8"?>
<LinearLayout xmlns:android="http://schemas.android.com/apk/res/android"
    android:layout_width="fill_parent"
    android:layout_height="fill_parent"
    android:orientation="vertical" >

<Button
    android:id="@+id/btn_webbrowser"
    android:layout_width="fill_parent"
    android:layout_height="wrap_content"
    android:text="Web Browser"
    android:onClick="onClickWebBrowser" />

<Button
    android:id="@+id/btn_makecalls"
    android:layout_width="fill_parent"
    android:layout_height="wrap_content"
    android:text="Make Calls"
    android:onClick="onClickMakeCalls"  />

<Button
    android:id="@+id/btn_showMap"
    android:layout_width="fill_parent"
    android:layout_height="wrap_content"
    android:text="Show Map"
    android:onClick="onClickShowMap" />

</LinearLayout>
```

(3) 在 IntentsActivity.java 文件中添加下列粗体显示的语句：

```java
package net.learn2develop.Intents;

import android.app.Activity;
import android.content.Intent;
import android.net.Uri;
import android.os.Bundle;
import android.view.View;

public class IntentsActivity extends Activity {

    int request_Code = 1;

    /** Called when the activity is first created. */
```

```java
@Override
public void onCreate(Bundle savedInstanceState) {
    super.onCreate(savedInstanceState);
    setContentView(R.layout.main);
}

public void onClickWebBrowser(View view) {
    Intent i = new
            Intent(android.content.Intent.ACTION_VIEW,
                    Uri.parse("http://www.amazon.com"));
    startActivity(i);
}

public void onClickMakeCalls(View view) {
    Intent i = new
            Intent(android.content.Intent.ACTION_DIAL,
                    Uri.parse("tel:+651234567"));
    startActivity(i);
}

public void onClickShowMap(View view) {
    Intent i = new
            Intent(android.content.Intent.ACTION_VIEW,
                    Uri.parse("geo:37.827500,-122.481670"));
    startActivity(i);
}
}
```

(4) 按 F11 键在 Android 模拟器上调试应用程序。

(5) 单击 Web Browser 按钮在模拟器中加载 Browser 应用程序。在图 2-23 中，内置的 Browser 应用程序正在显示网站 www.amazon.com。

(6) 单击 Make Calls 按钮，加载 Phone 应用程序，如图 2-24 所示。

图 2-23　　　　　　　　　　图 2-24

(7) 类似地，单击 Show Map 按钮加载 Maps 应用程序，如图 2-25 所示。

 注意：为了显示 Maps 应用程序，需要在支持 Google APIs 的 AVD 上运行它。

示例说明

本例中，您看到了如何使用 Intent 类来调用 Android 中的一些内置应用程序(例如 Maps、Phone、Contacts 和 Browser)。

在 Android 中，意图通常是成对出现的：动作(action)和数据(data)。动作描述了要执行什么，例如编辑一个条目、查看一个条目的内容等。数据则指定了受影响的对象，例如 Contacts 数据库中的某个人。这一数据被指定为一个 Uri 对象。

动作的一些示例如下：

- ACTION_VIEW
- ACTION_DIAL
- ACTION_PICK

数据的一些示例如下：

图 2-25

- www.google.com
- tel:+651234567
- geo:37.827500, -122.481670
- content://contacts

 注意：2.4.2 节将介绍可以定义并在活动中使用的数据类型。

动作和数据对共同描述了要执行的操作。例如，要拨一个电话号码，使用 ACTION_DIAL/tel:+651234567 对；要显示存储于手机中的联系人列表，使用 ACTION_VIEW/content://contacts 对；要从联系人列表中选择一个联系人，使用 ACTION_PICK/content://contacts 对。

在第 1 个按钮中，创建了一个 Intent 对象，然后给它的构造函数传递了两个参数——动作和数据：

```
Intent i = new
        Intent(android.content.Intent.ACTION_VIEW,
            Uri.parse("http://www.amazon.com"));
startActivity(i);
```

这里的动作由 android.content.Intent.ACTION_VIEW 常量表示。使用 Uri 类的 parse() 方法将一个 URL 字符串转换为一个 Uri 对象。

android.content.Intent.ACTION_VIEW 常量实际上指的是"android.intent.action.VIEW

动作，所以前述代码可以重写为：

```
Intent i = new
        Intent("android.intent.action.VIEW",
                Uri.parse("http://www.amazon.com"));
startActivity(i);
```

前面的代码片段还可以按如下方式改写：

```
Intent i = new
        Intent("android.intent.action.VIEW");
i.setData(Uri.parse("http://www.amazon.com"));
startActivity(i);
```

在这里，我们使用 setData()方法单独设置了数据。

对于第 2 个按钮，我们通过在数据部分传入一个电话号码来拨出一个特定的号码：

```
Intent i = new
        Intent(android.content.Intent.ACTION_DIAL,
                Uri.parse("tel:+651234567"));
startActivity(i);
```

这时，拨号程序将显示被呼叫的号码。用户仍旧要按拨号按钮来拨出这个号码。如果想无须用户干预而直接拨出号码，则要修改动作如下：

```
Intent i = new
        Intent(android.content.Intent.ACTION_CALL,
                Uri.parse("tel:+651234567"));
startActivity(i);
```

注意：如果想让应用程序直接呼叫特定号码，需要为应用程序添加 android.permission.CALL_PHONE 权限。

如果仅仅只是显示拨号程序，而不指定任何号码，只要像下面这样省略数据部分即可：

```
Intent I = new
        Intent(android.content.Intent.ACTION_DIAL);
startActivity(i);
```

第 3 个按钮使用 Action_VIEW 常量显示了一个地图：

```
Intent i = new
        Intent(android.content.Intent.ACTION_VIEW,
                Uri.parse("geo:37.827500,-122.481670"));
startActivity(i);
```

这里，要使用 geo 模式来代替 http。

2.4.1 理解意图对象

到目前为止，您已经了解了 Intent 对象调用其他活动的作用。现在正好可以回顾一下并对 Intent 对象如何施展它的魔力获得更详细的认识。

首先，我们可以通过传递一个动作给一个 Intent 对象的构造函数来调用另一个活动：

```
startActivity(new Intent("net.learn2develop.SecondActivity"));
```

动作(本例中是 net.learn2develop.SecondActivity)也被称为组件名称，用来标识所要调用的目标活动/应用程序。也可以通过指定存在于项目中的活动的类名修改组件的名称，如下面的代码所示：

```
startActivity(new Intent(this, SecondActivity.class));
```

还可以通过传入一个动作常量和数据来创建一个 Intent 对象，例如：

```
Intent i = new
        Intent(android.content.Intent.ACTION_VIEW,
            Uri.parse("http://www.amazon.com"));
startActivity(i);
```

动作部分定义了您要干什么，而数据部分包含了目标活动执行的数据。也可以使用 setData()方法传递数据给 Intent 对象：

```
Intent i = new
        Intent("android.intent.action.VIEW");
i.setData(Uri.parse("http://www.amazon.com"));
```

在本例中，您表示要查看指定 URL 的 Web 页面。Android 操作系统会查找所有可以满足您要求的活动。这一过程被称为意图解析(intent resolution)。2.4.2 节将详细讨论您的活动是如何成为其他活动的目标的。

对于某些意图是无须指定数据的。例如，要从 Contacts 应用程序中选择一个联系人，可指定该动作，然后使用 setType()方法表明 MIME 类型：

```
Intent i = new
        Intent(android.content.Intent.ACTION_PICK);
i.setType(ContactsContract.Contacts.CONTENT_TYPE);
```

注意：第 7 章将讨论如何在自己的应用程序内使用 Contacts 应用程序。

setType()方法显式指定了 MIME 数据类型来表明要返回的数据类型。ContactsContract.Contacts.CONTENT_TYPE 的 MIME 类型是 vnd.android.cursor.dir/contact。

除了指定动作、数据和类型外，Intent 对象还可以指定类别。将活动按类别分组为逻辑单元，可以实现 Android 对活动的进一步筛选。下一节将更详细地讨论类别。

总的来说，Intent 对象可以包含以下信息：
- 动作
- 数据
- 类型
- 类别

2.4.2 使用意图筛选器

之前，您已经看到了一个活动是如何使用 Intent 对象调用另一个活动的。为了使其他活动调用您的活动，需要在 AndroidManifest.xml 文件的<intent-filter>元素中指定动作和类别，如下所示：

```
<intent-filter >
    <action android:name="net.learn2develop.SecondActivity" />
    <category android:name="android.intent.category.DEFAULT" />
</intent-filter>
```

这是一个活动使用 net.learn2develop.SecondActivity 动作调用另一个活动的非常简单的示例。下面的"试一试"则提供了一个更复杂的示例。

试一试　　更详细地指定意图筛选器

(1) 使用先前创建的 Intents 项目，给该项目添加一个新类，并命名为 MyBrowserActivity。在 res/layout 文件夹下再新增一个 XML 文件，命名为 browser.xml。

(2) 在 AndroidManifest.xml 文件中添加下列粗体显示的语句：

```
<?xml version="1.0" encoding="utf-8"?>
<manifest xmlns:android="http://schemas.android.com/apk/res/android"
    package="net.learn2develop.Intents"
    android:versionCode="1"
    android:versionName="1.0" >

    <uses-sdk android:minSdkVersion="14" />
    <uses-permission android:name="android.permission.CALL_PHONE"/>
    <uses-permission android:name="android.permission.INTERNET"/>
    <application
        android:icon="@drawable/ic_launcher"
        android:label="@string/app_name" >
        <activity
            android:label="@string/app_name"
            android:name=".IntentsActivity" >
            <intent-filter >
                <action android:name="android.intent.action.MAIN" />
                <category android:name="android.intent.category.LAUNCHER" />
            </intent-filter>
        </activity>
        <activity android:name=".MyBrowserActivity"
            android:label="@string/app_name">
```

```xml
            <intent-filter>
                <action android:name="android.intent.action.VIEW" />
                <action android:name="net.learn2develop.MyBrowser" />
                <category android:name="android.intent.category.DEFAULT" />
                <data android:scheme="http" />
            </intent-filter>
        </activity>

    </application>

</manifest>
```

(3) 在 main.xml 文件中添加下列粗体显示的代码:

```xml
<?xml version="1.0" encoding="utf-8"?>
<LinearLayout xmlns:android="http://schemas.android.com/apk/res/android"
    android:layout_width="fill_parent"
    android:layout_height="fill_parent"
    android:orientation="vertical" >

<Button
    android:id="@+id/btn_webbrowser"
    android:layout_width="fill_parent"
    android:layout_height="wrap_content"
    android:text="Web Browser"
    android:onClick="onClickWebBrowser" />

<Button
    android:id="@+id/btn_makecalls"
    android:layout_width="fill_parent"
    android:layout_height="wrap_content"
    android:text="Make Calls"
    android:onClick="onClickMakeCalls"  />

<Button
    android:id="@+id/btn_showMap"
    android:layout_width="fill_parent"
    android:layout_height="wrap_content"
    android:text="Show Map"
    android:onClick="onClickShowMap" />

<Button
    android:id="@+id/btn_launchMyBrowser"
    android:layout_width="fill_parent"
    android:layout_height="wrap_content"
    android:text="Launch My Browser"
    android:onClick="onClickLaunchMyBrowser" />

</LinearLayout>
```

(4) 在 IntentsActivity.java 文件中添加下列粗体显示的语句:

```java
package net.learn2develop.Intents;

import android.app.Activity;
import android.content.Intent;
import android.net.Uri;
import android.os.Bundle;
import android.view.View;

public class IntentsActivity extends Activity {

    int request_Code = 1;

    /** Called when the activity is first created. */
    @Override
    public void onCreate(Bundle savedInstanceState) { ¡- }

    public void onClickWebBrowser(View view) { ¡- }

    public void onClickMakeCalls(View view) { ... }

    public void onClickShowMap(View view) { ... }

    public void onClickLaunchMyBrowser(View view) {
        Intent i = new
                Intent("net.learn2develop.MyBrowser");
        i.setData(Uri.parse("http://www.amazon.com"));
        startActivity(i);
    }

}
```

(5) 在 browser.xml 文件中添加下列粗体显示的语句：

```xml
<?xml version="1.0" encoding="utf-8"?>
<LinearLayout xmlns:android="http://schemas.android.com/apk/res/android"
    android:orientation="vertical"
    android:layout_width="fill_parent"
    android:layout_height="fill_parent" >
<WebView
    android:id="@+id/WebView01"
    android:layout_width="wrap_content"
    android:layout_height="wrap_content" />
</LinearLayout>
```

(6) 在 MyBrowserActivity.java 文件中添加下列粗体显示的语句：

```java
package net.learn2develop.Intents;

import android.app.Activity;
import android.net.Uri;
import android.os.Bundle;
```

第 2 章 活动、碎片和意图

```java
import android.webkit.WebView;
import android.webkit.WebViewClient;

public class MyBrowserActivity extends Activity {
    @Override
    public void onCreate(Bundle savedInstanceState) {
        super.onCreate(savedInstanceState);
        setContentView(R.layout.browser);

        Uri url = getIntent().getData();
        WebView webView = (WebView) findViewById(R.id.WebView01);
        webView.setWebViewClient(new Callback());
        webView.loadUrl(url.toString());
    }

    private class Callback extends WebViewClient {
        @Override
        public boolean shouldOverrideUrlLoading (WebView view, String url) {
            return(false);
        }
    }
}
```

(7) 按 F11 键在 Android 模拟器上调试应用程序。

(8) 单击 Launch my Browser 按钮，将会看到显示着 Amazon.com 的 Web 页面的一个新活动(如图 2-26 所示)。

示例说明

在本例中，创建了一个名为 MyBrowserActivity 的新活动。首先需要在 AndroidManifest.xml 文件中声明它。

```xml
<activity android:name=".MyBrowserActivity"
        android:label="@string/app_name">
    <intent-filter>
        <action android:name="android.intent.
          action.VIEW" />
        <action android:name="net.learn2develop.
          MyBrowser" />
        <category android:name="android.intent.
          category.DEFAULT" />
        <data android:scheme="http" />
    </intent-filter>
</activity>
```

图 2-26

在<intent-filter>元素中，声明了活动的两个动作、一个类别以及一个数据。这意味着所有其他活动可以使用 android.intent.action.VIEW 或 net.learn2develop.MyBrowser 动作来调用这个活动。对于所有希望别人使用 startActivity()或 startActivityForResult()方法来调用的活动，它们需要具有 android.intent.category.DEFAULT 类别。如果没有的话，您的活动将不能被其他活动调用。<data>元素指定了活动期望的数据类型。在本例中，它期望的数据要以 http://前缀打头。

先前的意图筛选器还可以按如下方式改写：

```xml
<activity android:name=".MyBrowserActivity"
          android:label="@string/app_name">
    <intent-filter>
        <action android:name="android.intent.action.VIEW" />
        <category android:name="android.intent.category.DEFAULT" />
        <data android:scheme="http" />
    </intent-filter>
    <intent-filter>
        <action android:name="net.learn2develop.MyBrowser" />
        <category android:name="android.intent.category.DEFAULT" />
        <data android:scheme="http" />
    </intent-filter>
</activity>
```

按这种方式编写意图筛选器可以对一个意图筛选器中的动作、类别以及数据进行逻辑分组，使其更加具有可读性。

现在如果使用带有如下数据的 ACTION_VIEW 动作，Android 将显示一个选择(如图 2-27 所示)：

```java
Intent I = new
    Intent(android.content.Intent.ACTION_VIEW,
        Uri.parse("http://www.amazon.com"));
```

可以在使用 Browser 应用程序还是当前正在构建的 Intents 应用程序间进行选择。

注意，当有多个活动匹配 Intent 对象时，会出现 Complete action using 对话框。通过使用 Intent 类的 createChooser()方法来对该对话框进行自定义，如下所示：

```java
Intent i = new
    Intent(android.content.Intent.ACTION
    _VIEW,
        Uri.parse("http://www.amazon.com"));
startActivity(Intent.createChooser(i, "Open
URL using..."));
```

图 2-27

这段代码将对话框的标题改为 Open URL using …，如图 2-28 所示。注意，现在没有了 Use by default for this action 选项。

使用 createChooser()的另一个好处是，当没有活动与您的 Intent 对象匹配时，应用程序不会崩溃，而是会显示如图 2-29 所示的消息。

第 2 章 活动、碎片和意图

图 2-28

图 2-29

2.4.3 添加类别

可以在意图筛选器中使用<category>元素对活动进行分类。假设已经在 AndroidManifest.xml 文件中添加了下列<category>元素：

```xml
<?xml version="1.0" encoding="utf-8"?>
<manifest xmlns:android="http://schemas.android.com/apk/res/android"
    package="net.learn2develop.Intents"
    android:versionCode="1"
    android:versionName="1.0" >

    <uses-sdk android:minSdkVersion="14" />
    <uses-permission android:name="android.permission.CALL_PHONE"/>
    <uses-permission android:name="android.permission.INTERNET"/>
    <application
        android:icon="@drawable/ic_launcher"
        android:label="@string/app_name" >
        <activity
            android:label="@string/app_name"
            android:name=".IntentsActivity" >
            <intent-filter >
                <action android:name="android.intent.action.MAIN" />
                <category android:name="android.intent.category.LAUNCHER" />
            </intent-filter>
        </activity>

        <activity android:name=".MyBrowserActivity"
              android:label="@string/app_name">
            <intent-filter>
                <action android:name="android.intent.action.VIEW" />
                <action android:name="net.learn2develop.MyBrowser" />
                <category android:name="android.intent.category.DEFAULT" />
                <category android:name="net.learn2develop.Apps" />
                <data android:scheme="http" />
            </intent-filter>
        </activity>
```

```
        </application>
</manifest>
```

在这种情况下，下列代码将调用 MyBrowserActivity 活动：

```
Intent i = new
        Intent(android.content.Intent.ACTION_VIEW,
            Uri.parse("http://www.amazon.com"));
i.addCategory("net.learn2develop.Apps");
startActivity(Intent.createChooser(i, "Open URL using..."));
```

使用 addCategory()方法将类别添加到 Intent 对象中。如果遗漏了 addCategory()语句，前述的代码仍旧会调用 MyBrowserActivity 活动，因为它仍旧和默认类别 android.intent.category.DEFAULT 匹配。

不过，如果指定了一个和意图筛选器中定义的类别不匹配的类别，活动将不会被调用：

```
Intent I = new
        Intent(android.content.Intent.ACTION_VIEW,
            Uri.parse("http://www.amazon.com"));
//i.addCategory("net.learn2develop.Apps");
//---this category does not match any in the intent-filter---
i.addCategory("net.learn2develop.OtherApps");
startActivity(Intent.createChooser(i, "Open URL using..."));
```

上述的类别(net.learn2develop.OtherApps)不匹配意图筛选器中的任何类别，所以如果不使用 Intent 类的 createChooser()方法，将引发一个运行时异常。

如果在 MyBrowserActivity 的意图筛选器中添加前述类别，先前的代码就可以运行了：

```
<activity android:name=".MyBrowserActivity"
        android:label="@string/app_name">
    <intent-filter>
        <action android:name="android.intent.action.VIEW" />
        <action android:name="net.learn2develop.MyBrowser" />
        <category android:name="android.intent.category.DEFAULT" />
        <category android:name="net.learn2develop.Apps" />
        <category android:name="net.learn2develop.OtherApps" />
        <data android:scheme="http" />
    </intent-filter>
</activity>
```

可以添加多个类别到一个 Intent 对象中，以下语句在 Intent 对象中添加了 net.learn2develop.SomeOtherApps 类别：

```
Intent i = new
        Intent(android.content.Intent.ACTION_VIEW,
            Uri.parse("http://www.amazon.com"));
//i.addCategory("net.learn2develop.Apps");
//---this category does not match any in the intent-filter---
```

```
        i.addCategory("net.learn2develop.OtherApps");
        i.addCategory("net.learn2develop.SomeOtherApps");
        startActivity(Intent.createChooser(i, "Open URL using..."));
```

由于意图筛选器没有定义 net.learn2develop.SomeOtherApps 类别,上述代码将不能调用 MyBrowerActivity 活动。为此,需要再一次添加 net.learn2develop.SomeOtherApps 类别到意图筛选器中。

从这个示例可以看出,在一个活动可以被调用之前,当使用一个带有类别的 Intent 对象时,所有添加到 Intent 对象的类别必须完全匹配意图筛选器中所定义的类别。

2.5 显示通知

到目前为止,您一直使用 Toast 类来给用户显示消息。Toast 类虽然是一个向用户显示警报的方便的方法,但它不能持久。它只是在屏幕上闪那么几秒钟后就消失掉了。用户如果不一直盯着屏幕,一旦其中包含了重要信息的话,就很容易错过。

对于重要的消息,要使用更加持久的方法。这种情况下,应当使用 NotificationManager 在设备顶端的状态栏(有时也称为通知栏)中显示一条持久化的信息。下面的"试一试"演示了如何做到这一点。

试一试 在状态栏上显示通知

Notifications.zip 代码文件可以在Wrox.com 上下载

(1) 打开 Eclipse,创建一个新的 Android 项目并命名为 Notifications。

(2) 在包中添加一个新的类文件 Notifications.zip。另外,在 res/ layout 文件夹下添加一个新的 notification.xml 文件。

(3) 按如下所示填充 notification.xml 文件:

```
<?xml version="1.0" encoding="utf-8"?>
<LinearLayout xmlns:android="http://schemas.android.com/apk/res/android"
    android:orientation="vertical"
    android:layout_width="fill_parent"
    android:layout_height="fill_parent" >
<TextView
    android:layout_width="fill_parent"
    android:layout_height="wrap_content"
    android:text="Here are the details for the notification..." />
</LinearLayout>
```

(4) 按如下所示填充 NotificationView.java 文件:

```
package net.learn2develop.Notifications;

import android.app.Activity;
```

```java
import android.app.NotificationManager;
import android.os.Bundle;

public class NotificationView extends Activity
{
    @Override
    public void onCreate(Bundle savedInstanceState)
    {
        super.onCreate(savedInstanceState);
        setContentView(R.layout.notification);

        //---look up the notification manager service---
        NotificationManager nm = (NotificationManager)
            getSystemService(NOTIFICATION_SERVICE);

        //---cancel the notification that we started---
        nm.cancel(getIntent().getExtras().getInt("notificationID"));
    }
}
```

(5) 在 AndroidManifest.xml 文件中添加下列粗体显示的语句：

```xml
<?xml version="1.0" encoding="utf-8"?>
<manifest xmlns:android="http://schemas.android.com/apk/res/android"
    package="net.learn2develop.Notifications"
    android:versionCode="1"
    android:versionName="1.0" >

    <uses-sdk android:minSdkVersion="14" />
    <uses-permission android:name="android.permission.VIBRATE"/>

    <application
        android:icon="@drawable/ic_launcher"
        android:label="@string/app_name" >
        <activity
            android:label="@string/app_name"
            android:name=".NotificationsActivity" >
            <intent-filter >
                <action android:name="android.intent.action.MAIN" />

                <category android:name="android.intent.category.LAUNCHER" />
            </intent-filter>
        </activity>
        <activity android:name=".NotificationView"
            android:label="Details of notification">
            <intent-filter>
                <action android:name="android.intent.action.MAIN" />
                <category android:name="android.intent.category.DEFAULT" />
            </intent-filter>
        </activity>
    </application>
```

```
</manifest>
```

(6) 在 main.xml 文件中添加下列粗体显示的语句：

```xml
<?xml version="1.0" encoding="utf-8"?>
<LinearLayout xmlns:android="http://schemas.android.com/apk/res/android"
    android:layout_width="fill_parent"
    android:layout_height="fill_parent"
    android:orientation="vertical" >

<Button
    android:id="@+id/btn_displaynotif"
    android:layout_width="fill_parent"
    android:layout_height="wrap_content"
    android:text="Display Notification"
    android:onClick="onClick"/>

</LinearLayout>
```

(7) 最后，在 NotificationsActivity.java 文件中添加下列粗体显示的语句：

```java
package net.learn2develop.Notifications;

import android.app.Activity;
import android.app.Notification;
import android.app.NotificationManager;
import android.app.PendingIntent;
import android.content.Intent;
import android.os.Bundle;
import android.view.View;

public class NotificationsActivity extends Activity {
    int notificationID = 1;

    /** Called when the activity is first created. */
    @Override
    public void onCreate(Bundle savedInstanceState) {
        super.onCreate(savedInstanceState);
        setContentView(R.layout.main);
    }

    public void onClick(View view) {
        displayNotification();
    }

    protected void displayNotification()
    {
        //---PendingIntent to launch activity if the user selects
        // this notification---
        Intent i = new Intent(this, NotificationView.class);
```

```
            i.putExtra("notificationID", notificationID);

            PendingIntent pendingIntent =
                PendingIntent.getActivity(this, 0, i, 0);

            NotificationManager nm = (NotificationManager)
                getSystemService(NOTIFICATION_SERVICE);

            Notification notif = new Notification(
                R.drawable.ic_launcher,
                "Reminder: Meeting starts in 5 minutes",
                System.currentTimeMillis());

            CharSequence from = "System Alarm";
            CharSequence message = "Meeting with customer at 3pm...";

            notif.setLatestEventInfo(this, from, message, pendingIntent);

            //---100ms delay, vibrate for 250ms, pause for 100 ms and
            // then vibrate for 500ms---
            notif.vibrate = new long[] { 100, 250, 100, 500};
            nm.notify(notificationID, notif);
        }
```

}

(8) 按 F11 键在 Android 模拟器上调试应用程序。

(9) 单击 Display Notification 按钮，状态栏上将出现一个通知的滚动文本(在 Notification 对象的构造函数中设置)，如图 2-30 所示。

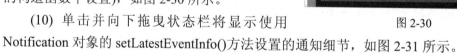

图 2-30

(10) 单击并向下拖曳状态栏将显示使用 Notification 对象的 setLatestEventInfo()方法设置的通知细节，如图 2-31 所示。

(11) 单击通知将显示 NotificationView 活动，如图 2-32 所示。同时也将把通知从状态栏上清除。

图 2-31

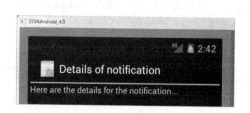

图 2-32

示例说明

要显示一个通知，首先要创建一个指向 NotificationView 类的 Intent 对象：

```
Intent i = new Intent(this, NotificationView.class);
i.putExtra("notificationID", notificationID);
```

当用户从通知列表中选择了一个通知时，这个意图将被用来启动另一个活动。在这个示例中，给 Intent 对象添加一个键/值对以便可以用来标记通知 ID，标识目标活动的通知。这个 ID 将在以后用来撤销这个通知。

还需要创建一个 PendingIntent 对象。PendingIntent 对象可以代表应用程序帮助您在后面某个时候执行一个动作，而不用考虑应用程序是否正在运行。在这里，按如下所示初始化它：

```
PendingIntent pendingIntent =
    PendingIntent.getActivity(this, 0, i, 0);
```

getActivity()方法检索一个 PendingIntent 对象并使用如下参数设置它：

- 上下文——应用程序上下文
- 请求码——用于意图的请求码
- 意图——用来启动目标活动的意图
- 标志——活动启动时使用的标志

然后，获取一个 NotificationManager 类的实例并创建一个 Notification 类的实例：

```
NotificationManager nm = (NotificationManager)
    getSystemService(NOTIFICATION_SERVICE);

Notification notif = new Notification(
    R.drawable.ic_launcher,
    "Reminder: Meeting starts in 5 minutes",
    System.currentTimeMillis());
```

当通知首次显示在状态栏上时，Notification 类使您能够指定通知的主要信息。Notification 构造函数的第二个参数在状态栏上设置了"滚动文本"(如图 2-33 所示)。

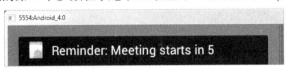

图 2-33

接下来，使用 setLatestEventInfo()方法来设置通知的详细内容：

```
CharSequence from = "System Alarm";
CharSequence message = "Meeting with customer at 3pm...";

notif.setLatestEventInfo(this, from, message, pendingIntent);

//---100ms delay, vibrate for 250ms, pause for 100 ms and
// then vibrate for 500ms---
```

```
notif.vibrate = new long[] { 100, 250, 100, 500};
```

上述代码还将通知设置为震动手机。最后，使用notify()方法来显示通知。

```
nm.notify(notificationID, notif);
```

当用户单击通知时，NotificationView 活动就会启动。这里，使用 NotificationManager 对象的 cancel()方法并传递给它通知的 ID(通过 Intent 对象传递)来取消这个通知：

```
//---look up the notification manager service---
NotificationManager nm = (NotificationManager)
    getSystemService(NOTIFICATION_SERVICE);
//---cancel the notification that we started---
nm.cancel(getIntent().getExtras().getInt("notificationID"));
```

2.6 本章小结

本章首先详细介绍了活动和碎片是如何工作的以及用于显示它们的各种形式。您还了解了如何使用活动来显示对话框窗口。

本章第二部分阐述了 Android 中的一个非常重要的概念——意图。意图是使不同活动连接起来的"胶水"，也是为 Android 平台进行开发需要了解的一个关键概念。

练 习

1. 如果有两个或多个活动具有相同的意图筛选器动作名称，那将会发生什么？
2. 写一段代码来调用内置的 Browser 应用程序。
3. 在意图筛选器中，可以指定哪些组成部分？
4. Toast 类和 NotificationManager 类的区别是什么？
5. 列举在活动中添加碎片的两种方法。
6. 说明碎片和活动的一个关键区别。

练习答案参见附录 C。

本章主要内容

主 题	关 键 概 念
创建活动	所有活动必须在 AndroidManifest.xml 文件中声明
活动的关键生命周期	当活动启动时，总是调用 onStart()和 onResume()事件。当活动被停止或转入后台时，总是调用 onPause()事件
以对话框形式显示活动	使用 showDialog()方法并实现 onCreateDialog()方法
碎片	碎片是可以在活动中添加或移除的"微活动"
以编程方式操作碎片	在运行时添加、移除或替换碎片时，需要使用 FragmentManager 和 FragmentTransaction 类

(续表)

主　题	关　键　概　念
碎片的生命周期	类似于活动的生命周期——在 onPause() 事件中保存碎片的状态，在下面的某个事件中还原其状态：onCreate()、onCreateView() 或 onActivityCreated()
意图	连接不同活动的"胶水"
意图筛选器	可以使您指定应当如何调用活动的"筛选器"
调用活动	使用 startActivity() 或 startActivityForResult() 方法
传递数据给一个活动	使用 Bundle 对象
Intent 对象中的组成部分	Intent 对象包含动作、数据、类型和类别
显示通知	使用 NotificationManager 类
PendingIntent 对象	PendingIntent 对象可以代表应用程序帮助您在后面某个时候执行一个动作，而不用考虑应用程序是否正在运行

第 3 章

Android 用户界面

本章将介绍以下内容:
- 可以用来布局视图的多种视图组(ViewGroup)
- 如何适应和管理屏幕方向的变化
- 如何以编程方式创建用户界面
- 如何侦听用户界面通知

在第 2 章中,您已经学习了 Activity 类和它的生命周期,了解了活动就是用户用来和应用程序交互的一个手段。然而,活动本身并没有在屏幕上呈现。相反,它需要使用视图和视图组来绘制屏幕。本章将学习有关在 Android 中创建用户界面的详细内容,以及用户是如何与之交互的。此外,还将学习如何处理 Android 设备屏幕方向的变化。

3.1 了解屏幕的构成

在第 2 章中,您已经知道了 Android 应用程序的基本单元是活动。活动显示了应用程序的用户界面,它可以包含按钮、标签、文本框等小部件。通常情况下,使用一个 XML 文件(例如,位于 res/layout 文件夹下的 main.xml 文件)来定义用户界面,类似如下所示:

```
<?xml version="1.0" encoding="utf-8"?>
<LinearLayout xmlns:android="http://schemas.android.com/apk/res/android"
    android:layout_width="fill_parent"
    android:layout_height="fill_parent"
    android:orientation="vertical" >

    <TextView
        android:layout_width="fill_parent"
        android:layout_height="wrap_content"
        android:text="@string/hello" />
```

```
</LinearLayout>
```

运行时,在 Activity 类的 onCreate()方法处理程序中使用 Activity 类的 setContentView()
方法加载 XML 用户界面:

```
@Override
    public void onCreate(Bundle savedInstanceState) {
        super.onCreate(savedInstanceState);
        setContentView(R.layout.main);
    }
```

在编译过程中,XML 文件中的每一个元素被编译成相应的 Android GUI 类,这些类使用方法来表示属性。在此文件被加载后,Android 系统就会创建活动的用户界面。

> 注意:尽管使用 XML 文件构建用户界面通常比较容易,但有时您还需要在运行时动态地构建用户界面(例如,编写游戏时)。因此,完全用代码创建用户界面也是可行的方法。本章稍后将为您提供一个示例,演示是如何实现这一点的。

3.1.1 视图和视图组

活动包含了视图和视图组。视图是一个可以在屏幕上显现的小部件,例如按钮、标签和文本框。视图派生自基类 android.view.View。

> 注意:第 4 章和第 5 章将讨论 Android 中各种常见的视图。

一个或多个视图可以组合成一个视图组。视图组(其本身就是一种特殊的视图类型)提供了一种布局,您可以按该布局定制视图的外观和顺序。视图组的例子包括 LinearLayout 和 FrameLayout。视图组派生于基类 android.view.ViewGroup。

Android 支持以下视图组:

- `LinearLayout`
- `AbsoluteLayout`
- `TableLayout`
- `RelativeLayout`
- `FrameLayout`
- `ScrollView`

在后面的章节中将对这每一个视图组进行详细讨论。注意,在实践中将不同类型的布局组合起来创建想要的用户界面是很常见的。

3.1.2 LinearLayout

LinearLayout 是以单行或单列的形式排列视图的。子视图可以水平或垂直地排列。要了解 LinearLayout 是如何工作的,请考虑 main.xml 文件中通常包含的以下元素:

```
<?xml version="1.0" encoding="utf-8"?>
<LinearLayout xmlns:android="http://schemas.android.com/apk/res/android"
    android:layout_width="fill_parent"
    android:layout_height="fill_parent"
    android:orientation="vertical" >

<TextView
    android:layout_width="fill_parent"
    android:layout_height="wrap_content"
    android:text="@string/hello" />
</LinearLayout>
```

在 main.xml 文件中,可注意到根元素<LinearLayout>及其包含的<TextView>元素,<LinearLayout>元素用来控制其所包含的视图的显示顺序。

每个视图和视图组都有一组公共属性,其中一些如表 3-1 所示。

表 3-1 视图和视图组使用的公共属性

属　　性	描　　述
layout_width	指定视图或视图组的宽度
layout_height	指定视图或视图组的高度
layout_marginTop	指定视图或视图组顶边额外的空间
layout_marginBottom	指定视图或视图组底边额外的空间
layout_marginLeft	指定视图或视图组左边额外的空间
layout_marginRight	指定视图或视图组右边额外的空间
layout_gravity	指定如何定位子视图
layout_weight	指定在布局中应该给视图分配多少额外空间
layout_x	指定视图或视图组的 x 坐标
layout_y	指定视图或视图组的 y 坐标

 注意:以上某些属性只有当一个视图在特定的视图组中时才适用。例如,layout_weight 和 layout_gravity 属性只适用于视图位于 LinearLayout 或 TableLayout 视图组中的情况。

举例来说,使用 fill_parent 常量可以让<TextView>元素的宽度填满其父元素(本例中指屏幕)的整个宽度。wrap_content 常量表明了其高度就是其内容(本例中指其包含的文本)的高度。如果不打算让<TextView>视图占据一整行,可以将其 layout_width 属性设置为

wrap_content,如下所示:

```xml
<TextView
    android:layout_width="wrap_content"
    android:layout_height="wrap_content"
    android:text="@string/hello" />
```

这将把视图的宽度设为此视图所包含的文本的宽度。

考虑以下布局:

```xml
<?xml version="1.0" encoding="utf-8"?>
<LinearLayout xmlns:android="http://schemas.android.com/apk/res/android"
    android:layout_width="fill_parent"
    android:layout_height="fill_parent"
    android:orientation="vertical" >

<TextView
    android:layout_width="100dp"
    android:layout_height="wrap_content"
    android:text="@string/hello" />

<Button
    android:layout_width="160dp"
    android:layout_height="wrap_content"
    android:text="Button"
    android:onClick="onClick" />

</LinearLayout>
```

度量单位

当指定 Android 用户界面上元素的大小时,应知道以下度量单位:

dp——与密度无关的像素(density-independent pixel)。1dp 相当于 160dpi 的屏幕上的 1 像素。当在布局中指定视图尺寸时,推荐将 dp 作为度量单位。160dpi 是 Android 假定的基准密度。当指的是与密度无关的像素时,可以使用 dp 或 dip。

sp——与比例无关的像素(scale-independent pixel)。与 dp 类似,推荐用于指定字体大小。

pt——磅。1 磅等于 1/72 英寸(基于屏幕的物理尺寸)。

px——像素。对应于屏幕上的实际像素。不建议使用这一单位,因为您的用户界面在不同屏幕尺寸的设备上可能不能正确呈现。

这里,将 TextView 和 Button 视图的宽度都设置成了一个绝对值。在本例中,TextView 的宽度被设置为 100dp,Button 的宽度被设置为 160dp。在具有不同像素密度的不同屏幕上查看这些视图的效果时,理解 Android 如何识别具有不同尺寸和密度的屏幕十分重要。

图 3-1 展现了 Nexus S 的屏幕。它有一个 4 英寸的屏幕(按对角线计算),屏幕宽度为 2.04 英寸。它的分辨率为 480(宽度)×800(高度)像素。将 480 像素分布到 2.04 英寸的宽度

上,结果就是像素密度大约为235dpi。

从图中可以看到,屏幕的像素密度会随着屏幕尺寸和分辨率发生变化。

Android定义并可识别4种屏幕密度:

- 低密度(*ldpi*)——120dpi
- 中等密度(*mdpi*)——160dpi
- 高密度(*hdpi*)——240dpi
- 超高密度(*xhdpi*)——320dpi

您的设备的屏幕密度会是上面列表中的一个。例如,Nexus S 被认为是一种 hdpi 设备,因为它的像素密度最接近240dpi。然而,HTC Hero 的屏幕尺寸为3.2英寸(按对角线计算),分辨率为320×480,因此其像素密度大约为180dpi。因为这个像素密度最接近160dpi,所以 HTC Hero 被认为是一种 mdpi 设备。

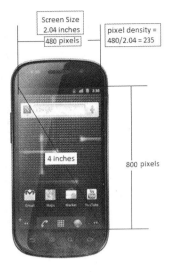

图 3-1

为了测试当 XML 文件中定义的视图显示在不同密度的屏幕上时效果如何,创建两个具有不同屏幕分辨率和 abstracted LCD 密度的 AVD。图 3-2 显示了一个分辨率为 480×800、LCD 密度为 235 的 AVD,它模拟了 Nexus S。

图 3-3 显示了另一个 AVD,它的分辨率为 320×480,LCD 密度为 180,模拟了 HTC Hero。

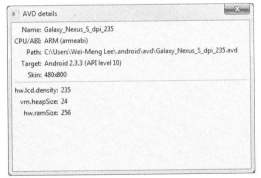

图 3-2   图 3-3

图 3-4 显示了在屏幕的像素密度为 235dpi 的模拟器中,视图看起来是什么样子。

图 3-5 显示了在屏幕的像素密度为 180dpi 的模拟器中,视图看起来是什么样子。

使用 dp 单位确保了无论屏幕密度如何,视图总是会以正确的比例显示,这是因为 Android 会根据屏幕的密度自动缩放视图的尺寸。以 Button 为例。如果它在一个 180dpi 的屏幕上显示(180dpi 的屏幕会被当做 160dpi 的屏幕进行处理),则宽度将是 160 像素。但是,如果在一个 235dpi 的屏幕上显示(235dpi 的屏幕会被当做 240dpi 的屏幕处理),那么宽度将是 240 像素。

图 3-4   图 3-5

如何将 dp 转换为 px

将 dp 转换为 px(像素)的公式如下：

实际像素=dp*(dpi/160)，其中 dpi 可以是 120、160、240 或 320。

因此，当 Button 显示在一个 235dpi 的屏幕上时，其实际宽度是 160 * (240/160) = 240 px。当运行在 180dpi 的模拟器中(被视为 160dpi 的设备)时，其实际像素则是 160 * (160/160) = 160 px。在本例中，1dp 等于 1px。

为了证明这是正确的，可以使用 View 对象的 getWidth()方法获得其以像素数表示的宽度：

```
public void onClick(View view) {
    Toast.makeText(this,
        String.valueOf(view.getWidth()),
        Toast.LENGTH_LONG).show();
}
```

如果不使用 dp，而是使用像素(px)指定尺寸，又会怎样？

```
<TextView
    android:layout_width="100px"
    android:layout_height="wrap_content"
    android:text="@string/hello" />
<Button
    android:layout_width="160px"
    android:layout_height="wrap_content"
    android:text="Button"
    android:onClick="onClick"/>
```

图 3-6 显示了 Label 和 Button 在 235dpi 的屏幕上的外观。图 3-7 显示了相同的视图在 180dpi 屏幕上的外观。在本例中，因为所有的尺寸都是使用像素指定的，所以 Android 不会执行任何转换。一般来说，当屏幕尺寸相同时，如果使用像素指定视图的尺寸，那么与具有低 dpi 的屏幕的设备相比，视图在具有高 dpi 屏幕的设备上会显得更小。

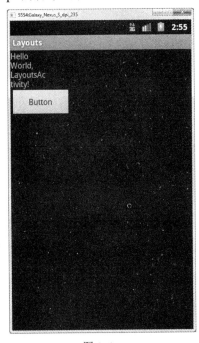

图 3-6

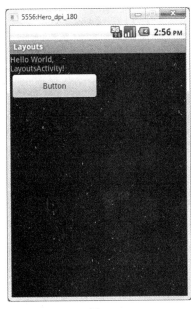

图 3-7

上面的例子还指定了布局的方向是垂直的：

```
<LinearLayout xmlns:android="http://schemas.
android.com/apk/res/android"
    android:layout_width="fill_parent"
    android:layout_height="fill_parent"
    android:orientation="vertical" >
```

默认的布局方向是横向的。因此，如果省略 android:orient-ation 属性，视图将显示如图 3-8 所示的效果。

在 LinearLayout 中，可以对其中包含的视图应用 layout_weight 和 layout_gravity 属性，可对 main.xml 文件作如下修改：

```
<LinearLayout
xmlns:android="http://schemas.android.com/ap
k/res/android"
    android:layout_width="fill_parent"
    android:layout_height="fill_parent"
    android:orientation="vertical" >
```

图 3-8

```
<Button
    android:layout_width="160dp"
    android:layout_height="wrap_content"
    android:text="Button"
    android:layout_gravity="left"
    android:layout_weight="1" />

<Button
    android:layout_width="160dp"
    android:layout_height="wrap_content"
    android:text="Button"
    android:layout_gravity="center"
    android:layout_weight="2" />

<Button
    android:layout_width="160dp"
    android:layout_height="wrap_content"
    android:text="Button"
    android:layout_gravity="right"
    android:layout_weight="3" />

</LinearLayout>
```

图 3-9 显示了视图的位置及其高度。layout-gravity 属性指定了视图应该朝着哪个方向移动，而 layout_weight 属性指定了可用空间的分布情况。在前面的示例中，3 个按钮分别占据了可用高度的 16.6%(1/(1+2+3)*100)、33.3%(2/(1+2+3)*100)和 50%(3/(1+2+3) *100)。

如果将 LinearLayout 的方向改为横向，就需要将每个视图的宽度改为 0dp，此时视图的显示情况如图 3-10 所示。

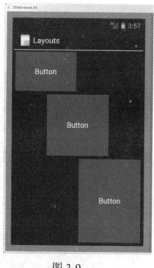

图 3-9

图 3-10

```
<LinearLayout xmlns:android="http://schemas.android.com/apk/res/android"
    android:layout_width="fill_parent"
    android:layout_height="fill_parent"
```

```xml
        android:orientation="horizontal" >

    <Button
        android:layout_width="0dp"
        android:layout_height="wrap_content"
        android:text="Button"
        android:layout_gravity="left"
        android:layout_weight="1" />

    <Button
        android:layout_width="0dp"
        android:layout_height="wrap_content"
        android:text="Button"
        android:layout_gravity="center_horizontal"
        android:layout_weight="2" />

    <Button
        android:layout_width="0dp"
        android:layout_height="wrap_content"
        android:text="Button"
        android:layout_gravity="right"
        android:layout_weight="3" />

</LinearLayout>
```

3.1.3 AbsoluteLayout

AbsoluteLayout 可用于指定其子元素的确切位置。考虑下列 main.xml 文件中定义的用户界面：

```xml
<AbsoluteLayout
    android:layout_width="fill_parent"
    android:layout_height="fill_parent"
    xmlns:android="http://schemas.android.com/apk/res/android" >
<Button
    android:layout_width="188dp"
    android:layout_height="wrap_content"
    android:text="Button"
    android:layout_x="126px"
    android:layout_y="361px" />
<Button
    android:layout_width="113dp"
    android:layout_height="wrap_content"
    android:text="Button"
    android:layout_x="12px"
    android:layout_y="361px" />
</AbsoluteLayout>
```

图 3-11 展示了使用 android_layout_x 和 android_layout_y 属性将两个 Button 视图(在一台 180dpi 的 AVD 上进行测试)定位在指定的位置处。

然而，当在高分辨率的屏幕上查看活动时，AbsoluteLayout 存在一个问题(如图 3-12 所示)。因此，从 Android 1.5 以后，AbsoluteLayout 已经被弃用了(尽管当前版本仍旧支持它)。鉴于不能保证在 Android 的未来版本中是否支持此视图组，应该在用户界面中避免使用 AbsoluteLayout，而使用本章描述的其他布局。

图 3-11

图 3-12

3.1.4 TableLayout

TableLayout 以行和列的形式组织视图。使用<TableRow>元素来指定表中的某一行。每一行可以包含一个或多个视图。行内的每个视图构成一个单元格。每一列的宽度由此列中最大单元格的宽度来决定。

考虑下列 main.xml 文件的内容：

```
<TableLayout
    xmlns:android="http://schemas.android.com/apk/res/android"
    android:layout_height="fill_parent"
    android:layout_width="fill_parent" >
<TableRow>
    <TextView
        android:text="User Name:"
        android:width ="120dp"
        />
    <EditText
        android:id="@+id/txtUserName"
        android:width="200dp" />
</TableRow>
<TableRow>
    <TextView
        android:text="Password:"
        />
```

```xml
        <EditText
            android:id="@+id/txtPassword"
            android:password="true"
            />
    </TableRow>
    <TableRow>
        <TextView />
        <CheckBox android:id="@+id/chkRememberPassword"
            android:layout_width="fill_parent"
            android:layout_height="wrap_content"
            android:text="Remember Password"
            />
    </TableRow>
    <TableRow>
        <Button
            android:id="@+id/buttonSignIn"
            android:text="Log In" />
    </TableRow>
</TableLayout>
```

图 3-13 展示了以上代码在 Android 模拟器上呈现的效果。

可注意到，在上述例子中，TableLayout 中有 4 行 2 列。直接位于 Password TextView 之下的单元格用<TextView/>空元素填充。如果不这样做，Remember Password 复选框将显示在 Password TextView 之下，如图 3-14 中所示。

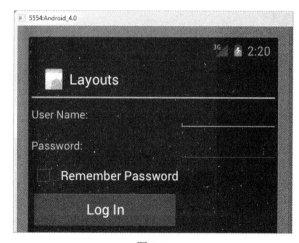

图 3-13   图 3-14

3.1.5 RelativeLayout

RelativeLayout 可用于指定子视图相对于彼此之间是如何定位的。考虑下列 main.xml

文件的内容：

```xml
<?xml version="1.0" encoding="utf-8"?>
<RelativeLayout
    android:id="@+id/RLayout"
    android:layout_width="fill_parent"
    android:layout_height="fill_parent"
    xmlns:android="http://schemas.android.com/apk/res/android" >

    <TextView
        android:id="@+id/lblComments"
        android:layout_width="wrap_content"
        android:layout_height="wrap_content"
        android:text="Comments"
        android:layout_alignParentTop="true"
        android:layout_alignParentLeft="true" />

    <EditText
        android:id="@+id/txtComments"
        android:layout_width="fill_parent"
        android:layout_height="170px"
        android:textSize="18sp"
        android:layout_alignLeft="@+id/lblComments"
        android:layout_below="@+id/lblComments"
        android:layout_centerHorizontal="true" />

    <Button
        android:id="@+id/btnSave"
        android:layout_width="125px"
        android:layout_height="wrap_content"
        android:text="Save"
        android:layout_below="@+id/txtComments"
        android:layout_alignRight="@+id/txtComments" />

    <Button
        android:id="@+id/btnCancel"
        android:layout_width="124px"
        android:layout_height="wrap_content"
        android:text="Cancel"
        android:layout_below="@+id/txtComments"
        android:layout_alignLeft="@+id/txtComments" />
</RelativeLayout>
```

可注意到，每一个嵌入 RelativeLayout 中的视图都有使它与其他视图对齐的属性。这些属性如下所示：

- layout_alignParentTop

- `layout_alignParentLeft`
- `layout_alignLeft`
- `layout_alignRight`
- `layout_below`
- `layout_centerHorizontal`

每一个属性的值是引用的视图的 ID。前面的 XML 用户界面创建的屏幕如图 3-15 所示。

3.1.6 FrameLayout

FrameLayout 是一个在屏幕上可以用来显示单个视图的占位符。添加到 FrameLayout 中的视图常常锚定在布局的左上方。考虑 main.xml 文件中的下列内容：

图 3-15

```xml
<?xml version="1.0" encoding="utf-8"?>
<RelativeLayout
    android:id="@+id/RLayout"
    android:layout_width="fill_parent"
    android:layout_height="fill_parent"
    xmlns:android="http://schemas.android.com/apk/res/android" >

    <TextView
        android:id="@+id/lblComments"
        android:layout_width="wrap_content"
        android:layout_height="wrap_content"
        android:text="Hello, Android!"
        android:layout_alignParentTop="true"
        android:layout_alignParentLeft="true" />

    <FrameLayout
        android:layout_width="wrap_content"
        android:layout_height="wrap_content"
        android:layout_alignLeft="@+id/lblComments"
        android:layout_below="@+id/lblComments"
        android:layout_centerHorizontal="true" >

    <ImageView
        android:src = "@drawable/droid"
        android:layout_width="wrap_content"
        android:layout_height="wrap_content" />

    </FrameLayout>
</RelativeLayout>
```

这里，RelativeLayout 中有一个 FrameLayout。在该 FrameLayout 中嵌入了一个 ImageView。用户界面如图 3-16 所示。

图 3-16

 注意：这个例子假设 res/drawable-mdpi 文件夹下有一个名为 droid.png 的图像。

如果在 FrameLayout 中添加另一个视图(如 Button 视图)，这个视图将覆盖先前的视图(如图 3-17 所示)：

```
<?xml version="1.0" encoding="utf-8"?>

<RelativeLayout
    android:id="@+id/RLayout"
    android:layout_width="fill_parent"
    android:layout_height="fill_parent"
    xmlns:android="http://schemas.android.com/apk/res/android"
    >
    <TextView
        android:id="@+id/lblComments"
        android:layout_width="wrap_content"
        android:layout_height="wrap_content"
        android:text="Hello, Android!"
        android:layout_alignParentTop="true"
        android:layout_alignParentLeft="true"
        />
    <FrameLayout
        android:layout_width="wrap_content"
        android:layout_height="wrap_content"
        android:layout_alignLeft="@+id/lblComments"
```

```
            android:layout_below="@+id/lblComments"
            android:layout_centerHorizontal="true" >
    <ImageView
        android:src = "@drawable/droid"
        android:layout_width="wrap_content"
        android:layout_height="wrap_content" />

    <Button
        android:layout_width="124dp"
        android:layout_height="wrap_content"
        android:text="Print Picture" />

    </FrameLayout>
</RelativeLayout>
```

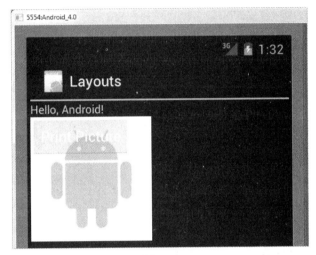

图 3-17

注意:虽然可以在 FrameLayout 中添加多个视图,但每一个视图都堆叠在前一个上面。这在想将一系列图像变成动画,使得每次只有一个视图可见的情况下很有用。

3.1.7 ScrollView

ScrollView 是一种特殊类型的 FrameLayout,因为它可以使用户滚动显示一个占据的空间大于物理显示的视图列表。ScrollView 只能包含一个子视图或视图组,通常是 LinearLayout。

注意:不要将 ScrollView 和 ListView(第 4 章将讨论)一起使用。ListView 用来显示一个相关信息的列表并针对大列表的处理进行了优化。

下面的 main.xml 的内容显示了一个包含 LinearLayout 的 ScrollView，而 LinearLayout 又包含了一些 Button 和 EditText 视图：

```xml
<ScrollView
    android:layout_width="fill_parent"
    android:layout_height="fill_parent"
    xmlns:android="http://schemas.android.com/apk/res/android" >

    <LinearLayout
        android:layout_width="fill_parent"
        android:layout_height="wrap_content"
        android:orientation="vertical" >
        <Button
            android:id="@+id/button1"
            android:layout_width="fill_parent"
            android:layout_height="wrap_content"
            android:text="Button 1" />
        <Button
            android:id="@+id/button2"
            android:layout_width="fill_parent"
            android:layout_height="wrap_content"
            android:text="Button 2" />
        <Button
            android:id="@+id/button3"
            android:layout_width="fill_parent"
            android:layout_height="wrap_content"
            android:text="Button 3" />
        <EditText
            android:id="@+id/txt"
            android:layout_width="fill_parent"
            android:layout_height="600dp" />
        <Button
            android:id="@+id/button4"
            android:layout_width="fill_parent"
            android:layout_height="wrap_content"
            android:text="Button 4" />
        <Button
            android:id="@+id/button5"
            android:layout_width="fill_parent"
            android:layout_height="wrap_content"
            android:text="Button 5" />
    </LinearLayout>
</ScrollView>
```

如果在 Android 模拟器中加载前面的代码，会看到如图 3-18 所示的效果。

图 3-18

因为 EditText 自动获得焦点，所以它会填充整个活动(因为高度被设为了 600dp)。为了防止它获得焦点，在<LinearLayout>元素中添加下面的两个属性：

```
<LinearLayout
    android:layout_width="fill_parent"
    android:layout_height="wrap_content"
    android:orientation="vertical"
    android:focusable="true"
    android:focusableInTouchMode="true" >
```

现在可以查看按钮并滚动视图列表了(如图 3-19 所示)。

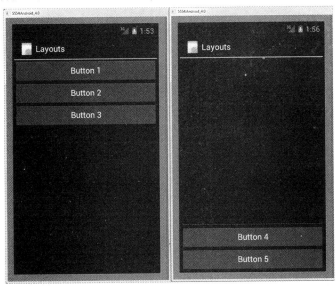

图 3-19

有时候可能想让 EditText 自动获得焦点,但是不想让软件输入面板(键盘)自动显示(在实际设备上会自动显示)。为了防止键盘显示,在 AndroidManifest.xml 文件的<activity>元素中添加下面的属性:

```
<activity
    android:label="@string/app_name"
    android:name=".LayoutsActivity"
    android:windowSoftInputMode="stateHidden" >
    <intent-filter >
        <action android:name="android.intent.action.MAIN" />
        <category android:name="android.intent.category.LAUNCHER" />
    </intent-filter>
</activity>
```

3.2 适应显示方向

现代智能手机的主要特征之一是它们切换屏幕方向的能力,Android 也不例外。Android 支持两种屏幕方向:纵向和横向。默认情况下,当改变 Android 设备的显示方向时,当前显示的活动将自动在新方向上重绘其内容。这是因为当显示方向上发生改变时,都会触发活动的 onCreate()方法。

注意:当改变 Android 设备的方向时,当前活动实际上是先被销毁,然后再重新创建。

然而,当重绘时,视图可能会按照其原始位置绘制(这取决于所选择的布局)。图 3-20 展示了之前提到的一个例子,分别以纵向和横向模式进行显示。

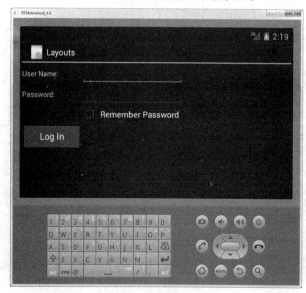

图 3-20

在横向模式下可以看到，屏幕右侧有大量的空余空间可以使用。而且，当屏幕方向被设置成横向时，任何位于屏幕底部的额外的视图将被隐藏。

通常，可以使用如下两种技术来处理屏幕方向的变化：
- 锚定——将视图锚定到屏幕的四条边是最容易的方法。当屏幕方向改变时，视图可以整齐地锚定于屏幕边缘。
- 调整大小和重新定位——尽管锚定和居中显示的技术简单，可以确保视图能处理屏幕方向变化，但最佳的技术还是根据当前屏幕方向重新调整每一个视图的大小。

3.2.1 锚定视图

使用 RelativeLayout 可以很容易实现锚定。考虑下列 main.xml 文件，其中包含了嵌入在<RelativeLayout>元素中的 5 个 Button 视图：

```xml
<RelativeLayout
    android:layout_width="fill_parent"
    android:layout_height="fill_parent"
    xmlns:android="http://schemas.android.com/apk/res/android">
    <Button
        android:id="@+id/button1"
        android:layout_width="wrap_content"
        android:layout_height="wrap_content"
        android:text="Top Left"
        android:layout_alignParentLeft="true"
        android:layout_alignParentTop="true" />
    <Button
        android:id="@+id/button2"
        android:layout_width="wrap_content"
        android:layout_height="wrap_content"
        android:text="Top Right"
        android:layout_alignParentTop="true"
        android:layout_alignParentRight="true" />
    <Button
        android:id="@+id/button3"
        android:layout_width="wrap_content"
        android:layout_height="wrap_content"
        android:text="Bottom Left"
        android:layout_alignParentLeft="true"
        android:layout_alignParentBottom="true" />
    <Button
        android:id="@+id/button4"
        android:layout_width="wrap_content"
        android:layout_height="wrap_content"
        android:text="Bottom Right"
        android:layout_alignParentRight="true"
        android:layout_alignParentBottom="true" />
    <Button
        android:id="@+id/button5"
        android:layout_width="fill_parent"
        android:layout_height="wrap_content"
```

```
            android:text="Middle"
            android:layout_centerVertical="true"
            android:layout_centerHorizontal="true" />
</RelativeLayout>
```

注意不同 Button 视图中所具有的以下属性：

- layout_alignParentLeft——将视图与其父视图的左侧对齐
- layout_alignParentRight——将视图与其父视图的右侧对齐
- layout_alignParentTop——将视图与其父视图的上部对齐
- layout_alignParentBottom——将视图与其父视图的底部对齐
- layout_centerVertical——使视图在其父视图中垂直居中
- layout_centerHorizontal——使视图在其父视图中水平居中

图 3-21 展示了以纵向模式观察到的活动的效果。

当屏幕方向改变为横向模式时，4 个按钮对齐到屏幕的四边。中间的按钮在屏幕中央显示，宽度完全拉伸到整个屏幕(如图 3-22 所示)。

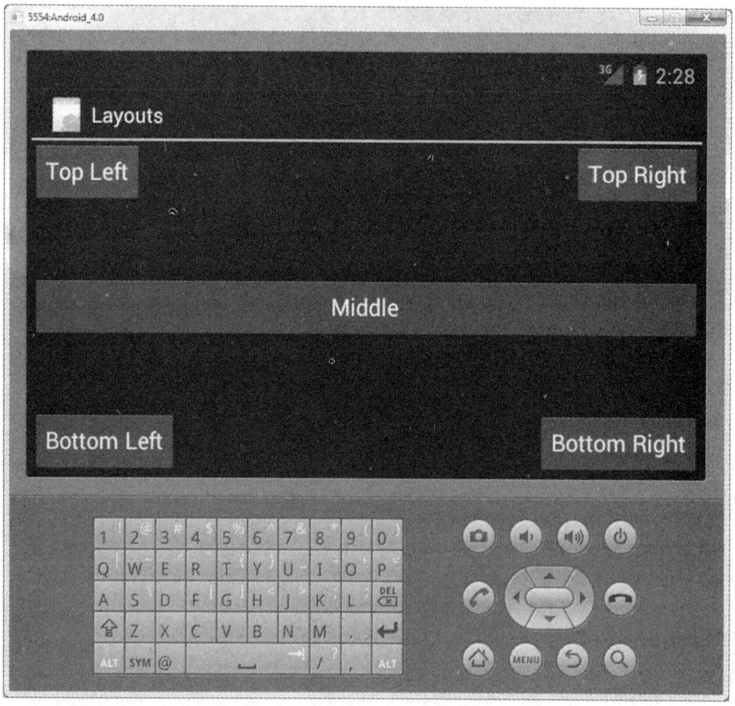

图 3-21

图 3-22

3.2.2 调整大小和重新定位

除了将视图锚定到屏幕的四边之外，基于屏幕方向定制用户界面的更简单的方法是为每个方向上的用户界面创建一个包含 XML 文件的单独的 res/layout 文件夹。为了支持横向模式，可以在 res 文件夹下创建一个新文件夹，并命名为 layout-land(表示横向)。图 3-23 显示了这样的含有 main.xml 文件的新文件夹。

基本上，包含在 layout 文件夹下的 main.xml 文件为活动定义了纵向模式的用户界面，而在 layout-land 文件夹下的 main.xml 文件定义了横向模式的用户界面。

图 3-21

layout 文件夹下的 main.xml 文件的内容如下所示：

```xml
<?xml version="1.0" encoding="utf-8"?>
<RelativeLayout
    android:layout_width="fill_parent"
    android:layout_height="fill_parent"
    xmlns:android="http://schemas.android.com/apk/res/android">
    <Button
        android:id="@+id/button1"
        android:layout_width="wrap_content"
        android:layout_height="wrap_content"
        android:text="Top Left"
        android:layout_alignParentLeft="true"
        android:layout_alignParentTop="true" />
    <Button
        android:id="@+id/button2"
        android:layout_width="wrap_content"
        android:layout_height="wrap_content"
        android:text="Top Right"
        android:layout_alignParentTop="true"
        android:layout_alignParentRight="true" />
    <Button
        android:id="@+id/button3"
        android:layout_width="wrap_content"
        android:layout_height="wrap_content"
        android:text="Bottom Left"
        android:layout_alignParentLeft="true"
        android:layout_alignParentBottom="true" />
    <Button
        android:id="@+id/button4"
        android:layout_width="wrap_content"
        android:layout_height="wrap_content"
        android:text="Bottom Right"
        android:layout_alignParentRight="true"
        android:layout_alignParentBottom="true" />
    <Button
```

```xml
        android:id="@+id/button5"
        android:layout_width="fill_parent"
        android:layout_height="wrap_content"
        android:text="Middle"
        android:layout_centerVertical="true"
        android:layout_centerHorizontal="true" />
</RelativeLayout>
```

下面显示了 layout-land 文件夹下的 main.xml 文件的内容(粗体显示的语句是横向模式下显示的额外视图):

```xml
<?xml version="1.0" encoding="utf-8"?>
<RelativeLayout
    android:layout_width="fill_parent"
    android:layout_height="fill_parent"
    xmlns:android="http://schemas.android.com/apk/res/android">
    <Button
        android:id="@+id/button1"
        android:layout_width="wrap_content"
        android:layout_height="wrap_content"
        android:text="Top Left"
        android:layout_alignParentLeft="true"
        android:layout_alignParentTop="true" />
    <Button
        android:id="@+id/button2"
        android:layout_width="wrap_content"
        android:layout_height="wrap_content"
        android:text="Top Right"
        android:layout_alignParentTop="true"
        android:layout_alignParentRight="true" />
    <Button
        android:id="@+id/button3"
        android:layout_width="wrap_content"
        android:layout_height="wrap_content"
        android:text="Bottom Left"
        android:layout_alignParentLeft="true"
        android:layout_alignParentBottom="true" />
    <Button
        android:id="@+id/button4"
        android:layout_width="wrap_content"
        android:layout_height="wrap_content"
        android:text="Bottom Right"
        android:layout_alignParentRight="true"
        android:layout_alignParentBottom="true" />
    <Button
        android:id="@+id/button5"
        android:layout_width="fill_parent"
        android:layout_height="wrap_content"
        android:text="Middle"
        android:layout_centerVertical="true"
```

```
        android:layout_centerHorizontal="true" />
    <Button
        android:id="@+id/button6"
        android:layout_width="180px"
        android:layout_height="wrap_content"
        android:text="Top Middle"
        android:layout_centerVertical="true"
        android:layout_centerHorizontal="true"
        android:layout_alignParentTop="true" />
    <Button
        android:id="@+id/button7"
        android:layout_width="180px"
        android:layout_height="wrap_content"
        android:text="Bottom Middle"
        android:layout_centerVertical="true"
        android:layout_centerHorizontal="true"
        android:layout_alignParentBottom="true" />
</RelativeLayout>
```

当活动以纵向模式加载时，将显示 5 个按钮，如图 3-24 所示。

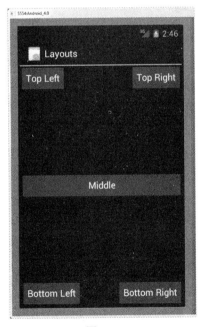

图 3-24

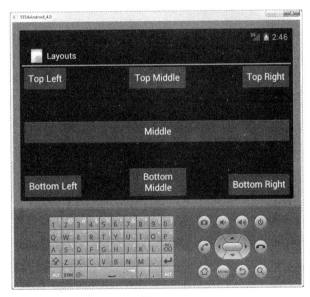

图 3-25

当活动以横向模式加载时，将出现 7 个按钮(如图 3-25 所示)，这证明当设备处于不同方向时，将加载不同的 XML 文件。

使用这种方法，当设备方向改变时，Android 将根据当前屏幕方向为活动自动加载相应的 XML 文件。

3.3 管理屏幕方向的变化

既然已经了解了如何实现两种技术以便适应屏幕方向的变化,那么让我们探究一下当设备改变方向时活动的状态会发生什么情况。

下面的"试一试"展示了当设备改变方向时活动的行为。

试一试 了解方向改变时活动的行为

Orientations.zip 代码文件可以在 Wrox.com 上下载

(1) 使用 Eclipse,创建一个名为 Orientations 的新项目。
(2) 在 main.xml 文件中添加下列粗体显示的语句:

```xml
<?xml version="1.0" encoding="utf-8"?>
<LinearLayout xmlns:android="http://schemas.android.com/apk/res/android"
    android:layout_width="fill_parent"
    android:layout_height="fill_parent"
    android:orientation="vertical" >

<EditText
    android:id="@+id/txtField1"
    android:layout_width="fill_parent"
    android:layout_height="wrap_content" />

<EditText
    android:layout_width="fill_parent"
    android:layout_height="wrap_content" />
</LinearLayout>
```

(3) 在 OrientationsActivity.java 文件中添加下列粗体显示的语句:

```java
package net.learn2develop.Orientations;

import android.app.Activity;
import android.os.Bundle;
import android.util.Log;

public class OrientationsActivity extends Activity {
    /** Called when the activity is first created. */
    @Override
    public void onCreate(Bundle savedInstanceState) {
        super.onCreate(savedInstanceState);
        setContentView(R.layout.main);
        Log.d("StateInfo", "onCreate");
    }

    @Override
```

```
    public void onStart() {
        Log.d("StateInfo", "onStart");
        super.onStart();
    }

    @Override
    public void onResume() {
        Log.d("StateInfo", "onResume");
        super.onResume();
    }

    @Override
    public void onPause() {
        Log.d("StateInfo", "onPause");
        super.onPause();
    }

    @Override
    public void onStop() {
        Log.d("StateInfo", "onStop");
        super.onStop();
    }

    @Override
    public void onDestroy() {
        Log.d("StateInfo", "onDestroy");
        super.onDestroy();
    }

    @Override
    public void onRestart() {
        Log.d("StateInfo", "onRestart");
        super.onRestart();
    }
```

(4) 按 F11 键在 Android 模拟器上调试应用程序。

(5) 在两个 EditText 视图中输入一些文本(如图 3-26 所示)。

(6) 按 Ctrl+F11 组合键改变 Android 模拟器的显示方向。图 3-27 显示了横向模式下的模拟器效果。注意，第 1 个 EditText 视图中的文本仍旧可见，而第 2 个 EditText 视图为空。

图 3-26

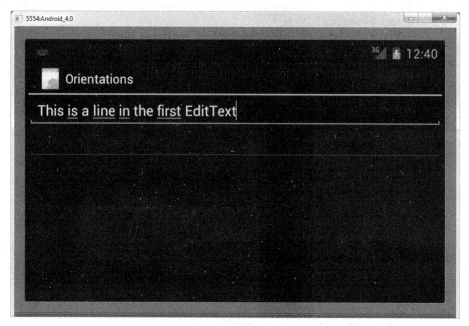

图 3-27

(7) 观察 LogCat 窗口中的输出(需要在 Eclipse 中切换到 Debug 透视图)，应该看到类似如下所示内容：

```
12-15 12:27:20.747: D/StateInfo(557): onCreate
12-15 12:27:20.747: D/StateInfo(557): onStart
12-15 12:27:20.747: D/StateInfo(557): onResume
...
12-15 12:39:37.846: D/StateInfo(557): onPause
12-15 12:39:37.846: D/StateInfo(557): onStop
12-15 12:39:37.866: D/StateInfo(557): onDestroy
12-15 12:39:38.206: D/StateInfo(557): onCreate
12-15 12:39:38.216: D/StateInfo(557): onStart
12-15 12:39:38.257: D/StateInfo(557): onResume
```

示例说明

从 LogCat 窗口中的输出内容可以明显看出，当设备改变方向时，活动先被销毁：

```
12-15 12:39:37.846: D/StateInfo(557): onPause
12-15 12:39:37.846: D/StateInfo(557): onStop
12-15 12:39:37.866: D/StateInfo(557): onDestroy
```

然后，它被重建：

```
12-15 12:39:38.206: D/StateInfo(557): onCreate
12-15 12:39:38.216: D/StateInfo(557): onStart
12-15 12:39:38.257: D/StateInfo(557): onResume
```

了解这一行为是重要的，因为需要确保采取必要的措施来保持方向改变之前活动的状态。例如，在活动中可能包含一些计算所需的值的变量。对于任意活动，应该在 onPause()

事件中保存任何您需要保存的状态，这一事件在每次活动改变方向时触发。下一节讲述保存状态信息的不同方法。

另一个需要理解的重要行为是当包含视图的活动被销毁时，只有那些在这个活动中被命名的视图(通过 android:id 属性)才能保持它们的状态。例如，在向 EditText 视图中输入一些文本的同时，用户可能会改变显示方向。出现这种情况时，EditText 视图中的任何文本将被保持并在活动重新创建时自动恢复。相反，如果没有使用 android:id 属性命名 EditText 视图，活动将无法保持视图中当前所含文本。

3.3.1 配置改变时保持状态信息

到目前为止，您已经学习了屏幕方向改变时会销毁和重建一个活动。记住，当重建一个活动时，活动的当前状态可能会丢失。当终止一个活动时，将触发以下两个方法中的一个或两个：

- onPause()——当一个活动被终止或转入后台时都会触发这一方法。
- onSaveInstanceState()——当一个活动将要被终止或转入后台时，也会触发这一方法(正如 onPause()方法一样)。然而，与 onPause()方法不同的是，当一个活动从栈中卸载时(例如用户按下了 Back 按钮)不会触发 onSaveInstanceState()方法，这是因为后面无须恢复其状态。

简而言之，要保持一个活动的状态，可以实现 onPause()方法，然后使用自己的方法来保存活动状态，如利用数据库、内部或外部的文件存储器等。

如果只是想保存活动状态用来在以后活动重建时(例如当设备改变方向时)进行恢复，那么更简单的方法是实现 onSaveInstanceState()方法。这个方法提供了 Bundle 对象作为一个参数，可以用它保存活动的状态。以下代码展示了在 onSaveInstanceState 方法中可以将字符串 ID 保存到 Bundle 对象中：

```
@Override
public void onSaveInstanceState(Bundle outState) {
    //---save whatever you need to persist---
    outState.putString("ID", "1234567890");
    super.onSaveInstanceState(outState);
}
```

当重建一个活动时，首先触发 onCreate()事件，随后是 onRestoreInstanceState()方法。此事件可以使您检索先前在 onSaveInstanceState 方法中通过其参数 Bundle 对象保存的状态：

```
@Override
public void onRestoreInstanceState(Bundle savedInstanceState) {
    super.onRestoreInstanceState(savedInstanceState);
    //---retrieve the information persisted earlier---
    String ID = savedInstanceState.getString("ID");
}
```

尽管可以使用 onSaveInstanceState()方法来保存状态信息，但要注意只能通过一个

Bundle 对象保存状态信息的局限性。如果需要保存更复杂的数据结构，这就不是一个合适的解决方法。

另一个可以使用的方法是 onRetainNonConfigurationInstance()方法。当一个活动由于配置改变(例如屏幕方向的变化、键盘是否可用等)将要被销毁时会触发这一方法。可以通过在该方法中返回来保存当前的数据，如下所示：

```
@Override
public Object onRetainNonConfigurationInstance() {
    //---save whatever you want here; it takes in an Object type---
    return("Some text to preserve");
}
```

注意：当屏幕方向改变时，这一变化是所谓配置改变的一部分。配置改变将销毁您的当前活动。

注意，此方法返回一个 Object 类型，它几乎允许返回任何数据类型。要提取保存的数据，可以使用 getLastNonConfigurationInstance()方法在 onCreate()方法中进行提取，如下所示：

```
@Override
public void onCreate(Bundle savedInstanceState) {
    super.onCreate(savedInstanceState);
    setContentView(R.layout.main);
    Log.d("StateInfo", "onCreate");
    String str = (String) getLastNonConfigurationInstance();
}
```

当需要临时保存一些数据，例如从 Web 服务中下载了数据，而用户改变了屏幕方向时，十分适合 onRetainNonConfigurationInstance()和 getLastNonConfigurationInstance()方法。在这种情况中，使用这两种方法保存数据要比再次下载数据高效得多。

3.3.2 检测方向改变

有时需要在运行时知道设备的当前方向。要确定这一点，可以使用 WindowManger 类。以下代码片段演示了如何以编程方式来检测活动的当前方向：

```
import android.view.Display;
import android.view.WindowManager;

    @Override
    public void onCreate(Bundle savedInstanceState) {
        super.onCreate(savedInstanceState);
        setContentView(R.layout.main);

        //---get the current display info---
        WindowManager wm = getWindowManager();
```

```
        Display d = wm.getDefaultDisplay();

        if (d.getWidth() > d.getHeight()) {
            //---landscape mode---
            Log.d("Orientation", "Landscape mode");
        }
        else {
            //---portrait mode---
            Log.d("Orientation", "Portrait mode");
        }
    }
```

getDefaultDisplay()方法返回一个表示设备屏幕的 Display 对象。然后，您可以获得其宽度和高度，进而推断出当前的屏幕方向。

3.3.3 控制活动的方向

有时，您也许想要保证应用程序只在一个特定的方向上显示。例如，您可能正在编写一个只能以横向模式呈现的游戏。在这种情况下，可以使用 Activity 类的 setRequestOrientation()方法以编程方式强制改变显示方向：

```
import android.content.pm.ActivityInfo;
    @Override
    public void onCreate(Bundle savedInstanceState) {
        super.onCreate(savedInstanceState);
        setContentView(R.layout.main);

        //---change to landscape mode---

        setRequestedOrientation(ActivityInfo.SCREEN_ORIENTATION_LANDSC
        APE);
    }
```

要改为纵向模式，可使用 ActivityInfo.SCREEN_ORIENTATION_PORTRAIT 常量。

除了使用 setRequestOrientation()方法，还可以在 AndroidManifest.xml 文件中的<activity>元素上使用 android:screenOrientation 属性，来将活动限制在一个特定方向上，如下所示：

```
<?xml version="1.0" encoding="utf-8"?>
<manifest xmlns:android="http://schemas.android.com/apk/res/android"
    package="net.learn2develop.Orientations"
    android:versionCode="1"
    android:versionName="1.0" >

    <uses-sdk android:minSdkVersion="14" />

    <application
        android:icon="@drawable/ic_launcher"
        android:label="@string/app_name" >
```

```
        <activity
            android:label="@string/app_name"
            android:name=".OrientationsActivity"
            android:screenOrientation="landscape" >
            <intent-filter >
                <action android:name="android.intent.action.MAIN" />
                <category android:name="android.intent.category.
                   LAUNCHER" />
            </intent-filter>
        </activity>
    </application>

</manifest>
```

上述的例子将活动限制在一个特定的方向上(此处是横向)，并且防止活动被销毁。也就是说，当设备方向改变时，活动不会被销毁并且也不会再次触发 onCreate()方法。

以下是在 android:screenOrientation 属性中可以指定的两个值：
- portrait——纵向模式
- sensor——基于加速计(默认)

3.4 使用 Action Bar

除了碎片，Android 3 和 Android 4 中引入的另一个新功能是 Action Bar。Action Bar 代替了传统的位于设备屏幕顶部的标题栏，它显示应用程序的图标和活动的名称。Action Bar 的右侧还可以有可选的动作项。图 3-28 显示了内置的 Email 应用程序在 Action Bar 中显示了应用程序图标、活动名称和一些动作项。下一节将详细讨论工作项。

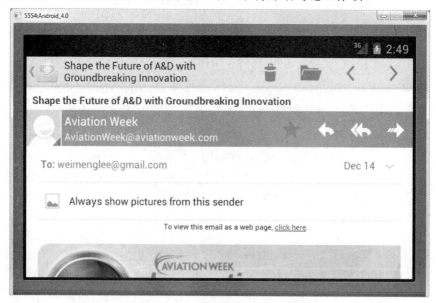

图 3-28

第 3 章 Android 用户界面

下面的"试一试"显示了如何以编程方式隐藏或显示 Action Bar。

试一试　　显示和隐藏 Action Bar

(1) 使用 Eclipse 创建一个新的 Android 项目,命名为 MyActionBar。

(2) 按 F11 键在 Android 模拟器上调试应用程序。应该会看到应用程序及位于其屏幕顶部的 Action Bar(包含应用程序图标和应用程序的名称 MyActionBar,如图 3-29 所示)。

图 3-29

(3) 为了隐藏 Action Bar,在 AndroidManifest.xml 文件中添加下面的粗体显示的代码:

```xml
<?xml version="1.0" encoding="utf-8"?>
<manifest xmlns:android="http://schemas.android.com/apk/res/android"
    package="net.learn2develop.MyActionBar"
    android:versionCode="1"
    android:versionName="1.0" >

    <uses-sdk android:minSdkVersion="14" />

    <application
        android:icon="@drawable/ic_launcher"
        android:label="@string/app_name" >
        <activity
            android:label="@string/app_name"
            android:name=".MyActionBarActivity"
            android:theme="@android:style/Theme.Holo.NoActionBar" >
            <intent-filter >
                <action android:name="android.intent.action.MAIN" />
                <category android:name="android.intent.category.LAUNCHER" />
            </intent-filter>
        </activity>
    </application>

</manifest>
```

(4) 在 Eclipse 中选择项目名称,然后按 F11 键在 Android 模拟器上再次调试应用程序。这一次,Action Bar 不会显示,如图 3-30 所示。

(5) 也可以以编程方式使用 ActionBar 类

图 3-30

移除 Action Bar。为此，首先需要删除前面的步骤中添加的 android:theme 属性。这一点很重要，否则下一步将导致应用程序引发一个异常。

(6) 按照如下代码修改 MyActionBarActivity.java 文件：

```java
package net.learn2develop.MyActionBar;

import android.app.ActionBar;
import android.app.Activity;
import android.os.Bundle;

public class MyActionBarActivity extends Activity {
    /** Called when the activity is first created. */
    @Override
    public void onCreate(Bundle savedInstanceState) {
        super.onCreate(savedInstanceState);
        setContentView(R.layout.main);

        ActionBar actionBar = getActionBar();
        actionBar.hide();
        //actionBar.show(); //---show it again---
    }
}
```

(7) 按 F11 键在模拟器上再次调试应用程序。Action Bar 将被隐藏。

示例说明

使用 android:theme 属性可以关闭活动对 Action Bar 的显示。将这个属性设为 @android:style/Theme.Holo.NoActionBar 会隐藏 Action Bar。或者，也可以通过编程方式，在运行时使用 getActionBar()方法获得对 Action Bar 的引用。调用 hide()方法会隐藏 Action Bar，调用 show()方法则会显示它。

注意，如果使用 android:theme 属性来关闭 Action Bar，在运行时调用 getActionBar()方法会返回 null。因此，使用 ActionBar 类以编程方式打开/关闭 Action Bar 总是更好的方法。

3.4.1 向 Action Bar 添加动作项

除了在 Action Bar 的左侧显示应用程序图标和活动名称以外，还可以在 Action Bar 上显示其他项，这些项叫做动作项。动作项是应用程序中经常执行的一些操作的快捷方式。例如，您可能正在构建一个 RSS 阅读器应用程序，此时 Refresh feed、Delete feed 和 Add new feed 就可以作为动作项。

下面的"试一试"演示了如何向 Action Bar 添加动作项。

试一试　　添加动作项

(1) 使用前一节创建的同一个 MyActionBar 项目，在 MyActionBarActivity.java 文件中

添加下面的粗体显示的代码：

```
package net.learn2develop.MyActionBar;

import android.app.Activity;
import android.os.Bundle;
import android.view.Menu;
import android.view.MenuItem;
import android.widget.Toast;

public class MyActionBarActivity extends Activity {
    /** Called when the activity is first created. */
    @Override
    public void onCreate(Bundle savedInstanceState) {
        super.onCreate(savedInstanceState);
        setContentView(R.layout.main);

        //ActionBar actionBar = getActionBar();
        //actionBar.hide();
        //actionBar.show(); //---show it again---
    }

    @Override
    public boolean onCreateOptionsMenu(Menu menu) {
        super.onCreateOptionsMenu(menu);
        CreateMenu(menu);
        return true;
    }

    @Override
    public boolean onOptionsItemSelected(MenuItem item)
    {
        return MenuChoice(item);
    }

    private void CreateMenu(Menu menu)
    {
        MenuItem mnu1 = menu.add(0, 0, 0, "Item 1");
        {
            mnu1.setIcon(R.drawable.ic_launcher);
            mnu1.setShowAsAction(MenuItem.SHOW_AS_ACTION_IF_ROOM);
        }
        MenuItem mnu2 = menu.add(0, 1, 1, "Item 2");
        {
            mnu2.setIcon(R.drawable.ic_launcher);
            mnu2.setShowAsAction(MenuItem.SHOW_AS_ACTION_IF_ROOM);
        }
        MenuItem mnu3 = menu.add(0, 2, 2, "Item 3");
```

```java
        {
            mnu3.setIcon(R.drawable.ic_launcher);
            mnu3.setShowAsAction(MenuItem.SHOW_AS_ACTION_IF_ROOM);
        }
        MenuItem mnu4 = menu.add(0, 3, 3, "Item 4");
        {
            mnu4.setShowAsAction(MenuItem.SHOW_AS_ACTION_IF_ROOM);
        }
        MenuItem mnu5 = menu.add(0, 4, 4, "Item 5");
        {
            mnu5.setShowAsAction(MenuItem.SHOW_AS_ACTION_IF_ROOM);
        }
    }

    private boolean MenuChoice(MenuItem item)
    {
        switch (item.getItemId()) {
        case 0:
            Toast.makeText(this, "You clicked on Item 1",
                Toast.LENGTH_LONG).show();
            return true;
        case 1:
            Toast.makeText(this, "You clicked on Item 2",
                Toast.LENGTH_LONG).show();
            return true;
        case 2:
            Toast.makeText(this, "You clicked on Item 3",
                Toast.LENGTH_LONG).show();
            return true;
        case 3:
            Toast.makeText(this, "You clicked on Item 4",
                Toast.LENGTH_LONG).show();
            return true;
        case 4:
            Toast.makeText(this, "You clicked on Item 5",
                Toast.LENGTH_LONG).show();
            return true;
        }
        return false;
    }
}
```

(2) 按 F11 键在 Android 模拟器上调试应用程序。观察 Action Bar 右侧的图标(如图 3-31 所示)。如果在模拟器中单击 MENU 按钮，会看到其余的菜单项，如图 3-32 所示。这叫做溢出菜单(overflow menu)。在没有 MENU 按钮的设备上，溢出菜单由一个带箭头的图标表示。图 3-33 显示了运行在 Asus Eee Pad Transformer(Android 3.2.1)上的相同应用程序。单击溢出菜单会显示其余菜单项。

(3) 单击每个菜单项会使 Toast 类显示所选菜单项的名称，如图 3-34 所示。

图 3-31

图 3-32

(4) 按 Control+F11 组合键将模拟器的显示方向改为横向。现在在 Action Bar 上可以看到 4 个动作项，3 个带图标，一个带文本，如图 3-35 所示。

图 3-33

图 3-34

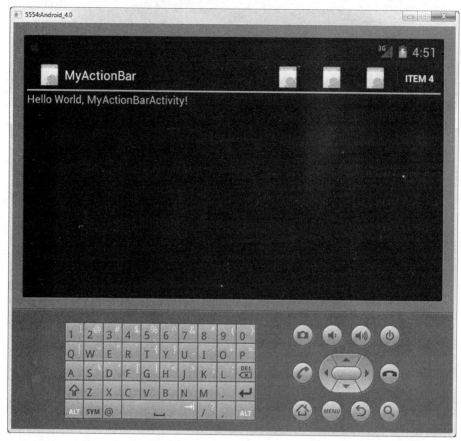

图 3-35

示例说明

通过调用活动的 onCreateOptionsMenu()方法，为 Action Bar 添加动作项：

```
@Override
public boolean onCreateOptionsMenu(Menu menu) {
    super.onCreateOptionsMenu(menu);
    CreateMenu(menu);
    return true;
}
```

在前面的示例中，调用了 CreateMenu()方法来显示一个菜单项列表：

```
private void CreateMenu(Menu menu)
{
    MenuItem mnu1 = menu.add(0, 0, 0, "Item 1");
    {
        mnu1.setIcon(R.drawable.ic_launcher);
        mnu1.setShowAsAction(MenuItem.SHOW_AS_ACTION_IF_ROOM);
    }
    MenuItem mnu2 = menu.add(0, 1, 1, "Item 2");
    {
        mnu2.setIcon(R.drawable.ic_launcher);
```

```
            mnu2.setShowAsAction(MenuItem.SHOW_AS_ACTION_IF_ROOM);
    }
    MenuItem mnu3 = menu.add(0, 2, 2, "Item 3");
    {
            mnu3.setIcon(R.drawable.ic_launcher);
            mnu3.setShowAsAction(MenuItem.SHOW_AS_ACTION_IF_ROOM);
    }
    MenuItem mnu4 = menu.add(0, 3, 3, "Item 4");
    {
            mnu4.setShowAsAction(MenuItem.SHOW_AS_ACTION_IF_ROOM);
    }
    MenuItem mnu5 = menu.add(0, 4, 4, "Item 5");
    {
            mnu5.setShowAsAction(MenuItem.SHOW_AS_ACTION_IF_ROOM);
    }
}
```

为了使菜单项显示为动作项，使用 SHOW_AS_ACTION_IF_ROOM 常量调用其 setShowAsAction() 方法。这告知 Android 设备如果 Action Bar 上有空间，将该菜单项显示为一个动作项。

当用户选择一个菜单项时，onOptionsItemSelected()方法将被调用：

```
@Override
public boolean onOptionsItemSelected(MenuItem item)
{
    return MenuChoice(item);
}
```

这里调用了自定义的 MenuChoice()方法来检查哪个菜单项被单击，然后输出一条消息：

```
private boolean MenuChoice(MenuItem item)
{
    switch (item.getItemId()) {
    case 0:
        Toast.makeText(this, "You clicked on Item 1",
            Toast.LENGTH_LONG).show();
        return true;
    case 1:
        Toast.makeText(this, "You clicked on Item 2",
            Toast.LENGTH_LONG).show();
        return true;
    case 2:
        Toast.makeText(this, "You clicked on Item 3",
            Toast.LENGTH_LONG).show();
        return true;
    case 3:
        Toast.makeText(this, "You clicked on Item 4",
            Toast.LENGTH_LONG).show();
        return true;
    case 4:
        Toast.makeText(this, "You clicked on Item 5",
```

```
            Toast.LENGTH_LONG).show();
        return true;
    }
    return false;
}
```

3.4.2 定制动作项和应用程序图标

在前面的示例中,显示的菜单项没有文本。如果想要为动作项同时显示图标和文本,可以使用"|"操作符和 MenuItem.SHOW_AS_ACTION_WITH_TEXT 常量:

```
MenuItem mnu1 = menu.add(0, 0, 0, "Item 1");
{
    mnu1.setIcon(R.drawable.ic_launcher);
    mnu1.setShowAsAction(
        MenuItem.SHOW_AS_ACTION_IF_ROOM |
        MenuItem.SHOW_AS_ACTION_WITH_TEXT);
}
```

这样,图标的旁边就会显示菜单项的文本,如图 3-36 所示。

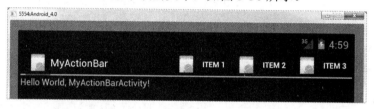

图 3-36

除了单击动作项,用户还可以单击 Action Bar 中的应用程序图标。此时,onOptionsItem-Selected()方法将被调用。为了确定被调用的应用程序图标,比较该项的 ID 和 android.R.id.home 常量:

```
private boolean MenuChoice(MenuItem item)
{
    switch (item.getItemId()) {
    case android.R.id.home:
        Toast.makeText(this,
            "You clicked on the Application icon",
            Toast.LENGTH_LONG).show();
        return true;

    case 0:
        Toast.makeText(this, "You clicked on Item 1",
            Toast.LENGTH_LONG).show();
        return true;
    case 1:
        //...
    }
    return false;
}
```

为使应用程序图标可被单击，需要调用 setDisplayHomeAsUpEnabled()方法：

```
@Override
public void onCreate(Bundle savedInstanceState) {
    super.onCreate(savedInstanceState);
    setContentView(R.layout.main);

    ActionBar actionBar=getActionBar();
    actionBar.setDisplayHomeAsUpEna
     bled(true);
    //actionBar.hide();
    //actionBar.show(); //---show it
     again---
}
```

图 3-37 显示了在应用程序图标的旁边出现了一个箭头按钮。

应用程序图标经常被应用程序用来返回到其主活动中。例如，应用程序中可能包含几个活动，此时可以使用应用程序图标作为一种快捷方式，让用户可以直接返回到应用程序的主活动中。为此，创建一个 Intent 对象并使用 Intent.FLAG_ACTIVITY_CLEAR_TOP 标签设置它总是一种很好的做法：

图 3-37

```
case android.R.id.home:
    Toast.makeText(this,
        "You clicked on the Application icon",
        Toast.LENGTH_LONG).show();

    Intent i = new Intent(this, MyActionBarActivity.class);
    i.addFlags(Intent.FLAG_ACTIVITY_CLEAR_TOP);
    startActivity(i);

    return true;
```

Intent.FLAG_ACTIVITY_CLEAR_TOP 标签确保了当用户单击 Action Bar 中的应用程序图标时，back stack 中的一系列活动被清除。这样，如果用户单击 Back 按钮，应用程序中的其他活动就不会再次显示。

3.5 以编程方式创建用户界面

到目前为止，您在本章中所看到的所有用户界面都是使用 XML 创建的。如前所述，除了使用 XML，还可以使用代码创建用户界面。如果用户界面需要在运行时动态生成，这个方法就很有用。例如，设想您正在构建一个电影票预订系统，您的应用程序将使用按钮显示每场电影的座位。这种情况下，就需要根据用户选择的电影来动态生成用户界面。

下面的"试一试"演示了在活动中动态构建用户界面所需的代码。

试一试 通过代码创建用户界面

UICode.zip 代码文件可以在Wrox.com 上下载

(1) 使用 Eclipse，创建一个名为 UICode 的新 Android 项目。
(2) 在 UICodeActivity.java 文件中，添加下列粗体显示的语句：

```java
package net.learn2develop.UICode;

import android.app.Activity;
import android.os.Bundle;
import android.view.ViewGroup.LayoutParams;
import android.widget.Button;
import android.widget.LinearLayout;
import android.widget.TextView;

public class UICodeActivity extends Activity {
    /** Called when the activity is first created. */
    @Override
    public void onCreate(Bundle savedInstanceState) {
        super.onCreate(savedInstanceState);
        //setContentView(R.layout.main);
        //---param for views---
        LayoutParams params =
            new LinearLayout.LayoutParams(
                LayoutParams.FILL_PARENT,
                LayoutParams.WRAP_CONTENT);

        //---create a layout---
        LinearLayout layout = new LinearLayout(this);
        layout.setOrientation(LinearLayout.VERTICAL);

        //---create a textview---
        TextView tv = new TextView(this);
        tv.setText("This is a TextView");
        tv.setLayoutParams(params);

        //---create a button---
        Button btn = new Button(this);
        btn.setText("This is a Button");
        btn.setLayoutParams(params);

        //---adds the textview---
        layout.addView(tv);
```

```
            //---adds the button---
            layout.addView(btn);

            //---create a layout param for the layout---
            LinearLayout.LayoutParams layoutParam =
                new LinearLayout.LayoutParams(
                        LayoutParams.FILL_PARENT,
                        LayoutParams.WRAP_CONTENT );

            this.addContentView(layout, layoutParam);
        }
    }
```

(3) 按 F11 键在 Android 模拟器上调试应用程序。图 3-38 显示活动被创建了。

图 3-38

示例说明

本例中,首先注释掉 setContentView()语句,这样就不从 main.xml 文件加载用户界面。
然后,创建一个 LayoutParams 对象来指定可以被别的视图(接下来将创建)使用的布局参数:

```
    //---param for views---
    LayoutParams params =
        new LinearLayout.LayoutParams(
            LayoutParams.FILL_PARENT,
            LayoutParams.WRAP_CONTENT);
```

还创建一个 LinearLayout 对象来包含活动中所有的视图:

```
    //---create a layout---
    LinearLayout layout = new LinearLayout(this);
    layout.setOrientation(LinearLayout.VERTICAL);
```

下一步,创建一个 TextView 视图和一个 Button 视图:

```
//---create a textview---
TextView tv = new TextView(this);
tv.setText("This is a TextView");
tv.setLayoutParams(params);

//---create a button---
Button btn = new Button(this);
btn.setText("This is a Button");
btn.setLayoutParams(params);
```

然后将它们添加到 LinearLayout 对象中：

```
//---adds the textview---
layout.addView(tv);

//---adds the button---
layout.addView(btn);
```

另外，创建一个被 LinearLayout 对象使用的 LayoutParams 对象：

```
//---create a layout param for the layout---
LinearLayout.LayoutParams layoutParam =
    new LinearLayout.LayoutParams(
        LayoutParams.FILL_PARENT,
        LayoutParams.WRAP_CONTENT );
```

最后，将 LinearLayout 对象添加到活动中：

```
this.addContentView(layout,layoutParam);
```

我们可以看出，使用代码创建用户界面是一个非常费力的工作。因此，只有在必要时才使用代码来动态生成用户界面。

3.6 侦听用户界面通知

用户和用户界面在两个层面上进行交互：活动层面和视图层面。在活动层面，Activity 类暴露了可以被重写的方法。一些常见的可以在活动中被重写的方法如下所示：

- onKeyDown——当一个键被按下并且没有被活动中的任何视图处理时调用
- onKeyUp——当一个键被释放并且没有被活动中的任何视图处理时调用
- onMenuItemSelected——当用户选择了面板的菜单时调用(见第 5 章)
- onMenuOpened——当用户打开了面板的菜单时调用(见第 5 章)

3.6.1 重写活动中定义的方法

为了演示活动是如何与用户交互的，下面的示例重写了活动的基类中定义的一些方法。

试一试　　重写活动的方法

> *UIActivity.zip 代码文件可以在 Wrox.com 上下载*

(1) 使用 Eclipse，创建一个新的 Android 项目，并命名为 UIActivity。
(2) 在 main.xml 文件中添加下列粗体显示的语句(替换 TextView)：

```xml
<?xml version="1.0" encoding="utf-8"?>
<LinearLayout xmlns:android="http://schemas.android.com/apk/res/android"
    android:layout_width="fill_parent"
    android:layout_height="fill_parent"
    android:orientation="vertical" >

    <TextView
        android:layout_width="214dp"
        android:layout_height="wrap_content"
        android:text="Your Name" />
    <EditText
        android:id="@+id/txt1"
        android:layout_width="214dp"
        android:layout_height="wrap_content" />
    <Button
        android:id="@+id/btn1"
        android:layout_width="106dp"
        android:layout_height="wrap_content"
        android:text="OK" />
    <Button
        android:id="@+id/btn2"
        android:layout_width="106dp"
        android:layout_height="wrap_content"
        android:text="Cancel" />

</LinearLayout>
```

(3) 在 UIActivityActivity.java 文件中添加下列粗体显示的语句：

```java
package net.learn2develop.UIActivity;

import android.app.Activity;
import android.os.Bundle;
import android.view.KeyEvent;
import android.widget.Toast;

public class UIActivityActivity extends Activity {
    /** Called when the activity is first created. */
    @Override
    public void onCreate(Bundle savedInstanceState) {
        super.onCreate(savedInstanceState);
        setContentView(R.layout.main);
```

}

```
@Override
public boolean onKeyDown(int keyCode, KeyEvent event)
{
    switch (keyCode)
    {
        case KeyEvent.KEYCODE_DPAD_CENTER:
            Toast.makeText(getBaseContext(),
                    "Center was clicked",
                    Toast.LENGTH_LONG).show();
            break;
        case KeyEvent.KEYCODE_DPAD_LEFT:
            Toast.makeText(getBaseContext(),
                    "Left arrow was clicked",
                    Toast.LENGTH_LONG).show();
            break;
        case KeyEvent.KEYCODE_DPAD_RIGHT:
            Toast.makeText(getBaseContext(),
                    "Right arrow was clicked",
                    Toast.LENGTH_LONG).show();
            break;
        case KeyEvent.KEYCODE_DPAD_UP:
            Toast.makeText(getBaseContext(),
                    "Up arrow was clicked",
                    Toast.LENGTH_LONG).show();
            break;
        case KeyEvent.KEYCODE_DPAD_DOWN:
            Toast.makeText(getBaseContext(),
                    "Down arrow was clicked",
                    Toast.LENGTH_LONG).show();
            break;
    }
    return false;
}
```

(4) 按 F11 键在 Android 模拟器上调试应用程序。

(5) 当活动加载后，在 EditText 中输入一些文本。下一步，单击方向键盘上的下箭头键。观察屏幕上显示的消息，如图 3-39 所示。

示例说明

当加载活动时，光标将在 EditText 视图中闪烁，因为它获得了焦点。

在 MainActivity 类中，按如下所示重写 Activity 基类的 onKeyDown()方法：

图 3-39

```java
@Override
public boolean onKeyDown(int keyCode, KeyEvent event)
{
    switch (keyCode)
    {
        case KeyEvent.KEYCODE_DPAD_CENTER:
            //...
            break;
        case KeyEvent.KEYCODE_DPAD_LEFT:
            //...
            break;
        case KeyEvent.KEYCODE_DPAD_RIGHT:
            //...
            break;
        case KeyEvent.KEYCODE_DPAD_UP:
            //...
            break;
        case KeyEvent.KEYCODE_DPAD_DOWN:
            //...
            break;
    }
    return false;
}
```

在 Android 中，当您按下设备上的任意一个键时，当前获得焦点的视图将试图处理生成的事件。本例中，当 EditText 获得焦点并且您按下了一个键时，EditText 视图将处理这一事件并在视图中将您刚刚输入的字符显示出来。然而，如果您按了上或下箭头键，EditText 视图不会对此作处理，相反会将这个事件传递给活动。在这种情况下，将调用 onKeyDown() 方法。本例中，您检查按下的键并显示一条消息表明该键被按下了。可以看到，现在焦点转移到了下一个视图上，即 OK 按钮。

有趣的是，如果 EditText 视图内已经有了一些文本，并且光标位于文本末尾，那么单击左箭头键不会触发 onKeyDown() 事件，而只是将光标向左移动一个字符。这是由于 EditText 视图已经处理了这一事件。如果按下的是右箭头键(当光标位于文本末尾时)，将调用 onKeyDown() 方法(因为现在 EditText 视图将不会处理这一事件)。这同样适用于光标位于 EditText 视图开始的情况：单击左箭头将触发 onKeyDown() 事件，而单击右箭头将只是将光标向右移动一个字符。

当 OK 按钮获得焦点时，按下方向键盘上的中间按钮。可以看到 Center was clicked 消息没有显示出来。这是由于 Button 视图本身处理了单击事件。因此，onKeyDown() 方法没有捕获这一事件。

还要注意，onKeyDown() 方法返回一个 boolean 类型的结果。当想告诉系统您已经处理完此事件并且系统不要再作进一步的处理时，应当返回 true。例如，考虑下列当每一个键匹配后返回 true 时的情况：

```java
@Override
public boolean onKeyDown(int keyCode, KeyEvent event)
{
    switch (keyCode)
    {
        case KeyEvent.KEYCODE_DPAD_CENTER:
            Toast.makeText(getBaseContext(),
                    "Center was clicked",
                    Toast.LENGTH_LONG).show();
            //break;
            return true;
        case KeyEvent.KEYCODE_DPAD_LEFT:
            Toast.makeText(getBaseContext(),
                    "Left arrow was clicked",
                    Toast.LENGTH_LONG).show();
             //break;
             return true;
        case KeyEvent.KEYCODE_DPAD_RIGHT:
            Toast.makeText(getBaseContext(),
                    "Right arrow was clicked",
                    Toast.LENGTH_LONG).show();
            //break;
            return true;
        case KeyEvent.KEYCODE_DPAD_UP:
            Toast.makeText(getBaseContext(),
                    "Up arrow was clicked",
                    Toast.LENGTH_LONG).show();

            //break;
            return true;
        case KeyEvent.KEYCODE_DPAD_DOWN:
            Toast.makeText(getBaseContext(),
                    "Down arrow was clicked",
                    Toast.LENGTH_LONG).show();
             //break;
             return true;
    }
    return false;
}
```

如果测试一下,就会发现现在不能使用箭头键在视图之间导航。

3.6.2 为视图注册事件

当用户和视图交互时,视图可以触发事件。例如,当用户触碰一个 Button 视图时,您需要响应这个事件以便能够采取适当的动作。要做到这一点,需要为视图显式地注册事件。

使用上一节讨论的同一个例子,我们知道那个活动有两个 Button 视图;因此,可以使用一个匿名类注册按钮单击事件,如下所示:

```java
package net.learn2develop.UIActivity;

import android.app.Activity;
import android.os.Bundle;
import android.view.KeyEvent;
import android.view.View;
import android.view.View.OnClickListener;
import android.widget.Button;
import android.widget.Toast;

public class UIActivityActivity extends Activity {
    /** Called when the activity is first created. */
    @Override
    public void onCreate(Bundle savedInstanceState) {
        super.onCreate(savedInstanceState);
        setContentView(R.layout.main);

        //---the two buttons are wired to the same event handler---
        Button btn1 = (Button)findViewById(R.id.btn1);
        btn1.setOnClickListener(btnListener);

        Button btn2 = (Button)findViewById(R.id.btn2);
        btn2.setOnClickListener(btnListener);
    }

    //---create an anonymous class to act as a button click listener---
    private OnClickListener btnListener = new OnClickListener()
    {
        public void onClick(View v)
        {
            Toast.makeText(getBaseContext(),
                    ((Button) v).getText() + " was clicked",
                    Toast.LENGTH_LONG).show();
        }
    };

    @Override
    public boolean onKeyDown(int keyCode, KeyEvent event)
    {
        //...
    }
}
```

现在，无论您按下的是 OK 按钮还是 Cancel 按钮，都会显示出相应的消息(如图 3-40 所示)，这证明该事件被正确地连接起来了。

除了为事件处理程序定义一个匿名类外，还可以定义一个匿名内部类来处理事件。下面的示例显示了如何为 EditText 视图处理 onFocusChange()方法。

图 3-40

```
import android.widget.EditText;

    @Override
    public void onCreate(Bundle savedInstanceState) {
        super.onCreate(savedInstanceState);
        setContentView(R.layout.main);

        //---the two buttons are wired to the same event handler---
        Button btn1 = (Button)findViewById(R.id.btn1);
        btn1.setOnClickListener(btnListener);

        Button btn2 = (Button)findViewById(R.id.btn2);
        btn2.setOnClickListener(btnListener);

        //---create an anonymous inner class to act as an onfocus listener---
        EditText txt1 = (EditText)findViewById(R.id.txt1);
        txt1.setOnFocusChangeListener(new View.OnFocusChangeListener()
        {
            @Override
            public void onFocusChange(View v, boolean hasFocus) {
                Toast.makeText(getBaseContext(),
                    ((EditText) v).getId() + " has focus - " + hasFocus,
                    Toast.LENGTH_LONG).show();
            }
        });
    }
```

 注意：对于本例而言，需要确保 onKeyDown()方法返回 false，而不是像前一节那样返回 true。

当 EditText 视图收到焦点时，屏幕上将输出一条消息，如图 3-41 所示。

图 3-41

通过使用匿名内部类,可以把两个 Button 视图的单击事件处理程序重写为如下所示:

```
//---the two buttons are wired to the same event handler---
Button btn1 = (Button)findViewById(R.id.btn1);
//btn1.setOnClickListener(btnListener);
btn1.setOnClickListener(new View.OnClickListener() {
    public void onClick(View v) {
        //---do something---
    }
});

Button btn2 = (Button)findViewById(R.id.btn2);
//btn2.setOnClickListener(btnListener);
btn2.setOnClickListener(new View.OnClickListener() {
    public void onClick(View v) {
        //---do something---
    }
});
```

应该使用哪个方法来处理事件?如果一个事件处理程序要处理多个视图,匿名类会很有帮助。如果一个视图有一个事件处理程序,匿名内部类方法(后一种方法)会很有帮助。

3.7 本章小结

本章学习了如何在 Android 中创建用户界面,还学习了可以用来在 Android 用户界面中定位视图的不同布局。由于 Android 设备支持多种屏幕方向,所以需要特别注意这一点,以确保您的用户界面能够适应屏幕方向的改变。

练 习

1. dp 单位和 px 单位有何不同?应该用哪一个来指定视图的尺寸?
2. 为什么不建议使用 AbsoluteLayout?
3. onPause()方法和 onSaveInstanceState()方法有何区别?
4. 列举 3 个可以重写来保存活动状态的方法。应该在哪些实例中使用各个方法?
5. 如何向 Action Bar 添加动作项?

练习答案参见附录 C。

本章主要内容

主 题	关 键 概 念
LinearLayout	以单行或单列的形式排列视图
AbsoluteLayout	可用于指定其子元素的确切位置
TableLayout	以行和列的形式组织视图
RelativeLayout	可用于指定子视图相对于彼此之间是如何定位的

(续表)

主　　题	关　键　概　念
FrameLayout	一个在屏幕上可以用来显示单个视图的占位符
ScrollView	一种特殊类型的 FrameLayout，因为它可以使用户滚动显示一个占据的空间大于物理显示的视图列表
度量单位	使用 dp 指定视图尺寸，使用 sp 指定字体大小
适应方向变化的两种方法	锚定与调整大小和重新定位
为不同方向使用不同的 XML 文件	纵向用户界面使用 layout 文件夹，横向用户界面使用 layout-land 文件夹
保持活动状态的 3 种方法	使用 onPause()事件 使用 onSaveInstanceState()事件 使用 onRetainNonConfigurationInstance()事件
获得当前设备的尺寸	使用 WindowManager 类的 getDefaultDisplay()方法
限制活动的方向	使用 setRequestOrientation()方法或者 AndroidManifest.xml 文件中的 android:screenOrientation 属性
Action Bar	取代了较早的 Android 版本中的标题栏
动作项	动作项显示在 Action Bar 的右侧，创建它们的方法与创建选项菜单很相似
应用程序图标	通常用于返回到应用程序的主活动。建立使用 Intent 对象和 Intent.FLAG_ACTIVITY_CLEAR_TOP 标志来实现这种功能

第 4 章

使用视图设计用户界面

本章将介绍以下内容：
- 如何使用 Android 中的基本视图设计用户界面
- 如何使用选取器视图显示项列表
- 如何使用列表视图显示项列表
- 如何使用特殊碎片

第 3 章中学习了可以用来在一个活动中定位视图的不同布局，还学习了可以用来适应不同屏幕分辨率和尺寸的相关技术。本章中，您将了解到可以用来为应用程序设计用户界面的各种视图。

特别地，将学习到以下视图组：
- 基本视图——常用的视图，如 TextView、EditText 和 Button 视图
- 选取器视图——可以使用户从一个列表中进行选择的视图，如 TimePicker 和 DatePicker 视图
- 列表视图——显示长的项列表的视图，如 ListView 和 SpinnerView 视图
- 专用碎片——执行某种功能的特殊碎片

随后的章节将涵盖本章未讨论到的其他视图，如模拟和数字时钟视图以及用于显示图形的其他视图等。

4.1 基本视图

首先，我们探究一些可以用于设计 Android 应用程序的用户界面的基本视图：
- `TextView`
- `EditText`
- `Button`
- `ImageButton`

- CheckBox
- ToggleButton
- RadioButton
- RadioGroup

这些基本视图可以用来显示文本信息以及执行一些基本的选择。下面的章节将详细研究所有这些视图。

4.1.1 TextView 视图

当创建一个新的 Android 项目时，Eclipse 总会创建一个包含一个<TextView>元素的 main.xml 文件(位于 res/layout 文件夹下)：

```xml
<?xml version="1.0" encoding="utf-8"?>
<LinearLayout xmlns:android="http://schemas.android.com/apk/res/android"
    android:layout_width="fill_parent"
    android:layout_height="fill_parent"
    android:orientation="vertical" >

    <TextView
        android:layout_width="fill_parent"
        android:layout_height="wrap_content"
        android:text="@string/hello" />

</LinearLayout>
```

TextView 视图用来向用户显示文本。这是最基本的视图，在开发 Android 应用程序时会频繁用到。如果想让用户可以编辑显示的文本，则应该使用 TextView 的子类——EditText，下一节将讨论它。

> **注意**：在某些平台上，TextView 常被称为标签视图。其唯一的目的就是在屏幕上显示文本。

4.1.2 Button、ImageButton、EditText、CheckBox、ToggleButton、RadioButton 和 RadioGroup 视图

除了最经常用到的 TextView 视图之外，还有其他一些您将频繁使用到的基础视图：

- Button——表示一个按钮的小部件。
- ImageButton——与 Button 视图类似，不过它还显示一个图像。
- EditText——TextView 视图的子类，还允许用户编辑其文本内容。
- CheckBox——具有两个状态的特殊按钮类型：选中或未选中。
- RadioGroup 和 RadioButton——RadioButton 有两个状态：选中或未选中。RadioGroup 用来把一个或多个 RadioButton 视图组合在一起，从而在该 RadioGroup 中只允许一

个 RadioButton 被选中。
- ToggleButton——用一个灯光指示器来显示选中/未选中状态。

下面的"试一试"揭示了这些视图工作原理的细节。

试一试　使用基本视图

BasicViews1.zip 代码文件可以在 Wrox.com 上下载

(1) 使用 Eclipse 创建一个 Android 项目，命名为 BasicView1。

(2) 在位于 res/layout 文件夹下的 main.xml 文件中添加下列粗体显示的元素：

```xml
<?xml version="1.0" encoding="utf-8"?>
<LinearLayout xmlns:android="http://schemas.android.com/apk/res/android"
    android:layout_width="fill_parent"
    android:layout_height="fill_parent"
    android:orientation="vertical" >

<Button android:id="@+id/btnSave"
    android:layout_width="fill_parent"
    android:layout_height="wrap_content"
    android:text="save" />

<Button android:id="@+id/btnOpen"
    android:layout_width="wrap_content"
    android:layout_height="wrap_content"
    android:text="Open" />

<ImageButton android:id="@+id/btnImg1"
    android:layout_width="fill_parent"
    android:layout_height="wrap_content"
    android:src="@drawable/ic_launcher" />

<EditText android:id="@+id/txtName"
    android:layout_width="fill_parent"
    android:layout_height="wrap_content" />

<CheckBox android:id="@+id/chkAutosave"
    android:layout_width="fill_parent"
    android:layout_height="wrap_content"
    android:text="Autosave" />

<CheckBox android:id="@+id/star"
    style="?android:attr/starStyle"
    android:layout_width="wrap_content"
    android:layout_height="wrap_content" />

<RadioGroup android:id="@+id/rdbGp1"
    android:layout_width="fill_parent"
```

```
        android:layout_height="wrap_content"
        android:orientation="vertical" >

    <RadioButton android:id="@+id/rdb1"
        android:layout_width="fill_parent"
        android:layout_height="wrap_content"
        android:text="Option 1" />

    <RadioButton android:id="@+id/rdb2"
        android:layout_width="fill_parent"
        android:layout_height="wrap_content"
        android:text="Option 2" />

</RadioGroup>

<ToggleButton android:id="@+id/toggle1"
    android:layout_width="wrap_content"
    android:layout_height="wrap_content" />

</LinearLayout>
```

(3) 要观察视图的效果，可在 Eclipse 中选择项目名称并按 F11 键进行调试。图 4-1 展示了在 Android 模拟器中显示的不同视图。

(4) 单击不同的视图并注意它们在外观和感觉上的变化。图 4-2 展示了视图的以下改变：
- 第 1 个 CheckBox 视图(Autosave)被选中。
- 第 2 个 CheckBox 视图(星形)被选中。
- 第 2 个 RadioButton(Option 2)被选中。
- ToggleButton 被打开。

图 4-1　　　　　　　　图 4-2

示例说明

到目前为止,所有的视图都是相对简单的——使用<LinearLayout>元素将它们一一列出,因此当在活动中显示时,它们堆叠在彼此之上。

对于第 1 个 Button,layout_width 属性被设置为 fill_parent,因此其宽度将占据整个屏幕的宽度:

```
<Button android:id="@+id/btnSave"
    android:layout_width="fill_parent"
    android:layout_height="wrap_content"
    android:text="save" />
```

对于第 2 个 Button,layout_width 属性被设置为 wrap_content,因此其宽度将是其所包含内容的宽度——具体来说,就是显示的文本(也就是"Open")的宽度:

```
<Button android:id="@+id/btnOpen"
    android:layout_width="wrap_content"
    android:layout_height="wrap_content"
    android:text="Open" />
```

ImageButton 显示了一个带有图像的按钮。图像通过 src 属性设置。本例中,使用曾用作应用程序图标的图像:

```
<ImageButton android:id="@+id/btnImg1"
    android:layout_width="fill_parent"
    android:layout_height="wrap_content"
    android:src="@drawable/ic_launcher" />
```

EditText 视图显示了一个矩形区域,用户可以向其中输入一些文本。layout_height 属性被设置为 wrap_content,这样,如果用户输入一个长的文本串,EditText 的高度将随着内容自动调整(如图 4-3 所示)。

```
<EditText android:id="@+id/txtName"
    android:layout_width="fill_parent"
    android:layout_height="wrap_content" />
```

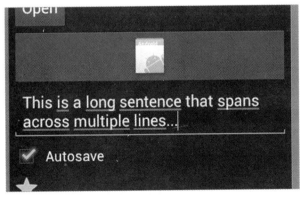

图 4-3

CheckBox 显示了一个用户可以通过轻点鼠标进行选中或取消选中的复选框:

```
<CheckBox android:id="@+id/chkAutosave"
    android:layout_width="fill_parent"
     android:layout_height="wrap_content"
     android:text="Autosave" />
```

如果您不喜欢 CheckBox 的默认外观，可以对其应用一个样式属性，使其显示为其他图像，如星形:

```
<CheckBox android:id="@+id/star"
     style="?android:attr/starStyle"
     android:layout_width="wrap_content"
     android:layout_height="wrap_content" />
```

style 属性的值的格式如下所示:

```
?[package:][type:]name
```

RadioGroup 包含了两个 RadioButton。这一点很重要，因为单选按钮通常用来表示多个选项以便用户选择。当选择了 RadioGroup 中的一个 RadioButton 时，其他所有 RadioButton 就自动取消选择:

```
<RadioGroup android:id="@+id/rdbGp1"
    android:layout_width="fill_parent"
    android:layout_height="wrap_content"
    android:orientation="vertical" >

    <RadioButton android:id="@+id/rdb1"
        android:layout_width="fill_parent"
        android:layout_height="wrap_content"
        android:text="Option 1" />

    <RadioButton android:id="@+id/rdb2"
        android:layout_width="fill_parent"
        android:layout_height="wrap_content"
        android:text="Option 2" />
```

`</RadioGroup>`

注意，RadioButton 是垂直排列的，一个位于另一个之上。如果想要水平排列，需要把 orientation 属性改为 horizontal。还需要确保 RadioButton 的 layout_width 属性被设置为 wrap_content:

```
<RadioGroup android:id="@+id/rdbGp1"
    android:layout_width="fill_parent"
    android:layout_height="wrap_content"
    android:orientation="horizontal" >
    <RadioButton android:id="@+id/rdb1"
        android:layout_width="wrap_content"
```

```
                android:layout_height="wrap_content"
                android:text="Option 1" />
    <RadioButton android:id="@+id/rdb2"
                android:layout_width="wrap_content"
                android:layout_height="wrap_content"
                android:text="Option 2" />
</RadioGroup>
```

图 4-4 显示了水平排列的 RadioButton。

图 4-4

ToggleButton 显示了一个矩形按钮，用户可以通过单击它来实现开和关的切换：

```
<ToggleButton android:id="@+id/toggle1"
    android:layout_width="wrap_content"
    android:layout_height="wrap_content" />
```

这个例子中，始终保持一致的一件事情是每个视图都有一个设置为特定值的 id 属性，如 Button 视图中所示：

```
<Button android:id="@+id/btnSave"
    android:layout_width="fill_parent"
    android:layout_height="wrap_content"
    android:text="@string/save" />
```

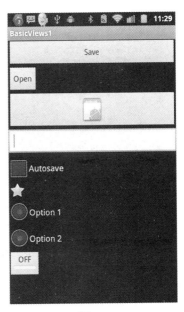

id 属性是视图的标识符，因此可以在以后使用 View.findViewById()或 Activity.findViewById()方法来检索它。

刚才看到的各种视图是在模拟 Android 4.0 智能手机的 Android 模拟器上进行测试的。当运行在较早版本的 Android 智能手机上时，它们会是什么样子？运行在 Android 平板电脑上时，又会是什么样子？

图 4-5 显示了将 AndroidManifest.xml 文件的 android:minSdkVersion 属性改为 10，并在运行 Android 2.3.6 的 Google Nexus S 上运行时，活动的外观：

```
<uses-sdk android:minSdkVersion="10" />
```

图 4-6 显示了将 AndroidManifest.xml 文件中的 android:minSdkVersion属性改为13,并在运行 Android 3.2.1 的 Asus Eee Pad Transformer 上运行时，活动的外观：

图 4-5

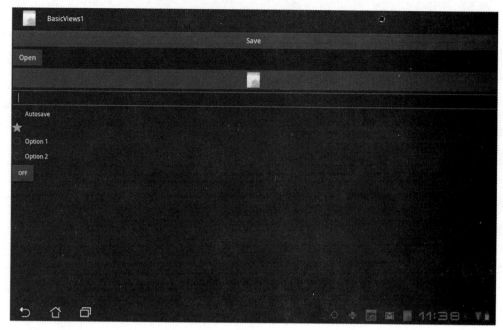

图 4-6

如果将 android:minSdkVersion 属性设为 8 或更小，然后在运行 Android 3.2.1 的 Asus Eee Pad Transformer 上运行，会看到额外的一个按钮，如图 4-7 所示。

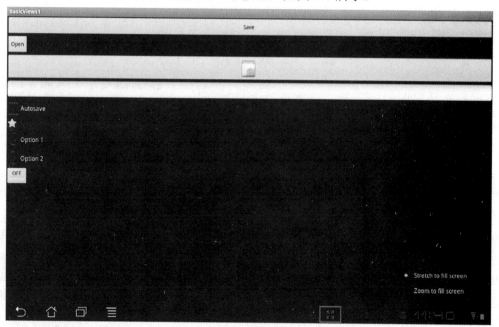

图 4-7

点击该按钮会出现两个选项，它们分别可以将活动拉伸到填充整个屏幕(默认选项)，或将活动缩放到填充整个屏幕，如图 4-8 所示。

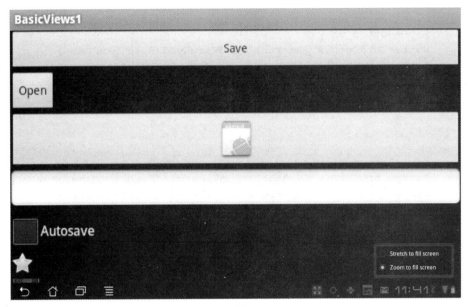

图 4-8

简言之,将最低 SDK 版本设为 8 或更低的应用程序可以在最初设计的屏幕尺寸上显示,也可以自动拉伸以填充屏幕(默认行为)。

现在,您已经了解了一个活动的多种视图的外观,下面的"试一试"将教您如何以编程方式控制它们。

试一试　处理视图事件

(1) 使用前面的"试一试"所创建的 BasicViews1 项目,修改 BasicViews1Activity.java 文件,添加下列粗体显示的语句:

```
package net.learn2develop.BasicViews1;

import android.app.Activity;
import android.os.Bundle;
import android.view.View;
import android.widget.Button;
import android.widget.CheckBox;
import android.widget.RadioButton;
import android.widget.RadioGroup;
import android.widget.RadioGroup.OnCheckedChangeListener;
import android.widget.Toast;
import android.widget.ToggleButton;

public class BasicViews1Activity extends Activity {
    /** Called when the activity is first created. */
    @Override
    public void onCreate(Bundle savedInstanceState) {
        super.onCreate(savedInstanceState);
        setContentView(R.layout.main);
```

```java
//---Button view---
Button btnOpen = (Button) findViewById(R.id.btnOpen);
btnOpen.setOnClickListener(new View.OnClickListener() {
    public void onClick(View v) {
        DisplayToast("You have clicked the Open button");
    }
});

//---Button view---
Button btnSave = (Button) findViewById(R.id.btnSave);
btnSave.setOnClickListener(new View.OnClickListener()
{
    public void onClick(View v) {
        DisplayToast("You have clicked the Save button");
    }
});

//---CheckBox---
CheckBox checkBox = (CheckBox) findViewById(R.id.chkAutosave);
checkBox.setOnClickListener(new View.OnClickListener()
{
    public void onClick(View v) {
        if (((CheckBox)v).isChecked())
            DisplayToast("CheckBox is checked");
        else
            DisplayToast("CheckBox is unchecked");
    }
});

//---RadioButton---
RadioGroup radioGroup = (RadioGroup) findViewById(R.id.rdbGp1);
radioGroup.setOnCheckedChangeListener(new OnCheckedChangeListener()
{
    public void onCheckedChanged(RadioGroup group, int checkedId) {
        RadioButton rb1 = (RadioButton) findViewById(R.id.rdb1);
        if (rb1.isChecked()) {
            DisplayToast("Option 1 checked!");
        } else {
            DisplayToast("Option 2 checked!");
        }
    }
});

//---ToggleButton---
ToggleButton toggleButton =
        (ToggleButton) findViewById(R.id.toggle1);
toggleButton.setOnClickListener(new View.OnClickListener()
{
    public void onClick(View v) {
        if (((ToggleButton)v).isChecked())
```

```
                    DisplayToast("Toggle button is On");
                else
                    DisplayToast("Toggle button is Off");
            }
        });
    }
    private void DisplayToast(String msg)
    {
        Toast.makeText(getBaseContext(), msg,
                Toast.LENGTH_SHORT).show();
    }
}
```

(2) 按 F11 键在 Android 模拟器中调试项目。

(3) 单击不同的视图，观察在 Toast 窗口中显示的消息。

示例说明

为了处理每一个视图所触发的事件，首先需要以编程方式定位在 onCreate()事件中所创建的视图。做法是使用 Acitivity 基类的 findViewById()方法，传入该视图的 ID。

```
//---Button view---
Button btnOpen = (Button) findViewById(R.id.btnOpen);
```

setOnClickListener()方法注册了一个在视图被单击时调用的回调函数：

```
btnOpen.setOnClickListener(new View.OnClickListener() {
    public void onClick(View v) {
        DisplayToast("You have clicked the Open button");
    }
});
```

当单击视图时，将调用 onClick()方法。

对于 CheckBox，为了确定其状态，必须把 onClick()方法的参数类型转换成一个 CheckBox，然后检查它的 isChecked()方法来确定其是否被选中：

```
CheckBox checkBox = (CheckBox) findViewById(R.id.chkAutosave);
checkBox.setOnClickListener(new View.OnClickListener()
{
    public void onClick(View v) {
        if (((CheckBox)v).isChecked())
            DisplayToast("CheckBox is checked");
        else
            DisplayToast("CheckBox is unchecked");
    }
});
```

对于 RadioButton，需要使用 RadioGroup 的 setOnCheckedChangeListener()方法注册一个回调函数，以便在该组中被选中的 RadioButton 发生变化时调用：

```
//---RadioButton---
RadioGroup radioGroup = (RadioGroup) findViewById(R.id.rdbGp1);
radioGroup.setOnCheckedChangeListener(new OnCheckedChangeListener()
{
    public void onCheckedChanged(RadioGroup group, int checkedId) {
        RadioButton rb1 = (RadioButton) findViewById(R.id.rdb1);
        if (rb1.isChecked()) {
            DisplayToast("Option 1 checked!");
        } else {
            DisplayToast("Option 2 checked!");
        }
    }
});
```

当选中一个 RadioButton 时,将触发 onCheckedChanged()方法。在这一过程中,找到那些单个的 RadioButton,然后调用它们的 isChecked()方法来确定是哪个 RadioButton 被选中。或者,onCheckedChanged()方法包含第 2 个参数,其中包含被选定 RadioButton 的唯一标识符。

ToggleButton 的工作方式与 CheckBox 类似。

到目前为止,为了处理视图上的事件,首先需要获得视图的一个引用,然后需要注册一个回调函数来处理事件。还有另外一种处理视图事件的方法。以 Button 为例,可以向其添加一个名为 onClick 的属性:

```
<Button android:id="@+id/btnSave"
    android:layout_width="fill_parent"
    android:layout_height="wrap_content"
    android:text="@string/save"
    android:onClick="btnSaved_clicked"/>
```

onClick 属性指定了按钮的单击事件。该属性的值就是事件处理程序的名称。因此,为处理按钮的单击事件,只需要创建一个名为 btnSaved_clicked 的方法,如下面的示例所示(注意该方法必须有一个 View 类型的参数):

```
public class BasicViews1Activity extends Activity {

    public void btnSaved_clicked (View view) {
        DisplayToast("You have clicked the Save button1");
    }

    /** Called when the activity is first created. */
    @Override
    public void onCreate(Bundle savedInstanceState) {
        super.onCreate(savedInstanceState);
        setContentView(R.layout.main);

        //...
    }
```

第 4 章　使用视图设计用户界面

```
private void DisplayToast(String msg)
{
    Toast.makeText(getBaseContext(), msg,
        Toast.LENGTH_SHORT).show();
}

}
```

如果与前面使用的方法进行比较，会发现这种方法更加简单。使用哪种方法取决于您自己，但本书中主要使用后面这种方法。

4.1.3　ProgressBar 视图

ProgressBar 视图提供了一些正在进行的任务的视觉反馈，如当您在后台执行一个任务时。例如，您可能正从 Web 上下载一些数据并需要更新用户的下载状态。在这种情况下，使用 ProgressBar 视图来完成这一任务是一个不错的选择。下面的活动演示了如何使用这个视图。

试一试　　使用 ProgressBar 视图

BasicViews2.zip 代码文件可以在 Wrox.com 上下载

(1) 打开 Eclipse，创建一个名为 BasicViews2 的 Android 项目。
(2) 修改位于 res/layout 文件夹下的 main.xml 文件，添加下列粗体显示的代码：

```
<?xml version="1.0" encoding="utf-8"?>
<LinearLayout xmlns:android="http://schemas.android.com/apk/res/android"
    android:layout_width="fill_parent"
    android:layout_height="fill_parent"
    android:orientation="vertical" >

<ProgressBar android:id="@+id/progressbar"
    android:layout_width="wrap_content"
    android:layout_height="wrap_content" />

</LinearLayout>
```

(3) 在 BasicViews2Activity.java 文件中添加下列粗体显示的语句：

```
package net.learn2develop.BasicViews2;

import android.app.Activity;
import android.os.Bundle;
import android.os.Handler;
import android.view.View;
import android.widget.ProgressBar;
```

```java
public class BasicViews2Activity extends Activity {
    static int progress;
    ProgressBar progressBar;
    int progressStatus = 0;
    Handler handler = new Handler();

    /** Called when the activity is first created. */
    @Override
    public void onCreate(Bundle savedInstanceState) {
        super.onCreate(savedInstanceState);
        setContentView(R.layout.main);

        progress = 0;
        progressBar = (ProgressBar) findViewById(R.id.progressbar);

        //---do some work in background thread---
        new Thread(new Runnable()
        {
            public void run()
            {
                //---do some work here---
                while (progressStatus < 10)
                {
                    progressStatus = doSomeWork();
                }

                //---hides the progress bar---
                handler.post(new Runnable()
                {
                    public void run()
                    {
                        //---0 - VISIBLE; 4 - INVISIBLE; 8 - GONE---
                        progressBar.setVisibility(View.GONE);
                    }
                });
            }

            //---do some long running work here---
            private int doSomeWork()
            {
                try {
                    //---simulate doing some work---
                    Thread.sleep(500);
                } catch (InterruptedException e)
                {
                    e.printStackTrace();
                }
                return ++progress;
            }
        }).start();
```

            }
        }

(4) 按 F11 键在 Android 模拟器中调试项目。图 4-9 显示了 ProgressBar 的动画。大约 5 秒钟后，它将消失。

图 4-9

示例说明

ProgressBar 视图的默认模式是不确定的——也就是说，它显示一个循环的动画。这种模式对于完成时间没有明确指示的任务是非常有用的，例如当您向一个 Web 服务发送一些数据并等待服务器的响应时。如果只是把<ProgressBar>元素放入 main.xml 文件中，它会不断地显示一个旋转的图标。当后台任务已经完成时，需要您来使它停止旋转。

已在 Java 文件中添加的代码显示了如何分配一个后台线程来模拟执行一些长时间运行的任务。要做到这一点，可以配合使用 Thread 类和一个 Runnable 对象。run()方法启动线程的执行，在这种情况下调用 doSomeWork()方法来模拟做一些工作。当模拟工作完成后(大约 5 秒种之后)，使用 Handler 对象给线程发送一条消息来取消 ProgressBar：

```java
//---do some work in background thread---
new Thread(new Runnable()
{
    public void run()
    {
        //---do some work here---
        while (progressStatus < 10)
        {
            progressStatus = doSomeWork();
        }

        //---hides the progress bar---
        handler.post(new Runnable()
        {
            public void run()
            {
                //---0 - VISIBLE; 4 - INVISIBLE; 8 - GONE---
                progressBar.setVisibility(View.GONE);
            }
        });
    }
```

```
            //---do some long running work here---
            private int doSomeWork()
            {
                try {
                    //---simulate doing some work---
                    Thread.sleep(500);
                } catch (InterruptedException e)
                {
                    e.printStackTrace();
                }
                return ++progress;
            }
    }).start();
```

当任务完成时,通过设置 ProgressBar 的 Visibility 属性为 View.GONE (值 8)来隐藏它。INVISIBLE 和 GONE 常量的区别在于 INVISIBLE 常量只是隐藏 ProgressBar(ProgressBar 仍旧在活动中占据空间)。GONE 常量则从活动中移除 ProgressBar 视图,它不再占据任何空间。

下面的"试一试"演示了如何改变 ProgressBar 的外观。

试一试　　定制 ProgressBar 视图

(1) 使用前面的"试一试"中所创建的 BasicViews2 项目,按如下所示修改 main.xml 文件:

```
<?xml version="1.0" encoding="utf-8"?>
<LinearLayout xmlns:android="http://schemas.android.com/apk/res/android"
    android:layout_width="fill_parent"
    android:layout_height="fill_parent"
    android:orientation="vertical" >

<ProgressBar android:id="@+id/progressbar"
    android:layout_width="wrap_content"
    android:layout_height="wrap_content"
    style="@android:style/Widget.ProgressBar.Horizontal" />

</LinearLayout>
```

(2) 修改 BasicViews2Activity.java 文件,添加下列粗体显示的语句:

```
package net.learn2develop.BasicViews2;

import android.app.Activity;
import android.os.Bundle;
import android.os.Handler;
import android.view.View;
import android.widget.ProgressBar;
```

```java
public class BasicViews2Activity extends Activity {
    static int progress;
    ProgressBar progressBar;
    int progressStatus = 0;
    Handler handler = new Handler();

    /** Called when the activity is first created. */
    @Override
    public void onCreate(Bundle savedInstanceState) {
        super.onCreate(savedInstanceState);
        setContentView(R.layout.main);

        progress = 0;
        progressBar = (ProgressBar) findViewById(R.id.progressbar);
        progressBar.setMax(200);

        //---do some work in background thread---
        new Thread(new Runnable()
        {
            public void run()
            {
                //---do some work here---
                while (progressStatus < 100)
                {
                    progressStatus = doSomeWork();

                    //---Update the progress bar---
                    handler.post(new Runnable()
                    {
                        public void run() {
                            progressBar.setProgress(progressStatus);
                        }
                    });
                }

                //---hides the progress bar---
                handler.post(new Runnable()
                {
                    public void run()
                    {
                        //---0 - VISIBLE; 4 - INVISIBLE; 8 - GONE---
                        progressBar.setVisibility(View.GONE);
                    }
                });
            }

            //---do some long running work here---
            private int doSomeWork()
            {
                try {
```

```
                //---simulate doing some work---
                Thread.sleep(500);
            } catch (InterruptedException e)
            {
                e.printStackTrace();
            }
            return ++progress;
        }
    }).start();
}
```

(3) 按 F11 键在 Android 模拟器中调试项目。

(4) 图 4-10 展示了正显示进度的 ProgressBar。当进度达到 50%时，ProgressBar 消失。

图 4-10

示例说明

为了使 ProgressBar 水平显示，只要设置其 style 属性为@android:style/Widget.ProgressBar.Horizontal:

```
<ProgressBar android:id="@+id/progressbar"
    android:layout_width="wrap_content"
    android:layout_height="wrap_content"
    style="@android:style/Widget.ProgressBar.Horizontal" />
```

为了显示进度，调用 setProgress()方法，传入一个表示进度的整数：

```
//---Update the progress bar---
    handler.post(new Runnable()
    {
        public void run() {
            progressBar.setProgress(progressStatus);
        }
    });
```

在本例中，设置了 ProgressBar 的范围为 0～200(通过 setMax()方法)。因此，ProgressBar 将在中途停止并消失(由于仅仅在 progressStatus 小于 100 时才持续调用 doSomeWork()方法)。为了确保 ProgressBar 只有当进度达到 100%时才消失，可以设置最大值为 100，或者修改 while 循环为当 progressStatus 达到 200 时停止，如下所示：

```
    //---do some work here---
```

```
        while (progressStatus < 200)
```

除了本例中为 ProgressBar 使用的水平样式，还可以使用以下常量：

- `Widget.ProgressBar.Horizontal`
- `Widget.ProgressBar.Small`
- `Widget.ProgressBar.Large`
- `Widget.ProgressBar.Inverse`
- `Widget.ProgressBar.Small.Inverse`
- `Widget.ProgressBar.Large.Inverse`

4.1.4 AutoCompleteTextView 视图

AutoCompleteTextView 是一种与 EditText 类似的视图(实际上，它是 EditText 的子类)，只不过它还在用户输入时自动显示完成建议的列表。下面的"试一试"展示了如何利用 AutoCompleteTextView 来自动协助用户完成文本输入。

试一试　　使用 AutoCompleteTextView

BasicViews3.zip 代码文件可以在 Wrox.com 上下载

(1) 打开 Eclipse，创建一个名为 BasicViews3 的 Android 项目。

(2) 按如下粗体显示内容修改位于 res/layout 文件夹下的 main.xml 文件：

```xml
<?xml version="1.0" encoding="utf-8"?>
<LinearLayout xmlns:android="http://schemas.android.com/apk/res/android"
    android:layout_width="fill_parent"
    android:layout_height="fill_parent"
    android:orientation="vertical" >

<TextView
    android:layout_width="fill_parent"
    android:layout_height="wrap_content"
    android:text="Name of President" />

<AutoCompleteTextView android:id="@+id/txtCountries"
    android:layout_width="fill_parent"
    android:layout_height="wrap_content" />

</LinearLayout>
```

(3) 在 BasicViews3Activity.java 文件中添加下列粗体显示的语句：

```java
package net.learn2develop.BasicViews3;

import android.app.Activity;
import android.os.Bundle;
```

```java
import android.widget.ArrayAdapter;
import android.widget.AutoCompleteTextView;

public class BasicViews3Activity extends Activity {
    String[] presidents = {
            "Dwight D. Eisenhower",
            "John F. Kennedy",
            "Lyndon B. Johnson",
            "Richard Nixon",
            "Gerald Ford",
            "Jimmy Carter",
            "Ronald Reagan",
            "George H. W. Bush",
            "Bill Clinton",
            "George W. Bush",
            "Barack Obama"
    };

    /** Called when the activity is first created. */
    @Override
    public void onCreate(Bundle savedInstanceState) {
        super.onCreate(savedInstanceState);
        setContentView(R.layout.main);

        ArrayAdapter<String> adapter = new ArrayAdapter<String>(this,
            android.R.layout.simple_dropdown
                _item_1line, presidents);
        AutoCompleteTextView textView =
            (AutoCompleteTextView)
                findViewById(R.id.txtCountries);

        textView.setThreshold(3);
        textView.setAdapter(adapter);
    }
}
```

(4) 按 F11 键在 Android 模拟器中调试应用程序。如图 4-11 所示，在向 AutoCompleteTextView 中输入时，会随之显示一个匹配名字的列表。

示例说明

在 BasicViews3Activity 类中，首先创建了一个包含一组总统名字的 String 数组：

```
String[] presidents = {
        "Dwight D. Eisenhower",
        "John F. Kennedy",
        "Lyndon B. Johnson",
```

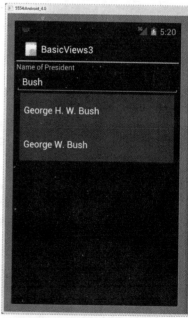

图 4-11

```
            "Richard Nixon",
            "Gerald Ford",
            "Jimmy Carter",
            "Ronald Reagan",
            "George H. W. Bush",
            "Bill Clinton",
            "George W. Bush",
            "Barack Obama"
    };
```

ArrayAdapter 对象管理将由 AutoCompleteTextView 显示的字符串数组。在先前的例子中，将 AutoCompleteTextView 设置为以 simple_dropdown_item_1line 模式显示：

```
ArrayAdapter<String> adapter = new ArrayAdapter<String>(this,
    android.R.layout.simple_dropdown_item_1line, presidents);
```

setThreshold()方法设置建议以下拉菜单形式出现前用户必须输入的最少字符个数：

```
textView.setThreshold(3);
```

为 AutoCompleteTextView 显示的建议列表从 ArrayAdapter 对象获得：

```
textView.setAdapter(adapter);
```

4.2 选取器视图

选择日期和时间是您在一个移动应用程序中需要执行的常见任务之一。Android 通过 TimePicker 和 DatePicker 视图来支持这一功能。下面的小节将阐述如何在活动中使用这些视图。

4.2.1 TimePicker 视图

TimePicker 视图可以使用户按 24 小时或 AM/PM 模式选择一天中的某个时间。下面的"试一试"展示了如何使用这一视图。

试一试　使用 TimePicker 视图

BasicViews4.zip 代码文件可以在 Wrox.com 上下载

(1) 打开 Eclipse，创建一个名为 BasicViews4 的 Android 项目。
(2) 修改位于 res/layout 文件夹下的 main.xml 文件，添加下列粗体显示的行：

```xml
<?xml version="1.0" encoding="utf-8"?>
<LinearLayout xmlns:android="http://schemas.android.com/apk/res/android"
    android:layout_width="fill_parent"
    android:layout_height="fill_parent"
    android:orientation="vertical" >
```

```xml
<TimePicker android:id="@+id/timePicker"
    android:layout_width="wrap_content"
    android:layout_height="wrap_content" />

<Button android:id="@+id/btnSet"
    android:layout_width="wrap_content"
    android:layout_height="wrap_content"
    android:text="I am all set!"
    android:onClick="onClick" />

</LinearLayout>
```

(3) 在 Eclipse 中选择该项目的名称并按 F11 键在 Android 模拟器中调试应用程序。图 4-12 显示了运用中的 TimePicker。除了单击加(+)和减(—)按钮外，还可以使用设备上的数字键盘来修改小时和分钟，单击 AM 按钮在 AM 和 PM 之间切换。

(4) 返回 Eclipse，在 BasicViews4Activity.java 文件中添加下列粗体显示的语句：

图 4-12

```java
package net.learn2develop.BasicViews4;

import android.app.Activity;
import android.os.Bundle;
import android.view.View;
import android.widget.TimePicker;
import android.widget.Toast;

public class BasicViews4Activity extends Activity {
    TimePicker timePicker;

    /** Called when the activity is first created. */
    @Override
    public void onCreate(Bundle savedInstanceState) {
        super.onCreate(savedInstanceState);
        setContentView(R.layout.main);

        timePicker = (TimePicker) findViewById(R.id.timePicker);
        timePicker.setIs24HourView(true);
    }

    public void onClick(View view) {
        Toast.makeText(getBaseContext(),
            "Time selected:" +
            timePicker.getCurrentHour() +
            ":" + timePicker.getCurrentMinute(),
            Toast.LENGTH_SHORT).show();
```

        }

    }

(5) 按 F11 键在 Android 模拟器上调试应用程序。这一次，TimePicker 将以 24 小时格式显示。单击 Button 将显示您在 TimePicker 中设置好的时间(如图 4-13 所示)。

示例说明

TimePicker 显示了一个可以让用户设置时间的标准用户界面。默认情况下，它以 AM/PM 格式显示时间。如果想要以 24 小时格式来显示，可以使用 setIs24HourView()方法。

为了以编程方式获得用户设置的时间，可以使用 getCurrentHour()和 getCurrentMinute()方法：

```
Toast.makeText(getBaseContext(),
        "Time selected:" +
        timePicker.getCurrentHour() +
        ":" + timePicker.getCurrent
        Minute(),
        Toast.LENGTH_SHORT).show();
```

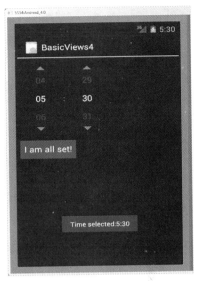

图 4-13

 注意：getCurrentHour()方法总是返回 24 小时格式的时间,也就是返回一个 0～23 之间的值。

虽然可以在一个活动中显示 TimePicker，但更好的方法是在一个对话框窗口中显示它。这样一旦设置好时间，TimePicker 就会消失，不再占据活动中的任何空间。下面的"试一试"展示了如何做到这一点。

试一试　使用对话框显示 TimePicker 视图

(1) 使用先前"试一试"中所创建的 BasicViews4 项目，按如下所示修改 BasicViews4Activity.java 文件：

```
package net.learn2develop.BasicViews4;

import java.text.SimpleDateFormat;
import java.util.Date;

import android.app.Activity;
import android.app.Dialog;
import android.app.TimePickerDialog;
import android.os.Bundle;
```

```java
import android.view.View;
import android.widget.TimePicker;
import android.widget.Toast;

public class BasicViews4Activity extends Activity {
    TimePicker timePicker;

    int hour, minute;
    static final int TIME_DIALOG_ID = 0;

    /** Called when the activity is first created. */
    @Override
    public void onCreate(Bundle savedInstanceState) {
        super.onCreate(savedInstanceState);
        setContentView(R.layout.main);

        timePicker = (TimePicker) findViewById(R.id.timePicker);
        timePicker.setIs24HourView(true);

        showDialog(TIME_DIALOG_ID);
    }

    @Override
    protected Dialog onCreateDialog(int id)
    {
        switch (id) {
        case TIME_DIALOG_ID:
            return new TimePickerDialog(
                    this, mTimeSetListener, hour, minute, false);
        }
        return null;
    }

    private TimePickerDialog.OnTimeSetListener mTimeSetListener =
    new TimePickerDialog.OnTimeSetListener()
    {
        public void onTimeSet(
                TimePicker view, int hourOfDay, int minuteOfHour)
        {
            hour = hourOfDay;
            minute = minuteOfHour;

            SimpleDateFormat timeFormat = new SimpleDateFormat("hh:mm aa");
            Date date = new Date(0,0,0, hour, minute);
            String strDate = timeFormat.format(date);

            Toast.makeText(getBaseContext(),
                    "You have selected " + strDate,
                    Toast.LENGTH_SHORT).show();
        }
```

第 4 章 使用视图设计用户界面

```
    };

    public void onClick(View view) {
        Toast.makeText(getBaseContext(),
            "Time selected:" +
                timePicker.getCurrentHour() +
                ":" + timePicker.getCurrentMinute(),
            Toast.LENGTH_SHORT).show();
    }

}
```

(2) 按 F11 键在 Android 模拟器中调试应用程序。当活动被加载时，可以看到 TimePicker 显示在一个对话框窗口内(如图 4-14 所示)。设置一个时间，然后单击 Set 按钮，将看到 Toast 窗口显示了您刚刚设置好的时间。

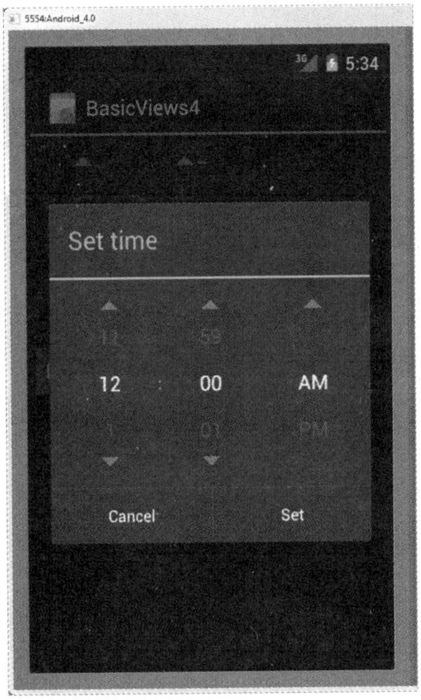

示例说明

为了显示一个对话框窗口，可以使用 showDialog() 方法，传入一个 ID 来标识对话框的源：

```
showDialog(TIME_DIALOG_ID);
```

当调用 showDialog()方法时，onCreateDialog()方法将被调用：

```
@Override
protected Dialog onCreateDialog(int id)
{
    switch (id) {
    case TIME_DIALOG_ID:
        return new TimePickerDialog(
            this, mTimeSetListener, hour, minute, false);
    }
    return null;
}
```

图 4-14

这里，创建了一个 TimePickerDialog 类的新实例，给它传递了当前上下文、回调函数、初始的小时和分钟，以及 TimePicker 是否以 24 小时格式显示。

当用户单击 TimePicker 对话框窗口中的 Set 按钮时，将调用 onTimeSet()方法：

```
private TimePickerDialog.OnTimeSetListener mTimeSetListener =
new TimePickerDialog.OnTimeSetListener()
{
    public void onTimeSet(
            TimePicker view, int hourOfDay, int minuteOfHour)
    {
        hour = hourOfDay;
```

```
            minute = minuteOfHour;

            SimpleDateFormat timeFormat = new SimpleDateFormat("hh:mm aa");
            Date date = new Date(0,0,0, hour, minute);
            String strDate = timeFormat.format(date);

            Toast.makeText(getBaseContext(),
                "You have selected " + strDate,
                Toast.LENGTH_SHORT).show();
        }
    };
```

这里，onTimeSet()方法将包含用户分别通过 hourOfDay 和 minuteOfHour 参数设置的小时和分钟。

4.2.2 DatePicker 视图

与 TimePicker 类似的另外一种视图就是 DatePicker。利用 DatePicker，可以使用户在活动中选择一个特定的日期。下面的"试一试"展示了如何使用 DatePicker。

试一试   使用 DatePicker 视图

(1) 使用前述"试一试"中创建的 BasicViews4 项目，按如下所示修改 main.xml 文件：

```xml
<?xml version="1.0" encoding="utf-8"?>
<LinearLayout xmlns:android="http://schemas.android.com/apk/res/android"
    android:layout_width="fill_parent"
    android:layout_height="fill_parent"
    android:orientation="vertical" >

<Button android:id="@+id/btnSet"
    android:layout_width="wrap_content"
    android:layout_height="wrap_content"
    android:text="I am all set!"
    android:onClick="onClick" />

<DatePicker android:id="@+id/datePicker"
    android:layout_width="wrap_content"
    android:layout_height="wrap_content" />

<TimePicker android:id="@+id/timePicker"
    android:layout_width="wrap_content"
    android:layout_height="wrap_content" />

</LinearLayout>
```

(2) 按 F11 键在 Android 模拟器上调试应用程序。图 4-15 显示了 DatePicker 视图(需要按 Ctrl+F11 组合键将模拟器的方向改为横向；纵向太窄，不能很好地显示 DatePicker)。

第 4 章 使用视图设计用户界面

图 4-15

(3) 返回 Eclipse，在 BasicViews4Activity.java 文件中添加下列粗体显示的语句：

```
package net.learn2develop.BasicViews4;

import java.text.SimpleDateFormat;
import java.util.Date;

import android.app.Activity;
import android.app.Dialog;
import android.app.TimePickerDialog;
import android.os.Bundle;
import android.view.View;
import android.widget.DatePicker;
import android.widget.TimePicker;
import android.widget.Toast;

public class BasicViews4Activity extends Activity {
    TimePicker timePicker;
    DatePicker datePicker;

    int hour, minute;
    static final int TIME_DIALOG_ID = 0;

    /** Called when the activity is first created. */
    @Override
    public void onCreate(Bundle savedInstanceState) {
        super.onCreate(savedInstanceState);
        setContentView(R.layout.main);
```

```java
        timePicker = (TimePicker) findViewById(R.id.timePicker);
        timePicker.setIs24HourView(true);

        // showDialog(TIME_DIALOG_ID);
        datePicker = (DatePicker) findViewById(R.id.datePicker);
    }

    @Override
    protected Dialog onCreateDialog(int id)
    {
        switch (id) {
        case TIME_DIALOG_ID:
            return new TimePickerDialog(
                    this, mTimeSetListener, hour, minute, false);
        }
        return null;
    }

    private TimePickerDialog.OnTimeSetListener mTimeSetListener =
    new TimePickerDialog.OnTimeSetListener()
    {
        public void onTimeSet(
                TimePicker view, int hourOfDay, int minuteOfHour)
        {
            hour = hourOfDay;
            minute = minuteOfHour;

            SimpleDateFormat timeFormat = new SimpleDateFormat("hh:mm aa");
            Date date = new Date(0,0,0, hour, minute);
            String strDate = timeFormat.format(date);

            Toast.makeText(getBaseContext(),
                    "You have selected " + strDate,
                    Toast.LENGTH_SHORT).show();
        }
    };

    public void onClick(View view) {
        Toast.makeText(getBaseContext(),
                "Date selected:" + (datePicker.getMonth() + 1) +
                "/" + datePicker.getDayOfMonth() +
                "/" + datePicker.getYear() + "\n" +
                "Time selected:" + timePicker.getCurrentHour() +
                ":" + timePicker.getCurrentMinute(),
                Toast.LENGTH_SHORT).show();
    }

}
```

(4) 按 F11 键在 Android 模拟器上调试应用程序。一旦设置了日期，单击 Button 将显示设置的日期，如图 4-16 所示。

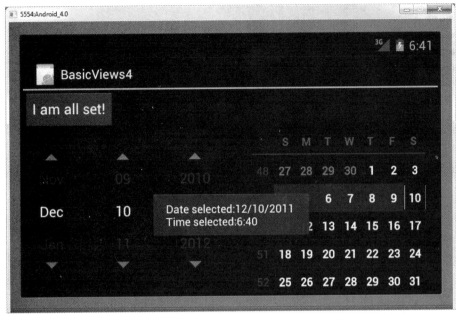

图 4-16

示例说明

与 TimePicker 类似，通过调用 getMonth()、getDayOfMonth()和 getYear()方法来分别获取月份、日子和年份：

```
"Date selected:" + (datePicker.getMonth() + 1) +
        "/" + datePicker.getDayOfMonth() +
        "/" + datePicker.getYear() + "\n" +
```

注意，getMonth()方法返回 0 代表一月、返回 1 代表二月，依次类推。因此，需要将此方法返回的结果加 1 来获得对应的月份数。

像 TimePicker 一样，也可以在对话框窗口中显示 DatePicker。下面的"试一试"将教您如何做到这一点。

试一试 使用对话框显示 DatePicker 视图

(1) 使用前述"试一试"中创建的 BasicViews4 项目，在 BasicViews4Activity.java 文件中添加下列粗体显示的语句：

```
package net.learn2develop.BasicViews4;

import java.text.SimpleDateFormat;
import java.util.Calendar;
import java.util.Date;

import android.app.Activity;
import android.app.DatePickerDialog;
```

```java
import android.app.Dialog;
import android.app.TimePickerDialog;
import android.os.Bundle;
import android.view.View;
import android.widget.DatePicker;
import android.widget.TimePicker;
import android.widget.Toast;

public class BasicViews4Activity extends Activity {
    TimePicker timePicker;
    DatePicker datePicker;

    int hour, minute;
    int yr, month, day;

    static final int TIME_DIALOG_ID = 0;
    static final int DATE_DIALOG_ID = 1;

    /** Called when the activity is first created. */
    @Override
    public void onCreate(Bundle savedInstanceState) {
        super.onCreate(savedInstanceState);
        setContentView(R.layout.main);

        timePicker = (TimePicker) findViewById(R.id.timePicker);
        timePicker.setIs24HourView(true);

        // showDialog(TIME_DIALOG_ID);
        datePicker = (DatePicker) findViewById(R.id.datePicker);

        //---get the current date---
        Calendar today = Calendar.getInstance();
        yr = today.get(Calendar.YEAR);
        month = today.get(Calendar.MONTH);
        day = today.get(Calendar.DAY_OF_MONTH);

        showDialog(DATE_DIALOG_ID);
    }

    @Override
    protected Dialog onCreateDialog(int id)
    {
        switch (id) {
        case TIME_DIALOG_ID:
            return new TimePickerDialog(
                    this, mTimeSetListener, hour, minute, false);
        case DATE_DIALOG_ID:
            return new DatePickerDialog(
                    this, mDateSetListener, yr, month, day);
        }
```

```java
            return null;
    }

    private DatePickerDialog.OnDateSetListener mDateSetListener =
    new DatePickerDialog.OnDateSetListener()
    {
        public void onDateSet(
                DatePicker view, int year, int monthOfYear, int dayOfMonth)
        {
            yr = year;
            month = monthOfYear;
            day = dayOfMonth;
            Toast.makeText(getBaseContext(),
                    "You have selected : " + (month + 1) +
                    "/" + day + "/" + year,
                Toast.LENGTH_SHORT).show();
        }
    };

    private TimePickerDialog.OnTimeSetListener mTimeSetListener =
    new TimePickerDialog.OnTimeSetListener()
    {
        public void onTimeSet(
                TimePicker view, int hourOfDay, int minuteOfHour)
        {
            hour = hourOfDay;
            minute = minuteOfHour;

            SimpleDateFormat timeFormat = new SimpleDateFormat("hh:mm aa");
            Date date = new Date(0,0,0, hour, minute);
            String strDate = timeFormat.format(date);

            Toast.makeText(getBaseContext(),
                    "You have selected " + strDate,
                    Toast.LENGTH_SHORT).show();
        }
    };

    public void onClick(View view) {
        Toast.makeText(getBaseContext(),
                "Date selected:" + (datePicker.getMonth() + 1) +
                "/" + datePicker.getDayOfMonth() +
                "/" + datePicker.getYear() + "\n" +
                "Time selected:" + timePicker.getCurrentHour() +
                ":" + timePicker.getCurrentMinute(),
                Toast.LENGTH_SHORT).show();
    }

}
```

(2) 按 F11 键在 Android 模拟器上调试应用程序。当活动加载时，可以看到 DatePicker

显示在一个对话框窗口中(如图 4-17 所示)。设定好一个日期并单击 Set 按钮。Toast 窗口将显示出您刚刚设置好的日期。

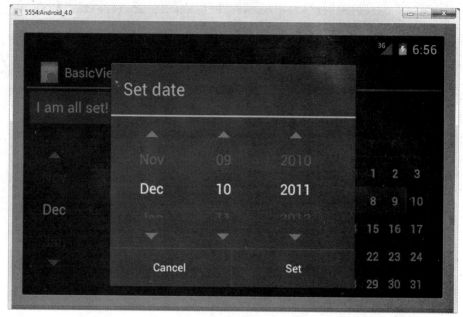

图 4-17

示例说明

DatePicker 和 TimePicker 的工作原理是一致的。当设置日期时，它将触发 onDateSet() 方法，从中可以获取由用户设定的日期：

```
public void onDateSet(
        DatePicker view, int year, int monthOfYear, int dayOfMonth)
{
    yr = year;
    month = monthOfYear;
    day = dayOfMonth;
    Toast.makeText(getBaseContext(),
        "You have selected : " + (month + 1) +
        "/" + day + "/" + year,
        Toast.LENGTH_SHORT).show();
}
```

注意，在显示对话框之前，需要初始化 3 个变量——yr、month 和 day：

```
//---get the current date---
Calendar today = Calendar.getInstance();
yr = today.get(Calendar.YEAR);
month = today.get(Calendar.MONTH);
day = today.get(Calendar.DAY_OF_MONTH);

showDialog(DATE_DIALOG_ID);
```

如果不这样做，当在运行时创建一个 DatePickerDialog 类的实例时，将发生非法参数异常(current should be >= start and <= end)。

4.3 使用列表视图显示长列表

列表视图是一种可以用来显示长的项列表的视图。在 Android 中，有两种列表视图：ListView 和 SpinnerView，两者都用于显示长的项列表。下面的"试一试"展示了这两种视图的使用。

4.3.1 ListView 视图

ListView 在一个垂直滚动列表中显示项列表。下面的"试一试"演示了如何使用 ListView 显示一个项列表。

试一试 使用 ListView 显示一个长的项列表

BasicViews5.zip 代码文件可以在Wrox.com 上下载

(1) 打开 Eclipse，创建一个名为 BasicViews5 的 Android 项目。
(2) 修改 BasicViews5Activity.java 文件，插入下列粗体显示的语句：

```java
package net.learn2develop.BasicViews5;

import android.app.ListActivity;
import android.os.Bundle;
import android.view.View;
import android.widget.ArrayAdapter;
import android.widget.ListView;
import android.widget.Toast;

public class BasicViews5Activity extends ListActivity {
    String[] presidents = {
            "Dwight D. Eisenhower",
            "John F. Kennedy",
            "Lyndon B. Johnson",
            "Richard Nixon",
            "Gerald Ford",
            "Jimmy Carter",
            "Ronald Reagan",
            "George H. W. Bush",
            "Bill Clinton",
            "George W. Bush",
            "Barack Obama"
    };

    /** Called when the activity is first created. */
```

```
@Override
public void onCreate(Bundle savedInstanceState) {
    super.onCreate(savedInstanceState);
    //---no need to call this---
    //setContentView(R.layout.main);

    setListAdapter(new ArrayAdapter<String>(this,
        android.R.layout.simple_list_item_1, presidents));
}

public void onListItemClick(
ListView parent, View v, int position, long id)
{
    Toast.makeText(this,
        "You have selected " + presidents[position],
        Toast.LENGTH_SHORT).show();
}

}
```

(3) 按 F11 键在 Android 模拟器上调试应用程序。图 4-18 展示了显示总统名字列表的活动。

(4) 单击一个列表项,将显示一个包含所选择项的消息。

示例说明

在本例中,首先要注意的是 BasicViews5Activity 类扩展了 ListActivity 类。ListActivity 类扩展了 Activity 类并且通过绑定到一个数据源来显示一个项列表。还要注意,无须修改 main.xml 文件来包含 ListView; ListActivity 类本身已经包含了一个 ListView。因此,在 onCreate() 方法中,不需要调用 setContentView() 方法来从 main.xml 文件中加载用户界面:

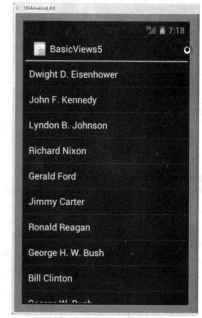

图 4-18

```
//---no need to call this---
//setContentView(R.layout.main);
```

在 onCreate()方法中,使用 setListAdapter()方法来用一个 ListView 以编程方式填充活动的整个屏幕。ArrayAdapter 对象管理将由 ListView 显示的字符串数组。在前面的例子中,将 ListView 设置为在 simple_list_item_1 模式下显示:

```
setListAdapter(new ArrayAdapter<String>(this,
    android.R.layout.simple_list_item_1, presidents));
```

当单击 ListView 中的一个列表项时,将触发 onListItemClick()方法:

```
public void onListItemClick(
```

```
    ListView parent, View v, int position, long id)
    {
        Toast.makeText(this,
            "You have selected " + presidents[position],
            Toast.LENGTH_SHORT).show();
    }
```

这里，只是使用 Toast 类来显示所选择的总统名字。

定制 ListView

ListView 是一个可以进一步定制的通用视图。下面的"试一试"展示了如何允许在 ListView 中选择多个项以及如何使之支持筛选功能。

试一试 在 ListView 中启用对筛选和多列表项的支持

(1) 打开前一节中创建的 BasicViews5 项目，在 BasicViews5Activity.java 文件中添加下列粗体显示的语句：

```
/** Called when the activity is first created. */
@Override
public void onCreate(Bundle savedInstanceState) {
    super.onCreate(savedInstanceState);

    //---no need to call this---
    //setContentView(R.layout.main);

    ListView lstView = getListView();
    //lstView.setChoiceMode(ListView.CHOICE_MODE_NONE);
    //lstView.setChoiceMode(ListView.CHOICE_MODE_SINGLE);
    lstView.setChoiceMode(ListView.CHOICE_MODE_MULTIPLE);
    lstView.setTextFilterEnabled(true);

    setListAdapter(new ArrayAdapter<String>(this,
        android.R.layout.simple_list_item_checked, presidents));
}
```

(2) 按 F11 键在 Android 模拟器上调试应用程序。现在，可以单击每个项以显示其旁边的勾号图标(如图 4-19 所示)。

示例说明

为了以编程方式获得对 ListView 对象的引用，可以使用能获取 ListActivity 的列表视图的 getListView()方法。想要以编程方式修改 ListView 的行为，就需要这么做。在此情况下，使用 setChoiceMode()方法来告诉 ListView 如何处理一个用户的单击。在本例中，将其设置为 ListView.CHOICE_MODE_MULTIPLE，这意味着用户可以选择多个项：

```
    ListView lstView = getListView();
```

```
//lstView.setChoiceMode(ListView.CHOIC
E_MODE_NONE);
//lstView.setChoiceMode(ListView.CHOIC
E_MODE_SINGLE);
lstView.setChoiceMode(ListView.CHOICE_
MODE_MULTIPLE);
```

ListView 的一个非常酷的功能是支持筛选。如果通过 setTextFilterEnabled()方法启用了筛选功能，用户将可以在键盘上输入并且 ListView 将自动筛选来匹配已经输入的内容：

```
lstView.setTextFilterEnabled(true);
```

图 4-20 显示了起作用的列表筛选功能。这里，列表中所有包含单词 john 的项将在结果列表中显示出来。

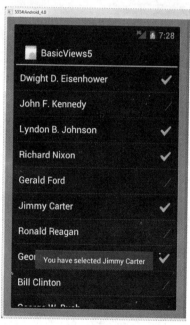

图 4-19

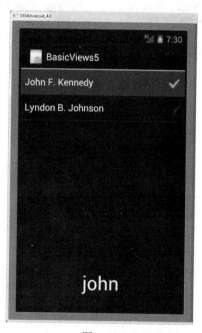

图 4-20

虽然本例中显示了总统名字列表存储在一个数组中，但在实际的应用中，建议从数据库中检索它们或至少将它们存储在 strings.xml 文件中。下面的"试一试"展示了这一点。

试一试　　将列表项存储在 strings.xml 文件中

（1）使用前一节创建的 BasicViews5 项目，在位于 res/values 文件夹下的 strings.xml 文件中添加下列粗体显示的行：

```
<?xml version="1.0" encoding="utf-8"?>
<resources>
    <string name="hello">Hello World, BasicViews5Activity!</string>
    <string name="app_name">BasicViews5</string>
    <string-array name="presidents_array">
        <item>Dwight D. Eisenhower</item>
```

```xml
        <item>John F. Kennedy</item>
        <item>Lyndon B. Johnson</item>
        <item>Richard Nixon</item>
        <item>Gerald Ford</item>
        <item>Jimmy Carter</item>
        <item>Ronald Reagan</item>
        <item>George H. W. Bush</item>
        <item>Bill Clinton</item>
        <item>George W. Bush</item>
        <item>Barack Obama</item>
    </string-array>
</resources>
```

(2) 按下列粗体显示内容修改 BasicViews5Activity.java 文件：

```java
public class BasicViews5Activity extends ListActivity {
    String[] presidents;

    /** Called when the activity is first created. */
    @Override
    public void onCreate(Bundle savedInstanceState) {
        super.onCreate(savedInstanceState);

        //---no need to call this---
        //setContentView(R.layout.main);

        ListView lstView = getListView();
        //lstView.setChoiceMode(ListView.CHOICE_MODE_NONE);
        //lstView.setChoiceMode(ListView.CHOICE_MODE_SINGLE);
        lstView.setChoiceMode(ListView.CHOICE_MODE_MULTIPLE);
        lstView.setTextFilterEnabled(true);

        presidents =
                getResources().getStringArray(R.array.presidents_array);

        setListAdapter(new ArrayAdapter<String>(this,
            android.R.layout.simple_list_item_checked, presidents));
    }

    public void onListItemClick(
    ListView parent, View v, int position, long id)
    {
        Toast.makeText(this,
            "You have selected " + presidents[position],
            Toast.LENGTH_SHORT).show();
    }
}
```

(3) 按 F11 键在 Android 模拟器上调试应用程序。您将会看到同前面的"试一试"中一样的名字列表。

示例说明

由于现在名字存储在 strings.xml 文件中，所以可以在这个 BasicViews5Activity.java 文件中使用 getResources()方法以编程方式来检索它：

```
presidents =
        getResources().getStringArray(R.array.presidents_array);
```

一般地，可以使用 getResources()方法以编程方式来检索与应用程序捆绑的资源。

这个示例演示了如何使 ListView 中的列表项可被选择。在选择过程的结尾，如何知道哪个项或哪些项被选中？下面的"试一试"演示了具体做法。

试一试　　检查哪些项被选中

(1) 再次使用 BasicView5 项目，在 main.xml 文件中添加下列粗体显示的代码：

```xml
<?xml version="1.0" encoding="utf-8"?>
<LinearLayout xmlns:android="http://schemas.android.com/apk/res/android"
    android:layout_width="fill_parent"
    android:layout_height="fill_parent"
    android:orientation="vertical" >

<Button
    android:id="@+id/btn"
    android:layout_width="fill_parent"
    android:layout_height="wrap_content"
    android:text="Show selected items"
    android:onClick="onClick"/>

<ListView
    android:id="@+id/android:list"
    android:layout_width="wrap_content"
    android:layout_height="wrap_content" />

</LinearLayout>
```

(2) 在 BasicViews5Activity.java 文件中添加下列粗体显示的代码：

```java
package net.learn2develop.BasicViews5;

import android.app.ListActivity;
import android.os.Bundle;
import android.view.View;
import android.widget.ArrayAdapter;
import android.widget.ListView;
import android.widget.Toast;

public class BasicViews5Activity extends ListActivity {
```

```java
    String[] presidents;

    /** Called when the activity is first created. */
    @Override
    public void onCreate(Bundle savedInstanceState) {
        super.onCreate(savedInstanceState);

        setContentView(R.layout.main);

        ListView lstView = getListView();
        //lstView.setChoiceMode(ListView.CHOICE_MODE_NONE);
        //lstView.setChoiceMode(ListView.CHOICE_MODE_SINGLE);
        lstView.setChoiceMode(ListView.CHOICE_MODE_MULTIPLE);
        lstView.setTextFilterEnabled(true);

        presidents =
            getResources().getStringArray(R.array.presidents_array);

        setListAdapter(new ArrayAdapter<String>(this,
            android.R.layout.simple_list_item_checked, presidents));
    }

    public void onListItemClick(
    ListView parent, View v, int position, long id)
    {
        Toast.makeText(this,
            "You have selected " + presidents[position],
            Toast.LENGTH_SHORT).show();
    }

    public void onClick(View view) {
        ListView lstView = getListView();

        String itemsSelected = "Selected items: \n";
        for (int i=0; i<lstView.getCount(); i++) {
            if (lstView.isItemChecked(i)) {
                itemsSelected += lstView.getItemAtPosition(i) + "\n";
            }
        }
        Toast.makeText(this, itemsSelected, Toast.LENGTH_LONG).show();
    }

}
```

(3) 按 F11 键在 Android 模拟器上调试应用程序。单击一些列表项，然后单击 Show selected items 按钮，如图 4-21 所示。所选名字的列表将会显示出来。

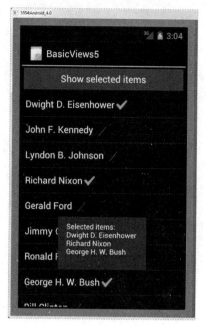

图 4-21

示例说明

在前一节的练习中,看到了如何填充一个占据整个活动的 ListView——在该例中,不需要向 main.xml 文件添加一个<ListView>元素。在本例中,看到了 ListView 可以部分填充一个活动。为此,需要添加一个<ListView>元素,并将其 id 属性设为@+id/android:list:

```
<ListView
    android:id="@+id/android:list"
    android:layout_width="wrap_content"
    android:layout_height="wrap_content" />
```

然后需要使用 setContentView()方法加载活动的内容(之前注释掉了):

```
setContentView(R.layout.main);
```

为了找出 ListView 中哪些项被选中,使用了 isItemChecked()方法:

```
ListView lstView = getListView();
String itemsSelected = "Selected items: \n";
for (int i=0; i<lstView.getCount(); i++) {
    if (lstView.isItemChecked(i)) {
        itemsSelected += lstView.getItemAtPosition(i) + "\n";
    }
}
Toast.makeText(this, itemsSelected, Toast.LENGTH_LONG).show();
```

getItemAtPosition()方法返回了指定位置的列表项的名称。

第 4 章 使用视图设计用户界面

> 注意：到目前为止，所有的例子都显示了如何在一个 ListActivity 内使用 ListView。这不是绝对必要的——您也可以在 Activity 内使用 ListView。在本例中，为了以编程方式引用 ListView，使用了 findViewByID()方法而不是 getListView()方法。<ListView>元素的 id 属性可以使用这种格式：@+id/<view_name>。

4.3.2 使用 Spinner 视图

ListView 在一个活动中显示一个长的项列表，但有时需要在用户界面上显示其他视图，因此没有额外的空间来显示像 ListView 这样的全屏视图。在这种情况下，应该使用 SpinnerView。SpinnerView 一次显示列表中的一项，并可以用户在其中进行选择。

下面的"试一试"展示了如何在活动中使用 SpinnerView。

试一试 使用 SpinnerView 一次显示一个项

BasicViews6.zip 代码文件可以在 Wrox.com 上下载

(1) 打开 Eclipse，创建一个名为 BasicViews6 的 Android 项目。

(2) 按如下所示修改位于 res/layout 文件夹下的 main.xml 文件：

```xml
<?xml version="1.0" encoding="utf-8"?>
<LinearLayout xmlns:android="http://schemas.android.com/apk/res/android"
    android:layout_width="fill_parent"
    android:layout_height="fill_parent"
    android:orientation="vertical" >

<Spinner
    android:id="@+id/spinner1"
    android:layout_width="wrap_content"
    android:layout_height="wrap_content"
    android:drawSelectorOnTop="true" />

</LinearLayout>
```

(3) 把下列粗体显示的行添加到位于 res/values 文件夹下的 strings.xml 文件中：

```xml
<?xml version="1.0" encoding="utf-8"?>
<resources>
    <string name="hello">Hello World, BasicViews6Activity!</string>
    <string name="app_name">BasicViews6</string>
    <string-array name="presidents_array">
        <item>Dwight D. Eisenhower</item>
        <item>John F. Kennedy</item>
        <item>Lyndon B. Johnson</item>
```

```xml
            <item>Richard Nixon</item>
            <item>Gerald Ford</item>
            <item>Jimmy Carter</item>
            <item>Ronald Reagan</item>
            <item>George H. W. Bush</item>
            <item>Bill Clinton</item>
            <item>George W. Bush</item>
            <item>Barack Obama</item>
    </string-array>
</resources>
```

(4) 在 BasicViews6Activity.java 文件中添加下列粗体显示的语句:

```java
package net.learn2develop.BasicViews6;

import android.app.Activity;
import android.os.Bundle;
import android.view.View;
import android.widget.AdapterView;
import android.widget.AdapterView.OnItemSelectedListener;
import android.widget.ArrayAdapter;
import android.widget.Spinner;
import android.widget.Toast;

public class BasicViews6Activity extends Activity {
    String[] presidents;

    /** Called when the activity is first created. */
    @Override
    public void onCreate(Bundle savedInstanceState) {
        super.onCreate(savedInstanceState);
        setContentView(R.layout.main);

        presidents =
                getResources().getStringArray(R.array.presidents_array);
        Spinner s1 = (Spinner) findViewById(R.id.spinner1);

        ArrayAdapter<String> adapter = new ArrayAdapter<String>(this,
                android.R.layout.simple_spinner_item, presidents);

        s1.setAdapter(adapter);
        s1.setOnItemSelectedListener(new OnItemSelectedListener()
        {
            @Override
            public void onItemSelected(AdapterView<?> arg0,
            View arg1, int arg2, long arg3)
            {
                int index = arg0.getSelectedItemPosition();
                Toast.makeText(getBaseContext(),
                    "You have selected item : " + presidents[index],
                    Toast.LENGTH_SHORT).show();
```

```
            }

            @Override
            public void onNothingSelected(AdapterView<?> arg0) { }
        });
    }
}
```

(5) 按 F11 键在 Android 模拟器上调试应用程序。单击 SpinnerView，可以看到弹出一个显示总统名字的列表(如图 4-22 所示)。单击一个列表项将显示一个消息，表明这个列表项被选择了。

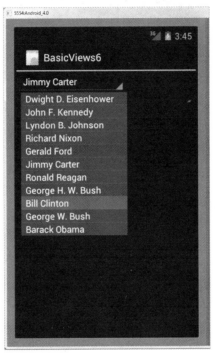

图 4-22

示例说明

上面的例子与 ListView 的工作原理很相像。需要实现的一个额外方法是 onNothingSelected()方法。当用户按下 Back 按钮时触发这一方法，撤销所显示的项列表。在这种情况下，没有任何项被选择，也不需要作任何处理。

除了在 ArrayAdapter 中以普通列表形式显示列表项之外，还可以使用单选按钮来显示它们。要做到这一点，需要修改 ArrayAdapter 类的构造函数中的第二个参数：

```
ArrayAdapter<String> adapter = new ArrayAdapter<String>(this,
    android.R.layout.simple_list_item_single_choice, presidents);
```

这样将使列表项以单选按钮列表形式显示(如图 4-23 所示)。

Android 4 编程入门经典——开发智能手机与平板电脑应用

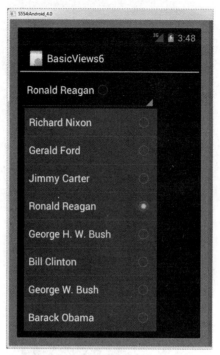

图 4-23

4.4 了解特殊碎片

在第 2 章,学习了 Android 3 中引入的碎片功能。使用碎片时,可以定制 Android 应用程序的用户界面,通过动态地重新排列碎片使其适应活动。这样就允许建立的应用程序在拥有不同的屏幕尺寸的设备上运行。

正如前面介绍过的,碎片是拥有自己的生命周期的"微活动"。为创建一个碎片,需要一个扩展 Fragment 基类的类。除了 Fragment 基类外,还可以扩展 Fragment 基类的其他一些子类,以创建更加特殊的碎片。下面将介绍 Fragment 的 3 个子类:ListFragment、DialogFragment 以及 PreferenceFragment。

4.4.1 使用 ListFragment

一个列表碎片就是一个包含 ListView 的碎片,它显示来自某个数据源(例如一个数组或一个 Cursor)的项目列表。列表碎片十分有用,因为经常需要用一个碎片包含一个项列表(例如一个 RSS 帖子的列表),用另一个碎片显示所选帖子的详细信息。为了创建一个碎片列表,必须扩展 ListFragment 基类。

下面"试一试"展示了启动一个碎片表的方法。

试一试 创建并使用一个列表碎片

ListFragmentExample.zip 代码文件可以在 wrox.com 上下载

(1) 利用 Eclipse 创建一个 Android 项目，并将其命名为 ListFragmentExample。
(2) 按照下列粗体显示的代码修改 main.xml 文件。

```xml
<?xml version="1.0" encoding="utf-8"?>
<LinearLayout xmlns:android="http://schemas.android.com/apk/res/android"
    android:layout_width="fill_parent"
    android:layout_height="fill_parent"
    android:orientation="horizontal" >

    <fragment
        android:name="net.learn2develop.ListFragmentExample.Fragment1"
        android:id="@+id/fragment1"
        android:layout_weight="0.5"
        android:layout_width="0dp"
        android:layout_height="200dp" />

    <fragment
        android:name="net.learn2develop.ListFragmentExample.Fragment1"
        android:id="@+id/fragment2"
        android:layout_weight="0.5"
        android:layout_width="0dp"
        android:layout_height="300dp" />

</LinearLayout>
```

(3) 将一个 XML 文件添加到 res/layout 文件夹中并将其命名为 fragment1.xml。
(4) 使用如下代码填充 fragment1.xml 文件:

```xml
<?xml version="1.0" encoding="utf-8"?>
<LinearLayout xmlns:android="http://schemas.android.com/apk/res/android"
    android:orientation="vertical"
    android:layout_width="fill_parent"
    android:layout_height="fill_parent">
    <ListView
        android:id="@id/android:list"
        android:layout_width="match_parent"
        android:layout_height="match_parent"
        android:layout_weight="1"
        android:drawSelectorOnTop="false"/>
</LinearLayout>
```

(5) 将一个 Java Class 文件添加到包里并把它命名为 Fragment1。
(6) 使用如下代码填充 Fragment1.java 文件:

```
package net.learn2develop.ListFragmentExample;
```

```java
import android.app.ListFragment;
import android.os.Bundle;
import android.view.LayoutInflater;
import android.view.View;
import android.view.ViewGroup;
import android.widget.ArrayAdapter;
import android.widget.ListView;
import android.widget.Toast;

public class Fragment1 extends ListFragment {
    String[] presidents = {
        "Dwight D. Eisenhower",
        "John F. Kennedy",
        "Lyndon B. Johnson",
        "Richard Nixon",
        "Gerald Ford",
        "Jimmy Carter",
        "Ronald Reagan",
        "George H. W. Bush",
        "Bill Clinton",
        "George W. Bush",
        "Barack Obama"
    };

    @Override
    public View onCreateView(LayoutInflater inflater,
    ViewGroup container, Bundle savedInstanceState) {
        return inflater.inflate(R.layout.fragment1, container, false);
    }

    @Override
    public void onCreate(Bundle savedInstanceState) {
        super.onCreate(savedInstanceState);
        setListAdapter(new ArrayAdapter<String>(getActivity(),
            android.R.layout.simple_list_item_1, presidents));
    }

    public void onListItemClick(ListView parent, View v,
    int position, long id)
    {
        Toast.makeText(getActivity(),
            "You have selected " + presidents[position],
            Toast.LENGTH_SHORT).show();
    }

}
```

(7) 按 F11 键在 Android 模拟器上调试应用程序,图 4-24 显示两个列表碎片,它们显示了两个总统名字列表。

(8) 单击两个 ListView 视图中的任意一个选项，会显示相应的消息(如图 4-25 所示)。

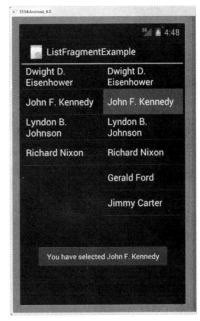

图 4-24　　　　　　　　图 4-25

示例说明

通过将一个 ListView 元素添加给碎片，首先为该碎片创建了一个 XML 文件：

```
<?xml version="1.0" encoding="utf-8"?>
<LinearLayout xmlns:android="http://schemas.android.com/apk/res/android"
    android:orientation="vertical"
    android:layout_width="fill_parent"
    android:layout_height="fill_parent">
    <ListView
        android:id="@id/android:list"
        android:layout_width="match_parent"
        android:layout_height="match_parent"
        android:layout_weight="1"
        android:drawSelectorOnTop="false"/>
</LinearLayout>
```

为了创建一个列表碎片，碎片所用的 Java 类必须扩展 ListFragement 基类：

```
Pubic class Fragment1 extends ListFragment{
}
```

接下来声明一个数组，用于包含活动中的总统名字列表：

```
String[] presidents = {
    "Dwight D. Eisenhower",
    "John F. Kennedy",
    "Lyndon B. Johnson",
```

```
            "Richard Nixon",
            "Gerald Ford",
            "Jimmy Carter",
            "Ronald Reagan",
            "George H. W. Bush",
            "Bill Clinton",
            "George W. Bush",
            "Barack Obama"
    };
```

在 onCreate()事件中,利用 setListAdapter()方法以编程方式将数组的内容填充到 ListView。ArrayAdapter 对象管理将被 ListView 显示的字符串数组。在本例中将 ListView 设为在 simple_list_item_1 模式下显示。

```
    @Override
    public void onCreate(Bundle savedInstanceState) {
        super.onCreate(savedInstanceState);
        setListAdapter(new ArrayAdapter<String>(getActivity(),
            android.R.layout.simple_list_item_1, presidents));
    }
```

每当单击 ListView 的一个列表项时,就会触发 onListItemClick()方法:

```
    public void onListItemClick(ListView parent, View v,
    int position, long id)
    {
        Toast.makeText(getActivity(),
            "You have selected " + presidents[position],
            Toast.LENGTH_SHORT).show();
    }
```

最后,将两个碎片添加到活动中,注意每个碎片的高度:

```xml
<?xml version="1.0" encoding="utf-8"?>
<LinearLayout xmlns:android="http://schemas.android.com/apk/res/android"
    android:layout_width="fill_parent"
    android:layout_height="fill_parent"
    android:orientation="horizontal" >

<fragment
    android:name="net.learn2develop.ListFragmentExample.Fragment1"
    android:id="@+id/fragment1"
    android:layout_weight="0.5"
    android:layout_width="0dp"
    android:layout_height="200dp" />

<fragment
    android:name="net.learn2develop.ListFragmentExample.Fragment1"
    android:id="@+id/fragment2"
    android:layout_weight="0.5"
```

```
                android:layout_width="0dp"
                android:layout_height="300dp" />

</LinearLayout>
```

4.4.2 使用 DialogFragment

可创建的另一种碎片类型是对话框碎片。一个对话框碎片浮动在活动上方，并且以模态方式显示。当需要获得用户的响应，然后才能继续执行操作的时候，对话框碎片十分有用。为了创建一个对话框碎片，需要扩展 DialogFragment 基类。

下面的"试一试"展示了创建对话框碎片的方法。

试一试 创建并使用一个对话碎片

DialogFragmentExample.zip 代码文件可以在 Wrox.com 上下载

(1) 使用 Eclipse 创建一个 Android 项目，并把它命名为 DialogFragmentExample。
(2) 向包里添加一个 Java 类文件并将其命名为 Fragment1。
(3) 使用如下代码填充 Fragment1.java 文件。

```java
package net.learn2develop.DialogFragmentExample;

import android.app.AlertDialog;
import android.app.Dialog;
import android.app.DialogFragment;
import android.content.DialogInterface;
import android.os.Bundle;

public class Fragment1 extends DialogFragment {

    static Fragment1 newInstance(String title) {
        Fragment1 fragment = new Fragment1();
        Bundle args = new Bundle();
        args.putString("title", title);
        fragment.setArguments(args);
        return fragment;
    }

    @Override
    public Dialog onCreateDialog(Bundle savedInstanceState) {
        String title = getArguments().getString("title");
        return new AlertDialog.Builder(getActivity())
        .setIcon(R.drawable.ic_launcher)
        .setTitle(title)
        .setPositiveButton("OK",
                new DialogInterface.OnClickListener() {
            public void onClick(DialogInterface dialog,
                    int whichButton) {
```

```
                    ((DialogFragmentExampleActivity)
                            getActivity()).doPositiveClick();
                }
            })
            .setNegativeButton("Cancel",
                    new DialogInterface.OnClickListener() {
                public void onClick(DialogInterface dialog,
                        int whichButton) {
                    ((DialogFragmentExampleActivity)
                            getActivity()).doNegativeClick();
                }
            }).create();
    }

}
```

(4) 使用下列粗体显示的代码填充 DialogFragmentExampleActivity.java 文件。

```
package net.learn2develop.DialogFragmentExample;

import android.app.Activity;
import android.os.Bundle;
import android.util.Log;

public class DialogFragmentExampleActivity extends Activity {
    /** Called when the activity is first created. */
    @Override
    public void onCreate(Bundle savedInstanceState) {
        super.onCreate(savedInstanceState);
        setContentView(R.layout.main);

        Fragment1 dialogFragment = Fragment1.newInstance(
                "Are you sure you want to do this?");
        dialogFragment.show(getFragmentManager(), "dialog");
    }

    public void doPositiveClick() {
        //---perform steps when user clicks on OK---
        Log.d("DialogFragmentExample", "User clicks on OK");
    }

    public void doNegativeClick() {
        //---perform steps when user clicks on Cancel---
        Log.d("DialogFragmentExample", "User clicks on Cancel");
    }

}
```

(5) 按 F11 键在 Android 模拟器上调试应用程序。图 4-26 所示的碎片在一个警告对话框中显示。单击 OK 按钮或 Cancel 按钮，观察显示的消息。

示例说明

为创建一个对话碎片，首先 Java 类要扩展 DialogFragment 基类：

```java
public class Fragment1 extends DialogFragment{
}
```

在本例中，创建了一个警告对话框，这是一个显示一条消息及可选按钮的对话框窗口。在 Fragment1 类中，定义了 newInstance()方法：

```java
static Fragment1 newInstance(String title) {
    Fragment1 fragment = new Fragment1();
    Bundle args = new Bundle();
    args.putString("title", title);
    fragment.setArguments(args);
    return fragment;
}
```

newInstance()方法允许创建碎片的一个新实例，同时它接受一个指定警告对话框中要显示的字符串(title)的参数。title 随后存储在一个 Bundle 对象里供之后使用。

接下来定义了 onCreateDialog()方法，该方法在 onCreate()之后、onCreateView()之前调用：

```java
@Override
public Dialog onCreateDialog(Bundle savedInstanceState) {
    String title = getArguments().getString("title");
    return new AlertDialog.Builder(getActivity())
    .setIcon(R.drawable.ic_launcher)
    .setTitle(title)
    .setPositiveButton("OK",
            new DialogInterface.OnClickListener() {
        public void onClick(DialogInterface dialog,
                int whichButton) {
            ((DialogFragmentExampleActivity)
                    getActivity()).doPositiveClick();
        }
    })
    .setNegativeButton("Cancel",
            new DialogInterface.OnClickListener() {
        public void onClick(DialogInterface dialog,
                int whichButton) {
            ((DialogFragmentExampleActivity)
                    getActivity()).doNegativeClick();
        }
    }).create();
}
```

在这里，创建的警告对话框有两个按钮：OK 和 Cancel。要在该对话框中显示的字符串从保存在 Bundle 对象中的 title 参数中获取。

为了显示对话框碎片，创建它的一个实例并调用它的 show()方法：

```java
Fragment1 dialogFragment = Fragment1.newInstance(
```

```
    "Are you sure you want to do this?");
dialogFragment.show(getFragmentManager(), "dialog");
```

还需要实现两种方法：doPostiveClick()和 doNegativeClick()，分别用于处理用户单击 OK 按钮或 Cancel 按钮的情况。

```
public void doPositiveClick() {
    //---perform steps when user clicks on OK---
    Log.d("DialogFragmentExample", "User clicks on OK");
}

public void doNegativeClick() {
    //---perform steps when user clicks on Cancel---
    Log.d("DialogFragmentExample", "User clicks on Cancel");
}
```

4.4.3 使用 PreferenceFragment

Android 应用程序通常要提供首选项，以允许用户定制应用程序。例如，可以允许用户保存那些用于访问 Web 资源的登录凭据，或者保存源刷新频率的信息(比如在一个 RSS 阅读器应用程序中)等等。在 Android 中，可以使用 PreferenceActivity 基类为用户显示一个用于编辑首选项的活动。在 Android 3.0 和更高版本中，可以使用 PreferenceFragment 类实现相同的功能。

下面"试一试"显示了在 Android 3 和 Android 4 版本中创建并使用一个首选项碎片的方法。

试一试   创建并使用一个首选项碎片

PreferenceFragmentExample.zip 代码文件可以在 Wrox.com 上下载

(1) 使用 Eclipse 创建一个 Android 项目并把它命名为 PreferenceFragmentExample。

(2) 在 res 文件夹下创建一个新的 xml 文件夹，然后将一个新的 Android XML 文件添加到该 xml 文件夹里。将该 XML 文件命名为 preferences.xml(如图 4-27 所示)。

(3) 使用如下代码填充 preferences.xml 文件。

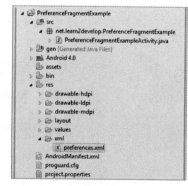

图 4-27

```xml
<?xml version="1.0" encoding="utf-8"?>
<PreferenceScreen
xmlns:android="http://schemas.android.com/apk/res/android">

    <PreferenceCategory android:title="Category 1">
        <CheckBoxPreference
            android:title="Checkbox"
            android:defaultValue="false"
            android:summary="True of False"
```

```xml
            android:key="checkboxPref" />
    </PreferenceCategory>

    <PreferenceCategory android:title="Category 2">
        <EditTextPreference
            android:name="EditText"
            android:summary="Enter a string"
            android:defaultValue="[Enter a string here]"
            android:title="Edit Text"
            android:key="editTextPref" />
        <RingtonePreference
            android:name="Ringtone Preference"
            android:summary="Select a ringtone"
            android:title="Ringtones"
            android:key="ringtonePref" />
        <PreferenceScreen
            android:title="Second Preference Screen"
            android:summary=
                "Click here to go to the second Preference Screen"
            android:key="secondPrefScreenPref">
            <EditTextPreference
                android:name="EditText"
                android:summary="Enter a string"
                android:title="Edit Text (second Screen)"
                android:key="secondEditTextPref" />
        </PreferenceScreen>
    </PreferenceCategory>

</PreferenceScreen>
```

(4) 将一个 Java 类文件添加到包里，并将其命名为 Fragment1。

(5) 使用如下代码填充 Fragment1.java 文件。

```java
package net.learn2develop.PreferenceFragmentExample;

import android.os.Bundle;
import android.preference.PreferenceFragment;

public class Fragment1 extends PreferenceFragment {
    @Override
    public void onCreate(Bundle savedInstanceState) {
        super.onCreate(savedInstanceState);

        //---load the preferences from an XML file---
        addPreferencesFromResource(R.xml.preferences);
    }
}
```

(6) 按照下列粗体显示的代码修改 PreferenceFragmentExampleActivity.java 文件。

```java
package net.learn2develop.PreferenceFragmentExample;

import android.app.Activity;
```

```java
import android.app.FragmentManager;
import android.app.FragmentTransaction;
import android.os.Bundle;

public class PreferenceFragmentExampleActivity extends Activity {
    /** Called when the activity is first created. */
    @Override
    public void onCreate(Bundle savedInstanceState) {
        super.onCreate(savedInstanceState);
        setContentView(R.layout.main);

        FragmentManager fragmentManager = getFragmentManager();
        FragmentTransaction fragmentTransaction =
            fragmentManager.beginTransaction();
        Fragment1 fragment1 = new Fragment1();
        fragmentTransaction.replace(android.R.id.content, fragment1);
        fragmentTransaction.addToBackStack(null);
        fragmentTransaction.commit();
    }
}
```

(7) 按 F11 键在 Android 模拟器上调试应用程序，图 4-28 展示了首选项碎片，它显示了用户可以修改的首选项列表。

(8) 当单击 Edit Text 首选项时，会显示一个弹出窗口(如图 4-29 所示)。

(9) 单击 Second Preference Screen 选项会使第二个首选项屏幕项显示出来(如图 4-30 所示)。

图 4-28　　　　　　　　图 4-29　　　　　　　　图 4-30

(10) 要想让首选项碎片消失，在模拟器上单击 Back 按钮。

(11) 如果查看 File Explore(在 DDMS 透视图中可用)，将可以定位到位于/data/data/net.learn2develop.preferenceFragmentExample/shAred_prefs/文件夹中的首选项文件(如图 4-31 所示)，用户所做的所有修改都保存在这个文件中。

第 4 章 使用视图设计用户界面

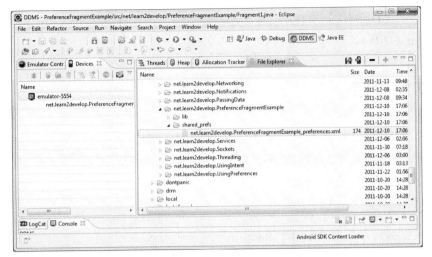

图 4-31

 注意：第 6 章将介绍如何检索保存在首选项文件中的值。

示例说明

为了在 Android 应用程序中创建一个首选项的列表，首先需要创建 preferences.xml 文件并将不同的 XML 元素填充到该文件中。这个 XML 文件定义了各种想保存在应用程序中的项。

为了创建首选项碎片，需要扩展 PreferenceFragment 基类：

```
public class Fragment1 extends preferenceFragment{
}
```

为在首选项碎片中加载首选项文件，可以使用 addPreferencesFromResource()方法：

```
@Override
public void onCreate(Bundle savedInstanceState) {
    super.onCreate(savedInstanceState);

    //---load the preferences from an XML file---
    addPreferencesFromResource(R.xml.preferences);
}
```

为在活动中显示首选项碎片，可以使用 FragmentManger 类和 FragmentTransaction 类：

```
FragmentManager fragmentManager = getFragmentManager();
FragmentTransaction fragmentTransaction =
    fragmentManager.beginTransaction();
Fragment1 fragment1 = new Fragment1();
fragmentTransaction.replace(android.R.id.content, fragment1);
fragmentTransaction.addToBackStack(null);
fragmentTransaction.commit();
```

205

需要使用 addToBackStack()方法将首选项碎片添加到 back stack，从而用户可以通过单击 Back 按钮关闭碎片。

4.5 本章小结

本章对在 Android 应用程序中经常会用到的一些视图作了概述。虽然不可能详细研究每一个视图，但这里所学习到的视图会为设计 Android 应用程序的用户界面提供一个良好的基础，而不用管其需求是什么。

练习

1. 如何以编程方式来确定一个 RadioButton 是否被选中？
2. 如何访问存储在 strings.xml 文件中的字符串资源？
3. 写一段代码来获取当前日期。
4. 列举在 Android 应用程序中可以使用的 3 种专用碎片，并描述它们的用法。

练习答案参见附录 C。

本章主要内容

主　题	关　键　概　念

TextView	```<TextView
	    android:layout_width="fill_parent"
	    android:layout_height="wrap_content"
	    android:text="@ string/hello"
	/>```

Button	```<Button android:id="@+id/btnSave"
	    android:layout_width="fill_parent"
	    android:layout_height="wrap_content"
	    android:text="Save" />```

ImageButton	```<ImageButton android:id="@+id/btnImg1"
	    android:layout_width="fill_parent"
	    android:layout_height="wrap_content"
	    android:src="@drawable/icon" />```

EditText	```<EditText android:id="@+id/txtName"
	    android:layout_width="fill_parent"
	    android:layout_height="wrap_content" />```

CheckBox	```<CheckBox android:id="@+id/chkAutosave"
	    android:layout_width="fill_parent"
	    android:layout_height="wrap_content"
	    android:text="Autosave" />```

(续表)

主 题	关 键 概 念
RadioGroup 和 RadioButton	```<RadioGroup android:id="@+id/rdbGp1"
android:layout_width="fill_parent"
android:layout_height="wrap_content"
android:orientation="vertical" >
<RadioButton android:id="@+id/rdb1"
android:layout_width="fill_parent"
android:layout_height="wrap_content"
android:text="Option 1" />
<RadioButton android:id="@+id/rdb2"
android:layout_width="fill_parent"
android:layout_height="wrap_content"
android:text="Option 2" />
</RadioGroup>```
ToggleButton	```<ToggleButton android:id="@+id/toggle1"
droid:layout_width="wrap_content"```
ProgressBar	```<ProgressBar android:id="@+id/progressbar"
android:layout_width="wrap_content"
android:layout_height="wrap_content" />```
AutoCompleteTextBox	```<AutoCompleteTextView
android:id="@+id/txtCountries"
android:layout_width="fill_parent"
android:layout_height="wrap_content" />```
TimePicker	```<TimePicker android:id="@+id/timePicker"
android:layout_width="wrap_content"
android:layout_height="wrap_content" />```
DatePicker	```<DatePicker android:id="@+id/datePicker"
android:layout_width="wrap_content"
android:layout_height="wrap_content" />```
Spinner	```<Spinner android:id="@+id/spinner1"
android:layout_width="wrap_content"
android:layout_height="wrap_content"
android:drawSelectorOnTop="true" />```
特殊碎片类型	```ListFragment, DialogFragment, and PreferenceFragment
android:src="@drawable/icon" />```

第 5 章

使用视图显示图片和菜单

本章将介绍以下内容：
- 如何使用 Gallery、ImageSwitcher、GridView 和 ImageView 视图显示图像
- 如何显示选项菜单和上下文菜单
- 如何使用 AnalogClock 和 DigitalClock 视图显示时间
- 如何使用 WebView 显示 Web 内容

在第 4 章中，我们已经学习了可以用来构建 Android 应用程序的用户界面的不同视图。本章将继续研究其他可用来创建健壮的、吸引人的应用程序的视图。

特别是，我们会将注意力转到可以用来显示图像的视图上。此外，还将学习如何在 Android 应用程序中创建选项和上下文菜单。本章结束时将讨论一些可用来显示当前时间和 Web 内容的很炫酷的视图。

5.1 使用图像视图显示图片

到目前为止，我们已经学习的所有视图都是用来显示文本信息的。要显示图像，可以使用 ImageView、Gallery、ImageSwitcher 和 GridView 视图。

下面将详细介绍每一个视图。

5.1.1 Gallery 和 ImageView 视图

Gallery 是一种用固定在中间位置的水平滚动列表显示列表项(如图像)的视图。图 5-1 展示了 Gallery 视图在显示一些图像时的外观效果：

下面的"试一试"展示了如何使用 Gallery 视图显示一组图像。

图 5-1

> 试一试　　使用 Gallery 视图

Gallery.zip 代码文件可以在 Wrox.com 上下载

(1) 打开 Eclipse，创建一个新的 Android 项目，将其命名为 Gallery。

(2) 按照下列粗体显示内容修改 main.xml 文件：

```xml
<?xml version="1.0" encoding="utf-8"?>
<LinearLayout xmlns:android="http://schemas.android.com/apk/res/android"
    android:layout_width="fill_parent"
    android:layout_height="fill_parent"
    android:orientation="vertical" >

<TextView
    android:layout_width="fill_parent"
    android:layout_height="wrap_content"
    android:text="Images of San Francisco" />

<Gallery
    android:id="@+id/gallery1"
    android:layout_width="fill_parent"
    android:layout_height="wrap_content" />

<ImageView
    android:id="@+id/image1"
    android:layout_width="320dp"
    android:layout_height="250dp"
    android:scaleType="fitXY" />

</LinearLayout>
```

(3) 右击 res/values 文件夹，选择 New | File，并将新文件命名为 attrs.xml。

(4) 在 attrs.xml 文件中输入如下内容：

```xml
<?xml version="1.0" encoding="utf-8"?>
<resources>
    <declare-styleable name="Gallery1">
        <attr name="android:galleryItemBackground" />
    </declare-styleable>
</resources>
```

(5) 准备一组图像，并将每一张图像依次命名为 pic1.png、pic2.png 等，如图 5-2 所示。

第 5 章　使用视图显示图片和菜单

图 5-2

 注意：可以从本书的支持网站 www.wrox.com 上下载这组图像。

(6) 将所有图像拖放到 res/drawable-mdpi 文件夹下(如图 5-3 所示)。当显示一个对话框时，选中 Copy files 选项并单击 OK 按钮。

 注意：本例中假设这一项目将在具有中等 DPI 屏幕分辨率的 AVD 上进行测试。在实际的项目中，需要确保每一个 drawable 文件夹都有一组(不同分辨率的)图像。

图 5-3

(7) 在 GalleryActivity.java 文件中添加下列粗体显示的语句：

```java
package net.learn2develop.Gallery;

import android.app.Activity;
import android.content.Context;
import android.content.res.TypedArray;
import android.os.Bundle;
import android.view.View;
import android.view.ViewGroup;
import android.widget.AdapterView;
import android.widget.AdapterView.OnItemClickListener;
import android.widget.BaseAdapter;
import android.widget.Gallery;
import android.widget.ImageView;
import android.widget.Toast;

public class GalleryActivity extends Activity {
    //---the images to display---
    Integer[] imageIDs = {
            R.drawable.pic1,
            R.drawable.pic2,
            R.drawable.pic3,
            R.drawable.pic4,
            R.drawable.pic5,
            R.drawable.pic6,
            R.drawable.pic7
    };

    /** Called when the activity is first created. */
    @Override
    public void onCreate(Bundle savedInstanceState) {
        super.onCreate(savedInstanceState);
        setContentView(R.layout.main);

        Gallery gallery = (Gallery) findViewById(R.id.gallery1);

        gallery.setAdapter(new ImageAdapter(this));
        gallery.setOnItemClickListener(new OnItemClickListener()
        {
            public void onItemClick(AdapterView parent, View v,
            int position, long id)
            {
                Toast.makeText(getBaseContext(),
                    "pic" + (position + 1) + " selected",
                     Toast.LENGTH_SHORT).show();
            }
        });
    }

    public class ImageAdapter extends BaseAdapter
    {
```

```java
    Context context;
    int itemBackground;
    
    public ImageAdapter(Context c)
    {
        context = c;
        //---setting the style---
        TypedArray a = obtainStyledAttributes(
                        R.styleable.Gallery1);
        itemBackground = a.getResourceId(
        R.styleable.Gallery1_android_galleryItemBackground,
        0);
        a.recycle();
    }
    
    //---returns the number of images---
    public int getCount() {
        return imageIDs.length;
    }
    
    //---returns the item---
    public Object getItem(int position) {
        return position;
    }
    
     //---returns the ID of an item---
    public long getItemId(int position) {
        return position;
    }
    
    //---returns an ImageView view---
    public View getView(int position, View convertView,
    ViewGroup parent) {
        ImageView imageView;
        if (convertView == null) {
            imageView = new ImageView(context);
            imageView.setImageResource(imageIDs[position]);
            imageView.setScaleType(
                ImageView.ScaleType.FIT_XY);
            imageView.setLayoutParams(
                new Gallery.LayoutParams(150, 120));
        } else {
            imageView = (ImageView) convertView;
        }
        imageView.setBackgroundResource(itemBackground);
        return imageView;
    }
}
}

}
```

(8) 按 F11 键在 Android 模拟器上调试应用程序。图 5-4 展示了显示一系列图像的 Gallery 视图。可以轻扫图像来显示整个系列的图像。单击一个图像时可以观察到 Toast 类将显示图像名称。

图 5-4

(9) 为了在 ImageView 中显示所选择的图像，在 GalleryActivity.java 文件中添加下列粗体显示的语句：

```
@Override
public void onCreate(Bundle savedInstanceState) {
    super.onCreate(savedInstanceState);
    setContentView(R.layout.main);

    Gallery gallery = (Gallery) findViewById(R.id.gallery1);

    gallery.setAdapter(new ImageAdapter(this));
    gallery.setOnItemClickListener(new OnItemClickListener()
    {
        public void onItemClick(AdapterView parent, View v,
        int position, long id)
        {
            Toast.makeText(getBaseContext(),
                "pic" + (position + 1) + " selected",
                Toast.LENGTH_SHORT).show();

            //---display the images selected---
            ImageView imageView =
                (ImageView) findViewById(R.id.image1);
            imageView.setImageResource(imageIDs[position]);
        }
    });
```

}

(11) 按 F11 键再次调试应用程序。这一次,将会看到在 ImageView 中显示了所选择的图像(如图 5-5 所示)。

图 5-5

示例说明

首先将 Gallery 和 ImageView 视图添加到 main.xml 文件中:

```
<Gallery
    android:id="@+id/gallery1"
    android:layout_width="fill_parent"
    android:layout_height="wrap_content" />

<ImageView
    android:id="@+id/image1"
    android:layout_width="320dp"
    android:layout_height="250dp"
    android:scaleType="fitXY" />
```

正如先前所述,Gallery 视图用来在一个水平滚动列表中显示一系列图像。ImageView 用来显示用户选定的图像。

需要显示的图像列表存储在 imageIDs 数组中:

```
//---the images to display---
Integer[] imageIDs = {
        R.drawable.pic1,
        R.drawable.pic2,
        R.drawable.pic3,
        R.drawable.pic4,
        R.drawable.pic5,
```

```
            R.drawable.pic6,
            R.drawable.pic7
    };
```

创建的 ImageAdapter 类扩展了 BaseAdapter 类，可以用来将一系列 ImageView 视图绑定到 Gallery 视图上。BaseAdapter 类作为联系 AdapterView 和为它提供数据的数据源之间的桥梁。下面列出了 AdapterView 的一些示例：

- ListView
- GridView
- Spinner
- Gallery

在 Android 中，BaseAdapter 类有以下几个子类：

- ListAdapter
- ArrayAdapter
- CursorAdapter
- SpinnerAdapter

对于 ImageAdapter 类，实现了下列粗体显示的方法：

```
public class ImageAdapter extends BaseAdapter {
    public ImageAdapter(Context c) { ... }

    //---returns the number of images---
    public int getCount() { ... }

    //---returns the item---
    public Object getItem(int position) { ... }

     //---returns the ID of an item---
    public long getItemId(int position) { ... }

    //---returns an ImageView view---
    public View getView(int position, View convertView,
    ViewGroup parent) { ... }
}
```

特别是，getView()方法返回指定位置的 View。在本例中，返回了一个 ImageView 对象。

当选定(也即单击)Gallery 视图中的一张图像时，将显示出所选定图像的位置(第一张图像是 0，第二张图像是 1，依次类推)并在 ImageView 中显示此图像。

```
        Gallery gallery = (Gallery) findViewById(R.id.gallery1);

        gallery.setAdapter(new ImageAdapter(this));
        gallery.setOnItemClickListener(new OnItemClickListener()
        {
            public void onItemClick(AdapterView<?> parent, View v,
            int position, long id)
```

```
        {
            Toast.makeText(getBaseContext(),
                "pic" + (position + 1) + " selected",
                Toast.LENGTH_SHORT).show();

            //---display the images selected---
            ImageView imageView =
                (ImageView) findViewById(R.id.image1);
            imageView.setImageResource(imageIDs[position]);
        }
    });
```

5.1.2 ImageSwitcher

前面一节演示了如何将 Gallery 视图与一个 ImageView 视图一起使用来显示一系列缩略图像, 以便当其中之一被选中时, 选定的图像在 ImageView 中显示。然而, 有时您并不想当用户在 Gallery 视图中选择一张图像时该图像显示得太突然——例如, 您也许希望在图像之间进行过渡时应用一些动画效果。这时, Gallery 视图就需要 ImageSwitcher 来配合使用。下面的 "试一试" 展示了使用方法。

试一试　使用 ImageSwitcher 视图

ImageSwitcher.zip 代码文件可以在Wrox.com 上下载

(1) 打开 Eclipse, 创建一个名为 ImageSwitcher 的 Android 项目。

(2) 修改 main.xml 文件, 添加下列粗体显示的语句:

```xml
<?xml version="1.0" encoding="utf-8"?>
<LinearLayout xmlns:android="http://schemas.android.com/apk/res/android"
    android:layout_width="fill_parent"
    android:layout_height="fill_parent"
    android:orientation="vertical" >

<TextView
    android:layout_width="fill_parent"
    android:layout_height="wrap_content"
    android:text="Images of San Francisco" />

<Gallery
    android:id="@+id/gallery1"
    android:layout_width="fill_parent"
    android:layout_height="wrap_content" />

<ImageSwitcher
    android:id="@+id/switcher1"
    android:layout_width="fill_parent"
    android:layout_height="fill_parent"
    android:layout_alignParentLeft="true"
```

```xml
        android:layout_alignParentRight="true"
        android:layout_alignParentBottom="true" />

</LinearLayout>
```

(3) 右击 res/values 文件夹，选择 New | File，并将文件命名为 attrs.xml。

(4) 在 attrs.xml 文件中输入如下内容：

```xml
<?xml version="1.0" encoding="utf-8"?>
<resources>
    <declare-styleable name="Gallery1">
        <attr name="android:galleryItemBackground" />
    </declare-styleable>
</resources>
```

(5) 将一组图像拖放到 res/drawable-mdpi 文件夹下(参见前一个图像示例)。当显示一个对话框时，选中 Copy files 选项并单击 OK 按钮。

(6) 在 ImageSwitcherActivity.java 文件中添加下列粗体显示的语句：

```java
package net.learn2develop.ImageSwitcher;

import android.app.Activity;
import android.content.Context;
import android.content.res.TypedArray;
import android.os.Bundle;
import android.view.View;
import android.view.ViewGroup;
import android.view.ViewGroup.LayoutParams;
import android.view.animation.AnimationUtils;
import android.widget.AdapterView;
import android.widget.AdapterView.OnItemClickListener;
import android.widget.BaseAdapter;
import android.widget.Gallery;
import android.widget.ImageSwitcher;
import android.widget.ImageView;
import android.widget.ViewSwitcher.ViewFactory;

public class ImageSwitcherActivity extends Activity implements ViewFactory
{
    //---the images to display---
    Integer[] imageIDs = {
            R.drawable.pic1,
            R.drawable.pic2,
            R.drawable.pic3,
            R.drawable.pic4,
            R.drawable.pic5,
            R.drawable.pic6,
            R.drawable.pic7
    };
```

```java
    private ImageSwitcher imageSwitcher;

/** Called when the activity is first created. */
@Override
public void onCreate(Bundle savedInstanceState) {
    super.onCreate(savedInstanceState);
    setContentView(R.layout.main);

    imageSwitcher = (ImageSwitcher) findViewById(R.id.switcher1);
    imageSwitcher.setFactory(this);
    imageSwitcher.setInAnimation(AnimationUtils.loadAnimation(this,
            android.R.anim.fade_in));
    imageSwitcher.setOutAnimation(AnimationUtils.loadAnimation(this,
            android.R.anim.fade_out));

    Gallery gallery = (Gallery) findViewById(R.id.gallery1);
    gallery.setAdapter(new ImageAdapter(this));
    gallery.setOnItemClickListener(new OnItemClickListener()
    {
        public void onItemClick(AdapterView<?> parent,
        View v, int position, long id)
        {
            imageSwitcher.setImageResource(imageIDs[position]);
        }
    });
}

public View makeView()
{
    ImageView imageView = new ImageView(this);
    imageView.setBackgroundColor(0xFF000000);
    imageView.setScaleType(ImageView.ScaleType.FIT_CENTER);
    imageView.setLayoutParams(new
            ImageSwitcher.LayoutParams(
                    LayoutParams.FILL_PARENT,
                    LayoutParams.FILL_PARENT));
    return imageView;
}

public class ImageAdapter extends BaseAdapter
{
    private Context context;
    private int itemBackground;

    public ImageAdapter(Context c)
    {
        context = c;

        //---setting the style---
        TypedArray a = obtainStyledAttributes(R.styleable.Gallery1);
        itemBackground = a.getResourceId(
                R.styleable.Gallery1_android_galleryItemBackground, 0);
```

```java
        a.recycle();
    }

    //---returns the number of images---
    public int getCount()
    {
        return imageIDs.length;
    }

    //---returns the item---
    public Object getItem(int position)
    {
        return position;
    }

    //---returns the ID of an item---
    public long getItemId(int position)
    {
        return position;
    }

    //---returns an ImageView view---
    public View getView(int position, View convertView, ViewGroup parent)
    {
        ImageView imageView = new ImageView(context);

        imageView.setImageResource(imageIDs[position]);
        imageView.setScaleType(ImageView.ScaleType.FIT_XY);
        imageView.setLayoutParams(new Gallery.LayoutParams(150, 120));
        imageView.setBackgroundResource(itemBackground);

        return imageView;
    }
}
```

(7) 按 F11 键在 Android 模拟器上调试应用程序。图 5-6 展示了 Gallery 和 ImageSwitcher 视图，有两个图像集合以及选定的图像。

示例说明

在这个示例中，首先需要注意的是 ImageSwitcherActivity 不但扩展了 Activity，而且还实现了 ViewFactory。为了使用 ImageSwitcher 视图，需要实现 ViewFactory 接口，它创建了可与 ImageSwitcher 视图一起使用的视图。因此，需要实现 makeView()方法：

图 5-6

```java
public View makeView()
{
    ImageView imageView = new ImageView(this);
    imageView.setBackgroundColor(0xFF000000);
    imageView.setScaleType(ImageView.ScaleType.FIT_CENTER);
    imageView.setLayoutParams(new
            ImageSwitcher.LayoutParams(
                    LayoutParams.FILL_PARENT,
                    LayoutParams.FILL_PARENT));
    return imageView;
}
```

这个方法创建一个新的 View 来添加到 ImageSwitcher 视图中，在本例中这是一个 ImageView。

就像前一节的 Gallery 示例，需要实现一个 ImageAdapter 类来将一系列 ImageView 视图绑定到 Gallery 视图上。

在 onCreate() 方法中，可以获得对 ImageSwitcher 视图的引用并设置动画，指定图像应该如何"淡入"和"淡出"视图。最后，当在 Gallery 视图中选定一张图像时，它将在 ImageSwitcher 视图中显示出来：

```java
@Override
public void onCreate(Bundle savedInstanceState) {
    super.onCreate(savedInstanceState);
    setContentView(R.layout.main);

    imageSwitcher = (ImageSwitcher) findViewById(R.id.switcher1);
    imageSwitcher.setFactory(this);
    imageSwitcher.setInAnimation(AnimationUtils.loadAnimation(this,
            android.R.anim.fade_in));
    imageSwitcher.setOutAnimation(AnimationUtils.loadAnimation(this,
            android.R.anim.fade_out));

    Gallery gallery = (Gallery) findViewById(R.id.gallery1);
    gallery.setAdapter(new ImageAdapter(this));
    gallery.setOnItemClickListener(new OnItemClickListener()
    {
        public void onItemClick(AdapterView<?> parent,
        View v, int position, long id)
        {
            imageSwitcher.setImageResource(imageIDs[position]);
        }
    });
}
```

本例中，当在 Gallery 视图中选定一个图像时，它将以"淡入"的方式显示出来。当选定下一个图像时，当前图像将淡出。如果想让图像从左边滑入，而在选择另一幅图像时再从右边滑出，可尝试下面的动画效果：

```
    imageSwitcher.setInAnimation(AnimationUtils.loadAnimation(this,
        android.R.anim.slide_in_left));
    imageSwitcher.setOutAnimation(AnimationUtils.loadAnimation(this,
        android.R.anim.slide_out_right));
```

5.1.3 GridView

GridView 在一个二维的滚动网格中来显示项。可以将 GridView 与一个 ImageView 配合使用来显示一系列图像。下面的"试一试"展示了如何做到这一点。

试一试 使用 GridView 视图

Grid.zip 代码文件可以在 Wrox.com 上下载

(1) 打开 Eclipse，创建一个名为 Grid 的 Android 项目。

(2) 将一组图像拖放到 res/drawable-mdpi 文件夹下(参见前一个图像示例)。当显示一个对话框时，选中 Copy files 选项并单击 OK 按钮。

(3) 在 main.xml 文件中输入以下内容：

```xml
<?xml version="1.0" encoding="utf-8"?>
<LinearLayout xmlns:android="http://schemas.android.com/apk/res/android"
    android:layout_width="fill_parent"
    android:layout_height="fill_parent"
    android:orientation="vertical" >

<GridView
    android:id="@+id/gridview"
    android:layout_width="fill_parent"
    android:layout_height="fill_parent"
    android:numColumns="auto_fit"
    android:verticalSpacing="10dp"
    android:horizontalSpacing="10dp"
    android:columnWidth="90dp"
    android:stretchMode="columnWidth"
    android:gravity="center" />

</LinearLayout>
```

(4) 在 GridActivity.java 文件中添加下列粗体显示的语句：

```java
package net.learn2develop.Grid;

import android.app.Activity;
import android.content.Context;
import android.os.Bundle;
import android.view.View;
import android.view.ViewGroup;
import android.widget.AdapterView;
import android.widget.AdapterView.OnItemClickListener;
```

```java
import android.widget.BaseAdapter;
import android.widget.GridView;
import android.widget.ImageView;
import android.widget.Toast;

public class GridActivity extends Activity {
    //---the images to display---
    Integer[] imageIDs = {
            R.drawable.pic1,
            R.drawable.pic2,
            R.drawable.pic3,
            R.drawable.pic4,
            R.drawable.pic5,
            R.drawable.pic6,
            R.drawable.pic7
    };

    /** Called when the activity is first created. */
    @Override
    public void onCreate(Bundle savedInstanceState) {
        super.onCreate(savedInstanceState);
        setContentView(R.layout.main);

        GridView gridView = (GridView) findViewById(R.id.gridview);
        gridView.setAdapter(new ImageAdapter(this));

        gridView.setOnItemClickListener(new OnItemClickListener()
        {
            public void onItemClick(AdapterView parent,
            View v, int position, long id)
            {
                Toast.makeText(getBaseContext(),
                        "pic" + (position + 1) + " selected",
                        Toast.LENGTH_SHORT).show();
            }
        });

    }

    public class ImageAdapter extends BaseAdapter
    {
        private Context context;

        public ImageAdapter(Context c)
        {
            context = c;
        }

        //---returns the number of images---
        public int getCount() {
```

```
        return imageIDs.length;
    }

    //---returns the item---
    public Object getItem(int position) {
        return position;
    }

    //---returns the ID of an item---
    public long getItemId(int position) {
        return position;
    }

    //---returns an ImageView view---
    public View getView(int position, View convertView,
    ViewGroup parent)
    {
        ImageView imageView;
        if (convertView == null) {
            imageView = new ImageView(context);
            imageView.setLayoutParams(new
                GridView.LayoutParams(85, 85));
            imageView.setScaleType(
                ImageView.ScaleType.CENTER_CROP);
            imageView.setPadding(5, 5, 5, 5);
        } else {
            imageView = (ImageView) convertView;
        }
            imageView.setImageResource(imageIDs[position]);
        return imageView;
    }
}

}
```

(5) 按 F11 键在 Android 模拟器中调试应用程序。图 5-7 展示了显示所有图像的 GridView。

图 5-7

示例说明

与 Gallery 和 ImageSwitcher 示例一样,实现 ImageAdapter 类并绑定到 GridView:

```java
GridView gridView = (GridView) findViewById(R.id.gridview);
gridView.setAdapter(new ImageAdapter(this));

gridView.setOnItemClickListener(new OnItemClickListener()
{
    public void onItemClick(AdapterView parent,
    View v, int position, long id)
    {
        Toast.makeText(getBaseContext(),
                "pic" + (position + 1) + " selected",
                Toast.LENGTH_SHORT).show();
    }
});
```

当选择了一个图像时,将显示一个 Toast 消息表明该图像已被选定。

在 getView()方法中,可以指定图像的大小,并通过为每幅图像设置内边距在 GridView 中对图像进行分隔:

```java
//---returns an ImageView view---
public View getView(int position, View convertView,
ViewGroup parent)
{
    ImageView imageView;
    if (convertView == null) {
        imageView = new ImageView(context);
        imageView.setLayoutParams(new
            GridView.LayoutParams(85, 85));
        imageView.setScaleType(
            ImageView.ScaleType.CENTER_CROP);
        imageView.setPadding(5, 5, 5, 5);
    } else {
        imageView = (ImageView) convertView;
    }
    imageView.setImageResource(imageIDs[position]);
    return imageView;
}
```

5.2 将菜单和视图一起使用

菜单用来显示在一个应用程序的主用户界面中不是直接可见的额外选项。在 Android 中有两种主要的菜单类型:

- 选项菜单——显示和当前活动相关的信息。在 Android 中,通过单击并按下 MENU

按钮来激活选项菜单。
- 上下文菜单——显示和活动中一个特定的视图相关的信息。在 Android 中，通过单击并按住视图来激活上下文菜单。

图 5-8 展示了 Browser 应用程序中的一个选项菜单的示例。当用户按下 MENU 按钮时，选项菜单将显示出来。所显示的菜单项随当前正在运行的活动而各异。

图 5-9 展示了当用户按住页面上的一幅图像时所显示的上下文菜单。显示的菜单项随当前选定的组件或视图而各异。为了激活上下文菜单，用户可在屏幕上选择一项，然后按住它。

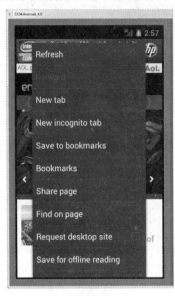
图 5-8

图 5-9

5.2.1 创建辅助方法

在深入学习和创建选项菜单以及上下文菜单之前，需要创建两个辅助方法。一个创建将在菜单内显示的项列表，另一个处理用户在菜单内选定一项时所触发的事件。

试一试　　创建菜单辅助方法

Menus.zip 代码文件可以在 Wrox.com 上下载

(1) 打开 Eclipse，创建一个名为 Menus 的 Android 项目。
(2) 在 MenusActivity.java 文件中添加下列粗体显示的语句：

```
package net.learn2develop.Menus;

import android.app.Activity;
import android.os.Bundle;
import android.view.Menu;
import android.view.MenuItem;
import android.widget.Toast;
```

```java
public class MenusActivity extends Activity {
    /** Called when the activity is first created. */
    @Override
    public void onCreate(Bundle savedInstanceState) {
        super.onCreate(savedInstanceState);
        setContentView(R.layout.main);
    }

    private void CreateMenu(Menu menu)
    {
        MenuItem mnu1 = menu.add(0, 0, 0, "Item 1");
        {
            mnu1.setAlphabeticShortcut('a');
            mnu1.setIcon(R.drawable.ic_launcher);
        }
        MenuItem mnu2 = menu.add(0, 1, 1, "Item 2");
        {
            mnu2.setAlphabeticShortcut('b');
            mnu2.setIcon(R.drawable.ic_launcher);
        }
        MenuItem mnu3 = menu.add(0, 2, 2, "Item 3");
        {
            mnu3.setAlphabeticShortcut('c');
            mnu3.setIcon(R.drawable.ic_launcher);
        }
        MenuItem mnu4 = menu.add(0, 3, 3, "Item 4");
        {
            mnu4.setAlphabeticShortcut('d');
        }
        menu.add(0, 4, 4, "Item 5");
        menu.add(0, 5, 5, "Item 6");
        menu.add(0, 6, 6, "Item 7");
    }

    private boolean MenuChoice(MenuItem item)
    {
        switch (item.getItemId()) {
        case 0:
            Toast.makeText(this, "You clicked on Item 1",
                Toast.LENGTH_LONG).show();
            return true;
        case 1:
            Toast.makeText(this, "You clicked on Item 2",
                Toast.LENGTH_LONG).show();
            return true;
        case 2:
            Toast.makeText(this, "You clicked on Item 3",
                Toast.LENGTH_LONG).show();
            return true;
        case 3:
```

```
            Toast.makeText(this, "You clicked on Item 4",
                Toast.LENGTH_LONG).show();
            return true;
        case 4:
            Toast.makeText(this, "You clicked on Item 5",
                Toast.LENGTH_LONG).show();
            return true;
        case 5:
            Toast.makeText(this, "You clicked on Item 6",
                Toast.LENGTH_LONG).show();
            return true;
        case 6:
            Toast.makeText(this, "You clicked on Item 7",
                Toast.LENGTH_LONG).show();
            return true;
    }
    return false;
}
```

}

示例说明

前面的示例创建了两个方法：CreateMenu()和 MenuChoice()。CreatMenu()方法接受一个 Menu 参数并向其添加一系列菜单项。

为了向菜单中添加菜单项，需要创建一个 MenuItem 类的实例并使用了 Menu 对象的 add()方法。

```
MenuItem mnu1 = menu.add(0, 0, 0, "Item 1");
{
    mnu1.setAlphabeticShortcut('a');
    mnu1.setIcon(R.drawable.ic_launcher);
}
```

add()方法的 4 个参数如下所示：

- groupId——菜单项所在的组的标识符，使用 0 表示一个菜单项不在一个组中
- itemId——唯一的菜单项 ID
- order——菜单项显示的顺序
- title——菜单项显示的文本

可以使用 setAlphabeticShortcut()方法来给菜单项分配快捷键，这样用户可以通过在键盘上按键来选择一个菜单项。setIcon()方法设置将在菜单项上显示的图像。

MenuChoice()方法接受一个 MenuItem 参数，并检查其 ID 来确定被选择的菜单项。然后，它显示一个 Toast 消息告诉用户哪一个菜单项被选中了。

5.2.2 选项菜单

现在准备修改应用程序，使得当用户在 Android 设备上按下 MENU 按钮时显示选项菜单。

第 5 章　使用视图显示图片和菜单

试一试　　**显示选项菜单**

(1) 使用前一节所创建的同一个项目，在 MenusActivity.java 文件中添加下列粗体显示的语句：

```
package net.learn2develop.Menus;

import android.app.Activity;
import android.os.Bundle;
import android.view.Menu;
import android.view.MenuItem;
import android.widget.Toast;

public class MenusActivity extends Activity {
    /** Called when the activity is first created. */
    @Override
    public void onCreate(Bundle savedInstanceState) {
        super.onCreate(savedInstanceState);
        setContentView(R.layout.main);
    }

    @Override
    public boolean onCreateOptionsMenu(Menu menu) {
        super.onCreateOptionsMenu(menu);
        CreateMenu(menu);
        return true;
    }

    @Override
    public boolean onOptionsItemSelected(MenuItem item)
    {
        return MenuChoice(item);
    }

    private void CreateMenu(Menu menu)
    {
        //...
    }

    private boolean MenuChoice(MenuItem item)
    {
        //...
    }
}
```

(2) 按 F11 键在 Android 模拟器中调试应用程序。图 5-10 展示了当单击 MENU 按钮时所弹出的选项菜单。要选择一个菜单项，可以单击单独的菜单项或者使用其快捷键(A 到 D；只对前 4 个菜单项有效)。注意虽然代码中为菜单项 1~3 显示了图标，但是这些图标并没有显示出来。

(3) 如果将 AndroidManifest.xml 文件的最低 SDK 属性改为小于或等于 10，然后在模拟器中重新运行应用程序,图标的显示将如图 5-11 所示。注意,第 5 个菜单项之后的所有菜单项被包含在名为 More 的菜单项中。单击 More 将显示其余的菜单项。

```
<uses-sdk android:minSdkVersion="10" />
```

　　　　图 5-10　　　　　　　　　　　　图 5-11

示例说明

　　为了显示活动的选项菜单,需要在活动中实现两个方法:onCreateOptionsMenu()和 onOptionsItemSelected()。当 MENU 按钮被按下时调用 onCreateOptionsMenu()方法。在这一事件中,调用了 CreateMenu()辅助方法来显示选项菜单。当选择了一个菜单项时,将调用 onOptionsItemSelected()方法。这时,调用 MenuChoice()方法来显示所选择的菜单项(并做任何想做的事)。

　　注意在不同版本的 Android 中选项按钮的外观和感觉。从 Honeycomb 开始,选项菜单项不再有图标,而是在一个可滚动的列表中显示所有菜单项。对于 Honeycomb 之前的 Android 版本,最多可以显示 5 个菜单项,多出的菜单项将被放到一个 More 菜单项中,该菜单项代表其余的所有菜单项。

5.2.3　上下文菜单

　　前一节讲述了在用户按下 MENU 按钮时是如何显示选项菜单的。除了选项菜单,还可以显示上下文菜单。上下文菜单通常和活动上的一个视图相关联,并在用户长按一个项时显示。例如,如果用户按住 Button 视图几秒钟,就会显示一个上下文菜单。

　　如果打算将上下文菜单和活动上的一个视图联系起来,需要调用那个视图的 setOnCreateContextMenuListener()方法。下面的"试一试"展示了如何使一个上下文菜单与一个 Button 视图进行关联。

第 5 章 使用视图显示图片和菜单

试一试　显示上下文菜单

Menus.zip 代码文件可以在 Wrox.com 上下载

(1) 使用前一节所创建的同一个项目，在 main.xml 文件中添加下列粗体显示的语句：

```xml
<?xml version="1.0" encoding="utf-8"?>
<LinearLayout xmlns:android="http://schemas.android.com/apk/res/android"
    android:layout_width="fill_parent"
    android:layout_height="fill_parent"
    android:orientation="vertical" >

<TextView
    android:layout_width="fill_parent"
    android:layout_height="wrap_content"
    android:text="@string/hello" />

<Button
    android:id="@+id/button1"
    android:layout_width="match_parent"
    android:layout_height="wrap_content"
    android:text="Click and hold on it" />

</LinearLayout>
```

(2) 在 MenusActivity.java 文件中添加下列粗体显示的语句：

```java
package net.learn2develop.Menus;

import android.app.Activity;
import android.os.Bundle;
import android.view.ContextMenu;
import android.view.ContextMenu.ContextMenuInfo;
import android.view.Menu;
import android.view.MenuItem;
import android.view.View;
import android.widget.Button;
import android.widget.Toast;

public class MenusActivity extends Activity {
    /** Called when the activity is first created. */
    @Override
    public void onCreate(Bundle savedInstanceState) {
        super.onCreate(savedInstanceState);
        setContentView(R.layout.main);

        Button btn = (Button) findViewById(R.id.button1);
        btn.setOnCreateContextMenuListener(this);
    }
```

231

```
@Override
public void onCreateContextMenu(ContextMenu menu, View view,
ContextMenuInfo menuInfo)
{
    super.onCreateContextMenu(menu, view, menuInfo);
    CreateMenu(menu);
}

@Override
public boolean onCreateOptionsMenu(Menu menu) {
    //...
}

@Override
public boolean onOptionsItemSelected(MenuItem item)
{
    return MenuChoice(item);
}

private void CreateMenu(Menu menu)
{
    //...
}

private boolean MenuChoice(MenuItem item)
{
    //...
}
}
```

注意：如果之前将AndroidManifest.xml文件的最低SDK属性改为10，那么一定要在下一步中调试应用程序之前将其改回14。

(3) 按F11键在Android模拟器上调试应用程序。图5-12展示了在长按Button视图时所显示的上下文菜单。

示例说明

在上述示例中，调用Button视图的setOnCreateContextMenuListener()方法将它和一个上下文菜单建立关联。

当用户长按Button视图时，onCreateContextMenu()方法将被调用。在这一方法中，通过调用CreateMenu()方法来显示上下文菜单。类似地，当上下文菜单内的一个项被选中时，onContextItemSelected()方法将被调用，在其中调用MenuChoice()方法来向用户显示一条消息。

图5-12

注意，菜单项的快捷键并未生效。要使之生效，需要调用 Menu 对象的 setQuertyMode()
方法，如下所示：

```java
private void CreateMenu(Menu menu)
{
    menu.setQwertyMode(true);
    MenuItem mnu1 = menu.add(0, 0, 0, "Item 1");
    {
        mnu1.setAlphabeticShortcut('a');
        mnu1.setIcon(R.drawable.ic_launcher);
    }
    //...
}
```

5.3 其他一些视图

除了目前已经了解的标准视图之外，Android SDK 还提供了其他一些可使应用程序变得更加有趣的视图。本节中，我们将学习以下视图：AnalogClock、DigitalClock 和 WebView。

5.3.1 AnalogClock 和 DigitalClock 视图

AnalogClock 视图显示一个有两根指针(一根分针，一根时针)的模拟时钟，而 DigitalClock 视图以数字形式显示时间。它们二者都显示的是系统时间，不允许用户用来显示一个特定时间(例如另一个时区的时间)。因此，如果您打算显示一个特定时区的时间，就必须构建一个自己的定制视图。

> **注意**：在 Android 中创建一个自己的定制视图已经超出了本书的范围。不过，如果您对这一领域感兴趣，可以在 Google 的 Android 文档中查看这一主题：
> http://developer.android.com/guide/topics/ui/custom-components.html。

使用 AnalogClock 和 DigitalClock 视图是很简单的。只需要在 XML 文件(例如 main.xml)中声明它们，如下所示：

```xml
<?xml version="1.0" encoding="utf-8"?>
<LinearLayout
    xmlns:android="http://schemas.android.com/apk/res/android"
    android:layout_width="fill_parent"
    android:layout_height="fill_parent"
    android:orientation="vertical" >

<AnalogClock
    android:layout_width="wrap_content"
    android:layout_height="wrap_content" />
```

```
<DigitalClock
    android:layout_width="wrap_content"
    android:layout_height="wrap_content" />

</LinearLayout>
```

图 5-13 展示了运用中的 AnalogClock 和 DigitalClock 视图。

5.3.2 WebView

WebView 可以使您在活动中嵌入一个 Web 浏览器。如果应用程序需要嵌入一些 Web 内容——如来自其他一些提供商的地图等——这个视图就很有用。下面的"试一试"展示了如何以编程方式加载一个 Web 页面的内容并在活动中显示出来。

图 5-13

试一试　　使用 WebView 视图

WebView.zip 代码文件可以在 Wrox.com 上下载

(1) 打开 Eclipse，创建一个名为 WebView 的 Android 新项目。
(2) 在 main.xml 文件中添加以下语句：

```
<?xml version="1.0" encoding="utf-8"?>
<LinearLayout xmlns:android="http://schemas.android.com/apk/res/android"
    android:layout_width="fill_parent"
    android:layout_height="fill_parent"
    android:orientation="vertical" >

<WebView android:id="@+id/webview1"
    android:layout_width="wrap_content"
    android:layout_height="wrap_content" />

</LinearLayout>
```

(3) 在 WebViewActivity.java 文件中添加下列粗体显示的语句：

```
package net.learn2develop.WebView;

import android.app.Activity;
import android.os.Bundle;
import android.webkit.WebSettings;
import android.webkit.WebView;

public class WebViewActivity extends Activity {
    /** Called when the activity is first created. */
```

```java
@Override
public void onCreate(Bundle savedInstanceState) {
    super.onCreate(savedInstanceState);
    setContentView(R.layout.main);

    WebView wv = (WebView) findViewById(R.id.webview1);

    WebSettings webSettings = wv.getSettings();
    webSettings.setBuiltInZoomControls(true);
    wv.loadUrl(
        "http://chart.apis.google.com/chart" +
        "?chs=300x225" +
        "&cht=v" +
        "&chco=FF6342,ADDE63,63C6DE" +
        "&chd=t:100,80,60,30,30,30,10" +
        "&chdl=A|B|C");
}
}
```

(4) 在 AndroidManifest.xml 文件中添加下面的权限：

```xml
<?xml version="1.0" encoding="utf-8"?>
<manifest xmlns:android="http://schemas.android.com/apk/res/android"
    package="net.learn2develop.WebView"
    android:versionCode="1"
    android:versionName="1.0" >

    <uses-sdk android:minSdkVersion="14" />
    <uses-permission android:name="android.permission.INTERNET"/>

    <application
        android:icon="@drawable/ic_launcher"
        android:label="@string/app_name" >
        <activity
          android:label="@string/app_name"
          android:name=".WebViewActivity" >
          <intent-filter >
                <action android:name="android.intent.action.MAIN" />

                <category android:name="android.intent.category.LAUNCHER" />
          </intent-filter>
        </activity>
    </application>

</manifest>
```

(5) 按 F11 键在 Android 模拟器上调试应用程序。图 5-14 展示了 WebView 的内容。

图 5-14

示例说明

为了利用 WebView 来加载一个 Web 页面，可以使用 loadUrl()方法并向其传入一个 URL，如下所示：

```
wv.loadUrl(
    "http://chart.apis.google.com/chart" +
    "?chs=300x225" +
    "&cht=v" +
    "&chco=FF6342,ADDE63,63C6DE" +
    "&chd=t:100,80,60,30,30,30,10" +
    "&chdl=A|B|C");
```

要显示内置的缩放控件，需要首先从 WebView 中获取 WebSettings 属性，然后调用它的 setBuiltInZoomControls()方法：

```
WebSettings webSettings = wv.getSettings();
webSettings.setBuiltInZoomControls(true);
```

图 5-15 展示了在 Android 模拟器上使用鼠标单击并拖拽 WebView 内容时显示的内置缩放控件。

 注意：尽管大多数 Android 设备支持多点触摸屏幕，但当在 Android 模拟器上测试应用程序时，可以使用内置的缩放控件对 Web 内容进行缩放。

有时，当加载一个会重定向的页面时(例如加载 www.wrox.com 会重定向到 www.wrox.com/wileyCDA)，WebView 将导致应用程序启动设备的 Browser 应用程序来加载所需的页面。在图 5-16 中，注意屏幕顶端的 URL 栏。

第 5 章 使用视图显示图片和菜单

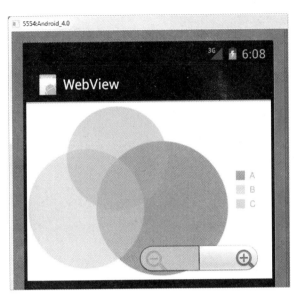

图 5-15

图 5-16

为了避免这种情况发生,需要实现 WebViewClient 类并重写 shouldOverrideUrlLoading()方法,如下面的示例所示:

```
package net.learn2develop.WebView;

import android.app.Activity;
import android.os.Bundle;
import android.webkit.WebSettings;
import android.webkit.WebView;
import android.webkit.WebViewClient;

public class WebViewActivity extends Activity {
    /** Called when the activity is first created. */
    @Override
    public void onCreate(Bundle savedInstanceState) {
        super.onCreate(savedInstanceState);
        setContentView(R.layout.main);

        WebView wv = (WebView) findViewById(R.id.webview1);

        WebSettings webSettings = wv.getSettings();
        webSettings.setBuiltInZoomControls(true);
        wv.setWebViewClient(new Callback());
        wv.loadUrl("http://www.wrox.com");
    }

    private class Callback extends WebViewClient {
        @Override
```

237

```
        public boolean shouldOverrideUrlLoading(
        WebView view, String url) {
            return(false);
        }
    }
}
```

图 5-17 展示了现在在 WebView 中正确加载了 Wrox.com 主页。

图 5-17

还可以使用 loadDataWithBaseURL()方法动态创建一个 HTML 字符串并加载到 WebView 中：

```
WebView wv = (WebView) findViewById(R.id.webview1);
final String mimeType = "text/html";
final String encoding = "UTF-8";
String html = "<H1>A simple HTML page</H1><body>" +
    "<p>The quick brown fox jumps over the lazy dog</p>" +
    "</body>";
wv.loadDataWithBaseURL("", html, mimeType, encoding, "");
```

图 5-18 展示了由 WebView 所显示的内容。

此外，如果在项目的 assets 文件夹下有一个 HTML 文件(如图 5-19 所示)，还可以使用 loadUrl()方法将其加载到 WebView 中：

```
WebView wv = (WebView) findViewById(R.id.webview1);
wv.loadUrl("file:///android_asset/Index.html");
```

第 5 章 使用视图显示图片和菜单

图 5-18

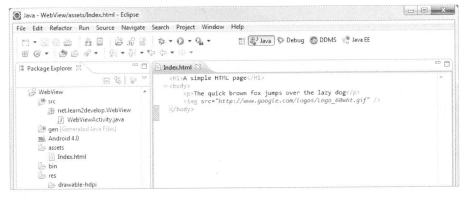

图 5-19

图 5-20 展示了 WebView 的内容。

图 5-20

5.4 本章小结

本章中，我们学习了可用来显示图像的各种视图：Gallery、ImageView、ImageSwitcher 和 GridView。此外，还了解了选项菜单和上下文菜单的不同，以及如何在应用程序中显示它们。最后，我们还学习了 AnalogClock 和 DigitalClock 视图(它们以图形化的方式显示当前时间)，以及用来显示 Web 页面内容的 WebView 视图。

练 习

1. ImageSwitcher 的作用是什么？
2. 说出在活动中实现选项菜单时需要重写的两个方法。
3. 说出在活动中实现上下文菜单时需要重写的两个方法。
4. 当在 WebView 中发生重定向时如何防止其启动设备的 Web 浏览器？

练习答案参见附录 C。

本章主要内容

主 题	关 键 概 念
使用 Gallery 视图	在水平滚动列表中显示一系列图像

Gallery	`<Gallery`
	`    android:id="@+id/gallery1"`
	`    android:layout_width="fill_parent"`
	`    android:layout_height="wrap_content" />`

ImageView	`<ImageView`
	`    android:id="@+id/image1"`
	`    android:layout_width="320px"`
	`    android:layout_height="250px"`
	`    android:scaleType="fitXY" />`

使用 ImageSwitcher 视图	当在图像间切换时应用动画效果

ImageSwitcher	`<ImageSwitcher`
	`    android:id="@+id/switcher1"`
	`    android:layout_width="fill_parent"`
	`    android:layout_height="fill_parent"`
	`    android:layout_alignParentLeft="true"`
	`    android:layout_alignParentRight="true"`
	`    android:layout_alignParentBottom="true" />`

使用 GridView	在一个二维的滚动网格中显示项

GridView	`<GridView`
	`    android:id="@+id/gridview"`
	`    android:layout_width="fill_parent"`
	`    android:layout_height="fill_parent"`
	`    android:numColumns="auto_fit"`
	`    android:verticalSpacing="10dp"`
	`    android:horizontalSpacing="10dp"`
	`    android:columnWidth="90dp"`

(续表)

主 题	关 键 概 念
GridView	`android:stretchMode="columnWidth"` `android:gravity="center" />`

AnalogClock	`<AnalogClock`
	`    android:layout_width="wrap_content"`
	`    android:layout_height="wrap_content" />`

DigitalClock	`<DigitalClock`
	`    android:layout_width="wrap_content"`
	`    android:layout_height="wrap_content" />`

WebView	`<WebView android:id="@+id/webview1"`
	`    android:layout_width="wrap_content"`
	`    android:layout_height="wrap_content" />`

第 6 章

数据持久化

本章将介绍以下内容：
- 如何使用 SharedPreferences 对象保存简单数据
- 使用 PreferenceActivity 类来允许用户修改首选项
- 如何写入和读取内部和外部存储器中的文件
- 如何创建并使用 SQLite 数据库

在本章中，我们将学习如何在 Android 应用程序中保持数据。由于用户通常希望在以后重新使用数据，因此持久化数据是应用程序开发中的一个重要主题。对于 Android 来说，持久化数据主要有 3 个基本的方法：
- 用来保存小块数据的轻量级的机制——共享首选项(shared preferences)
- 传统的文件系统
- 通过 SQLite 数据库支持的关系数据库管理系统

本章中讨论的技术可以使应用程序创建和访问它们自己的私有数据。在第 7 章，我们将学习如何跨应用程序共享数据。

6.1 保存和加载用户首选项

Android 提供了 SharedPreferences 对象来帮助我们保存简单的应用程序数据。例如，应用程序可能有一个允许用户指定在应用程序中显示的文本字体大小的选项。这时，应用程序需要记住用户对字体的设定值，以便该用户再次使用时，它可以正确设置字体大小。为了达到这一目的，可以有多种选择。将数据保存在一个文件中，但这需要执行一些文件管理例程，诸如向文件写数据、表明从文件中读取多少字符等。另外，如果有诸如文本尺寸、字体名称、首选的背景色等多条信息需要保存时，写文件的任务就会变得更加繁重。

替代写入文本文件的方法是使用数据库，但无论是从开发人员的角度还是考虑到应用程序运行时的性能，为存储简单数据而使用数据库都夸张了点。

不过，使用 SharedPreferences 对象的话，可以通过使用键/值对来保存所需的数据——为需要保存的数据指定一个键，然后其和其值将一起被自动保存到一个 XML 文件中。

6.1.1 使用活动访问首选项

在下面的"试一试"中，将学习如何使用 SharedPreferences 对象来存储应用程序数据。另外还将学到如何使用 Android 操作系统提供的一类特殊的活动，使用户能够直接修改存储的应用程序数据。

| 试一试 | 使用 SharedPreferences 对象保存数据 |

SharedPreferences.zip 代码文件可以在 Wrox.com 上下载

(1) 使用Eclipse创建一个Android项目，命名为UsingPreferences。

(2) 在 res 文件夹中创建一个新的子文件夹，命名为 xml。在这个新建的文件夹中添加一个文件，命名为 myapppreferences.xml，如图 6-1 所示。

图 6-1

(3) 在 myapppreferences.xml 文件中输入以下内容：

```xml
<?xml version="1.0" encoding="utf-8"?>
<PreferenceScreen
    xmlns:android="http://schemas.android.com/apk/res/android">
    <PreferenceCategory android:title="Category 1">
      <CheckBoxPreference
          android:title="Checkbox"
          android:defaultValue="false"
          android:summary="True or False"
          android:key="checkboxPref" />
    </PreferenceCategory>
    <PreferenceCategory android:title="Category 2">
      <EditTextPreference
          android:summary="Enter a string"
          android:defaultValue="[Enter a string here]"
          android:title="Edit Text"
          android:key="editTextPref" />
      <RingtonePreference
          android:summary="Select a ringtone"
          android:title="Ringtones"
          android:key="ringtonePref" />
      <PreferenceScreen
          android:title="Second Preference Screen"
          android:summary=
              "Click here to go to the second Preference Screen"
          android:key="secondPrefScreenPref" >
        <EditTextPreference
            android:summary="Enter a string"
```

```xml
            android:title="Edit Text (second Screen)"
            android:key="secondEditTextPref" />
    </PreferenceScreen>
  </PreferenceCategory>
</PreferenceScreen>
```

(4) 在包名下添加一个新的类文件,命名为 AppPreferenceActivity。

(5) 在 AppPreferenceActivity.java 文件中输入以下内容:

```java
package net.learn2develop.UsingPreferences;

import android.os.Bundle;
import android.preference.PreferenceActivity;

public class AppPreferenceActivity extends PreferenceActivity {
    @Override
    public void onCreate(Bundle savedInstanceState) {
        super.onCreate(savedInstanceState);
        //---load the preferences from an XML file---
        addPreferencesFromResource(R.xml.myapppreferences);
    }
}
```

(6) 在 AndroidManifest.xml 文件中,为 AppPreferenceActivity 类添加一个新条目:

```xml
<?xml version="1.0" encoding="utf-8"?>
<manifest xmlns:android="http://schemas.android.com/apk/res/android"
    package="net.learn2develop.UsingPreferences"
    android:versionCode="1"
    android:versionName="1.0" >

    <uses-sdk android:minSdkVersion="14" />

    <application
        android:icon="@drawable/ic_launcher"
        android:label="@string/app_name" >
        <activity
            android:label="@string/app_name"
            android:name=".UsingPreferencesActivity" >
            <intent-filter >
                <action android:name="android.intent.action.MAIN" />

                <category android:name="android.intent.category.LAUNCHER" />
            </intent-filter>
        </activity>
        <activity android:name=".AppPreferenceActivity"
                android:label="@string/app_name">
          <intent-filter>
            <action
                android:name="net.learn2develop.AppPreference
                Activity" />
```

```xml
            <category android:name="android.intent.category.DEFAULT" />
        </intent-filter>
      </activity>
   </application>
</manifest>
```

(7) 在 main.xml 文件中添加下列粗体显示的代码(替换现有的 TextView)：

```xml
<?xml version="1.0" encoding="utf-8"?>
<LinearLayout xmlns:android="http://schemas.android.com/apk/res/android"
    android:layout_width="fill_parent"
    android:layout_height="fill_parent"
    android:orientation="vertical" >

<Button
    android:id="@+id/btnPreferences"
    android:text="Load Preferences Screen"
    android:layout_width="fill_parent"
    android:layout_height="wrap_content"
    android:onClick="onClickLoad"/>

<Button
    android:id="@+id/btnDisplayValues"
    android:text="Display Preferences Values"
    android:layout_width="fill_parent"
    android:layout_height="wrap_content"
    android:onClick="onClickDisplay"/>

<EditText
    android:id="@+id/txtString"
    android:layout_width="fill_parent"
    android:layout_height="wrap_content" />

<Button
    android:id="@+id/btnModifyValues"
    android:text="Modify Preferences Values"
    android:layout_width="fill_parent"
    android:layout_height="wrap_content"
    android:onClick="onClickModify"/>

</LinearLayout>
```

(8) 在 UsingPreferencesActivity.java 文件中添加下列粗体显示的代码：

```java
package net.learn2develop.UsingPreferences;

import android.app.Activity;
import android.content.Intent;
import android.os.Bundle;
import android.view.View;
```

第 6 章　数据持久化

```
public class UsingPreferencesActivity extends Activity {
    /** Called when the activity is first created. */
    @Override
    public void onCreate(Bundle savedInstanceState) {
        super.onCreate(savedInstanceState);
        setContentView(R.layout.main);
    }

    public void onClickLoad(View view) {
        Intent i = new Intent("net.learn2develop.AppPreferenceActivity");
        startActivity(i);
    }
}
```

(9) 按 F11 键在 Android 模拟器上调试应用程序。单击 Load Preferences Screen 按钮可看到首选项界面，如图 6-2 所示。

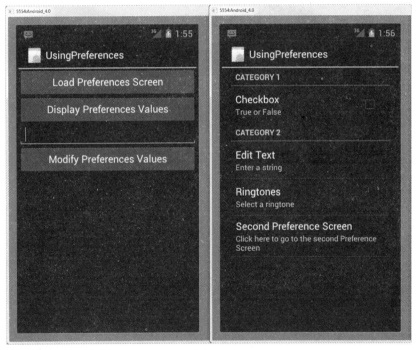

图 6-2

(10) 单击 Checkbox 项可以将复选框的值在选中与未选中之间切换。注意 Category 1 和 Category 2 这两个类别。单击 Edit Text 项，输入如图 6-3 所示的一些值。单击 OK 按钮关闭对话框。

(11) 单击 Ringtones 项，选择默认的铃声或静音模式，如图 6-4 所示。如果在真实的 Android 设备上测试应用程序，可以从一个更加丰富的铃声列表中进行选择。

(12) 单击 Second Preference Screen 项会进入下一个界面，如图 6-5 所示。

图 6-3　　　　　　　　图 6-4　　　　　　　　图 6-5

(13) 为返回前一个界面，单击 Back 按钮。为关闭首选项界面，需要再次单击 Back 按钮。

(14) 修改了至少一个首选项的值后，Android 模拟器的/data/data/net.learn2develop.UsingPreferences/shared_prefs 文件夹中会创建一个文件。为了验证这一点，在 Eclipse 中切换到 DDMS 透视图，观察 File Explorer 选项卡，如图 6-6 所示。您将看到一个名为 net.learn2develop.UsingPreferences_preferences.xml 的 XML 文件。

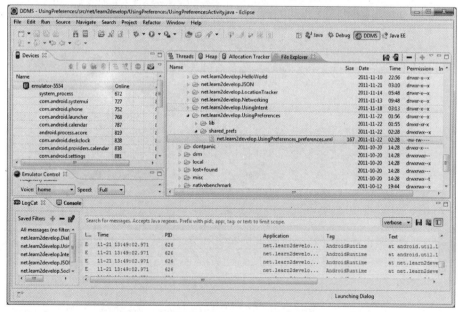

图 6-6

(15) 如果提取这个文件并观察其内容，可以看到如下所示的代码：

```xml
<?xml version='1.0' encoding='utf-8' standalone='yes'?>
<map>
<string name="editTextPref">[Enter a string here]</string>
<string name="ringtonePref"></string>
</map>
```

示例说明

首先创建了一个名为 myapppreferences.xml 的 XML 文件来存储想要为应用程序保存的首选项类型：

```xml
<?xml version="1.0" encoding="utf-8"?>
<PreferenceScreen
    xmlns:android="http://schemas.android.com/apk/res/android">
    <PreferenceCategory android:title="Category 1">
        <CheckBoxPreference
            android:title="Checkbox"
            android:defaultValue="false"
            android:summary="True or False"
            android:key="checkboxPref" />
    </PreferenceCategory>
    <PreferenceCategory android:title="Category 2">
        <EditTextPreference
            android:summary="Enter a string"
            android:defaultValue="[Enter a string here]"
            android:title="Edit Text"
            android:key="editTextPref" />
        <RingtonePreference
            android:summary="Select a ringtone"
            android:title="Ringtones"
            android:key="ringtonePref" />
        <PreferenceScreen
            android:title="Second Preference Screen"
            android:summary=
                "Click here to go to the second Preference Screen"
            android:key="secondPrefScreenPref" >
            <EditTextPreference
                android:summary="Enter a string"
                android:title="Edit Text (second Screen)"
                android:key="secondEditTextPref" />
        </PreferenceScreen>
    </PreferenceCategory>
</PreferenceScreen>
```

在前面的代码段中创建了以下内容：
- 两个首选项类别，用于分组不同的首选项类型。
- 两个复选框首选项，其键分别为 checkboxPref 和 secondEditTextPref。
- 一个铃声首选项，其键为 ringtonePref。
- 一个首选项界面，用于包含附加首选项。

android:key 属性指定了可以在代码中以编程方式引用以设置或检索该首选项的值的键。

为了使操作系统显示所有的首选项供用户编辑，创建了一个活动来扩展 PreferenceActivity 基类，然后调用 addPreferencesFromResource()方法来加载包含首选项的 XML 文件：

```
public class AppPreferenceActivity extends PreferenceActivity {
    @Override
    public void onCreate(Bundle savedInstanceState) {
        super.onCreate(savedInstanceState);
        //---load the preferences from an XML file---
        addPreferencesFromResource(R.xml.myapppreferences);
    }
}
```

PreferenceActivity 类是一种特殊类型的活动，向用户显示首选项的一个层次结构。为了显示首选项的活动，使用一个 Intent 对象调用它：

```
Intent i = new Intent("net.learn2develop.AppPreferenceActivity");
startActivity(i);
```

对首选项所做的所有更改会自动保存到应用程序的 shared_prefs 文件夹下的一个 XML 文件中。

6.1.2 通过编程检索和修改首选项值

在前一节看到，PreferenceActivity 类既允许开发人员方便地创建首选项，又允许用户在运行时修改首选项。为了在应用程序中使用这些首选项，需要使用 SharedPreferences 类。下面的"试一试"显示了具体做法。

试一试　检索和修改首选项

(1) 使用前一节创建的同一个项目，在 UsingPreferencesActivity.java 文件中添加下列粗体显示的代码：

```
package net.learn2develop.UsingPreferences;

import android.app.Activity;
import android.content.Intent;
import android.content.SharedPreferences;
import android.os.Bundle;
import android.view.View;
import android.widget.EditText;
import android.widget.Toast;

public class UsingPreferencesActivity extends Activity {
    /** Called when the activity is first created. */
    @Override
    public void onCreate(Bundle savedInstanceState) {
        super.onCreate(savedInstanceState);
        setContentView(R.layout.main);
```

```java
    }
    public void onClickLoad(View view) {
        Intent i = new Intent("net.learn2develop.AppPreferenceActivity");
        startActivity(i);
    }
    public void onClickDisplay(View view) {
        SharedPreferences appPrefs =
        getSharedPreferences("net.learn2develop.UsingPreferences_
        preferences",
                MODE_PRIVATE);
        DisplayText(appPrefs.getString("editTextPref", ""));
    }

    public void onClickModify(View view) {
        SharedPreferences appPrefs =
        getSharedPreferences("net.learn2develop.UsingPreferences_
           preferences",
                MODE_PRIVATE);
        SharedPreferences.Editor prefsEditor = appPrefs.edit();
        prefsEditor.putString("editTextPref",
                ((EditText)findViewById(R.id.txtString)).getText().
                    toString());
        prefsEditor.commit();
    }

    private void DisplayText(String str) {
        Toast.makeText(getBaseContext(), str, Toast.LENGTH_LONG).show();
    }
}
```

(2) 按F11键在Android模拟器上再次运行应用程序。这一次,单击Display Preferences Values 按钮将显示如图6-7所示的值。

(3) 在EditText视图中输入一个字符串,然后单击Modify Preferences Values按钮,如图6-8所示。

(4) 现在再次单击Display Preferences Values按钮。注意新值将被保存。

示例说明

在onClickDisplay()方法中,首先使用getSharedPreferences()方法来获得SharedPreferences类的一个实例。这是通过指定 XML 文件的名称实现的,在本例中 XML 文件的名称为net.learn2develop.UsingPreferences_preferences,其格式为<PackageName>_preferences。为了检索字符串首选项,使用getString()方法,向其传入想要检索的首选项的键:

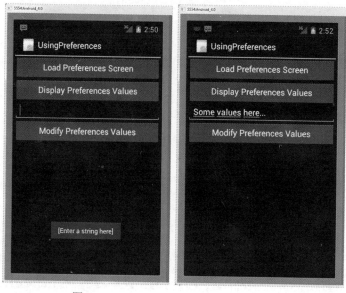

图 6-7　　　　　　　图 6-8

```
public void onClickDisplay(View view) {
    SharedPreferences appPrefs =
            getSharedPreferences("net.learn2develop.UsingPreferences
            _preferences",
                MODE_PRIVATE);
    DisplayText(appPrefs.getString("editTextPref", ""));
}
```

MODE_PRIVATE 常量表示只有创建首选项文件的应用程序能够打开首选项文件。

在 onClickModify()方法中，通过 SharedPreferences 对象的 edit()方法创建了一个 SharedPreferences.Editor 对象。为了修改字符串首选项的值，使用了 putString()方法。为了将更改保存到首选项文件中，使用了 commit()方法：

```
public void onClickModify(View view) {
    SharedPreferences appPrefs =
            getSharedPreferences("net.learn2develop.UsingPreferences
            _preferences",
                MODE_PRIVATE);
    SharedPreferences.Editor prefsEditor = appPrefs.edit();
    prefsEditor.putString("editTextPref",
            ((EditText) findViewById(R.id.txtString)).getText().
            toString());
    prefsEditor.commit();
}
```

6.1.3　修改首选项文件的默认名称

注意，保存在设备上的首选项文件的默认名称是 net.learn2develop.UsingPreferences_preferences.xml，包名作为前缀。但是在有些时候，为首选项文件命名一个特定的名字很有帮助。此时，可以执行以下操作。

在 AppPreferenceActivity.java 文件中添加下列粗体显示的代码：

```
package net.learn2develop.UsingPreferences;

import android.os.Bundle;
import android.preference.PreferenceActivity;
import android.preference.PreferenceManager;

public class AppPreferenceActivity extends PreferenceActivity {
    @Override
    public void onCreate(Bundle savedInstanceState) {
        super.onCreate(savedInstanceState);

        PreferenceManager prefMgr = getPreferenceManager();
        prefMgr.setSharedPreferencesName("appPreferences");

        //---load the preferences from an XML file---
        addPreferencesFromResource(R.xml.myapppreferences);
    }
}
```

这里使用 PreferenceManager 类将共享首选项文件的名称设为 appPreferences.xml。

按照如下所示修改 UsingPreferencesActivity.java 文件：

```
    public void onClickDisplay(View view) {
        /*
        SharedPreferences appPrefs =
                getSharedPreferences("net.learn2develop.UsingPreferences
                _preferences",
                    MODE_PRIVATE);
        */
        SharedPreferences appPrefs =
                getSharedPreferences("appPreferences", MODE_PRIVATE);

        DisplayText(appPrefs.getString("editTextPref", ""));
    }

    public void onClickModify(View view) {
        /*
        SharedPreferences appPrefs =
                getSharedPreferences("net.learn2develop.UsingPreferences
                _preferences",
                    MODE_PRIVATE);
        */
        SharedPreferences appPrefs =
                getSharedPreferences("appPreferences", MODE_PRIVATE);

        SharedPreferences.Editor prefsEditor = appPrefs.edit();
        prefsEditor.putString("editTextPref",
```

```
            ((EditText) findViewById(R.id.txtString)).getText().
                toString());
        prefsEditor.commit();
    }
```

重新运行应用程序并修改首选项时,注意现在创建了一个 **appPreferences.xml** 文件,如图 6-9 所示。

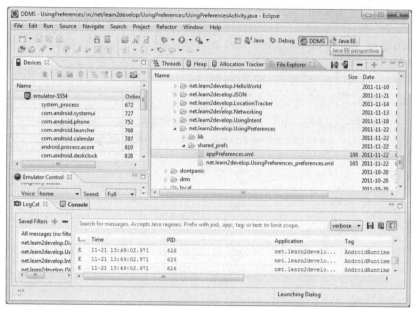

图 6-9

6.2 将数据持久化到文件中

SharedPreferences 对象允许您存储最适合以键/值对方式进行存储的数据,例如用户 ID、生日、性别、驾照号码等数据。然而,有时您还是希望利用传统的文件系统来存储数据。例如,您可能想存储一些诗歌的文本以便在您的应用程序中显示。在 Android 中,可以使用 java.io 包中的类来做到这一点。

6.2.1 保存到内部存储器

Android 应用程序中保存文件的第一个方法就是将其写入设备的内部存储器。下面的 "试一试"展示了如何将用户输入的一个字符串保存到设备的内部存储器中。

试一试　　将数据保存到内部存储器

Files.zip 代码文件可以在 Wrox.com 上下载

(1) 打开 Eclipse,创建一个 Android 项目,命名为 Files。

(2) 在 main.xml 文件中添加下列粗体显示的语句：

```xml
<?xml version="1.0" encoding="utf-8"?>
<LinearLayout xmlns:android="http://schemas.android.com/apk/res/android"
    android:layout_width="fill_parent"
    android:layout_height="fill_parent"
    android:orientation="vertical" >

<TextView
    android:layout_width="fill_parent"
    android:layout_height="wrap_content"
    android:text="Please enter some text" />

<EditText
    android:id="@+id/txtText1"
    android:layout_width="fill_parent"
    android:layout_height="wrap_content" />

<Button
    android:id="@+id/btnSave"
    android:text="Save"
    android:layout_width="fill_parent"
    android:layout_height="wrap_content"
    android:onClick="onClickSave" />

<Button
    android:id="@+id/btnLoad"
    android:text="Load"
    android:layout_width="fill_parent"
    android:layout_height="wrap_content"
    android:onClick="onClickLoad" />

</LinearLayout>
```

(3) 在 FilesActivity.java 文件中添加下列粗体显示的语句：

```java
package net.learn2develop.Files;

import java.io.FileInputStream;
import java.io.FileOutputStream;
import java.io.IOException;
import java.io.InputStreamReader;
import java.io.OutputStreamWriter;

import android.app.Activity;
import android.os.Bundle;
import android.view.View;
import android.widget.EditText;
import android.widget.Toast;

public class FilesActivity extends Activity {
```

```java
EditText textBox;
static final int READ_BLOCK_SIZE = 100;

/** Called when the activity is first created. */
@Override
public void onCreate(Bundle savedInstanceState) {
    super.onCreate(savedInstanceState);
    setContentView(R.layout.main);

    textBox = (EditText) findViewById(R.id.txtText1);
}

public void onClickSave(View view) {
    String str = textBox.getText().toString();
    try
    {
        FileOutputStream fOut =
                openFileOutput("textfile.txt",
                    MODE_WORLD_READABLE);
        OutputStreamWriter osw = new
                OutputStreamWriter(fOut);

        //---write the string to the file---
        osw.write(str);
        osw.flush();
        osw.close();

        //---display file saved message---
        Toast.makeText(getBaseContext(),
                "File saved successfully!",
                Toast.LENGTH_SHORT).show();

        //---clears the EditText---
        textBox.setText("");
    }
    catch (IOException ioe)
    {
        ioe.printStackTrace();
    }

}

public void onClickLoad(View view) {
    try
    {
        FileInputStream fIn =
                openFileInput("textfile.txt");
        InputStreamReader isr = new
                InputStreamReader(fIn);
```

```
            char[] inputBuffer = new char[READ_BLOCK_SIZE];
            String s = "";

            int charRead;
            while ((charRead = isr.read(inputBuffer))>0)
            {
                    //---convert the chars to a String---
                    String readString =
                            String.copyValueOf(inputBuffer, 0,
                            charRead);
                    s += readString;

                    inputBuffer = new char[READ_BLOCK_SIZE];
            }
            //---set the EditText to the text that has been
            // read---
            textBox.setText(s);

            Toast.makeText(getBaseContext(),
                    "File loaded successfully!",
                    Toast.LENGTH_SHORT).show();
        }
        catch (IOException ioe) {
            ioe.printStackTrace();
        }

    }

}
```

(4) 按 F11 键在 Android 模拟器上调试应用程序。

(5) 在 EditText 视图中输入一些文本(如图 6-10 所示)，然后单击 Save 按钮。

(6) 如果文件成功保存，将看到由 Toast 类显示出的消息"File saved successfully!"。EditText 视图中的文本将消失。

(7) 单击 Load 按钮，将又会看到字符串出现在 EditText 视图中。这表明文本被正确地保存了。

示例说明

图 6-10

为了将文本保存到文件中，要使用 FileOutputStream 类。openFileOutput()方法用指定的模式，打开一个指定的文件来写入。在本例中，使用 MODE_WORLD_READABLE 常量来表示此文件可被其他所有应用程序读取：

```
FileOutputStream fOut =
    openFileOutput("textfile.txt",
        MODE_WORLD_READABLE);
```

除了 MODE_WORLD_READABLE 常量之外，还可以从以下模式中进行选择：MODE_PRIVATE(文件只能被创建它的应用程序访问)、MODE_APPEND(附加到现有文件)和 MODE_WORLD_WRITEABLE(文件对于其他所有应用程序来说都是可写的)。

为了将字符流转换为字节流，要利用 OutputStreamWriter 类的一个实例，并给它传递一个 FileOutputStream 对象的实例：

```
OutputStreamWriter osw = new
        OutputStreamWriter(fOut);
```

然后使用其 write()方法将字符串写入到文件中。使用 flush()方法来保证所有字节都写入文件。最后，使用 close()方法来关闭文件：

```
//---write the string to the file---
osw.write(str);
osw.flush();
osw.close();
```

为了读取文件内容，FileInputStream 类和 InputStreamReader 类要配合使用：

```
FileInputStream fIn =
        openFileInput("textfile.txt");
InputStreamReader isr = new
        InputStreamReader(fIn);
```

由于不清楚要读取的文件的大小，因此按 100 个字符为一块将文件内容读到缓冲区(字符数组)中。然后将所读字符复制到一个 String 对象中：

```
char[] inputBuffer = new char[READ_BLOCK_SIZE];
String s = "";

int charRead;
while ((charRead = isr.read(inputBuffer))>0)
{
    //---convert the chars to a String---
    String readString =
            String.copyValueOf(inputBuffer, 0,
                    charRead);
    s += readString;

    inputBuffer = new char[READ_BLOCK_SIZE];
}
```

InputStreamReader 对象的 read()方法读取所读字符的个数，如果到达文件末尾，就返回-1。

当在 Android 模拟器上测试此应用程序时，可以使用 DDMS 透视图来验证应用程序的确将文件保存在了其 files 目录中(如图 6-11 所示；完整路径是/data/data/net.learn2develop. Files/files)。

第 6 章 数据持久化

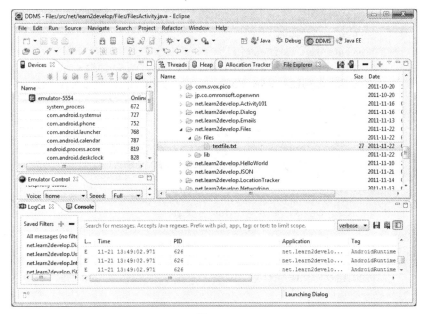

图 6-11

6.2.2 保存到外部存储器(SD 卡)

前面一节讲述了如何将文件保存到 Android 设备的内部存储器。有时,将文件保存到外部存储器(例如 SD 卡)也是很有用的,这是因为外部存储器容量大,而且很容易和其他用户共享文件(只要将 SD 卡移到别人的设备里就行了)。

使用前一节中创建的项目作为例子,为了将用户输入的文本保存在 SD 卡中,按粗体显示内容修改 Save 按钮的 onClick()方法:

```
import java.io.File;
import java.io.FileInputStream;
import java.io.FileOutputStream;
import java.io.IOException;
import java.io.InputStreamReader;
import java.io.OutputStreamWriter;

import android.app.Activity;
import android.os.Bundle;
import android.os.Environment;
import android.view.View;
import android.widget.EditText;
import android.widget.Toast;

    public void onClickSave(View view) {
        String str = textBox.getText().toString();
        try
        {
            //---SD Card Storage---
            File sdCard = Environment.getExternalStorageDirectory();
```

```
            File directory = new File (sdCard.getAbsolutePath() +
              "/MyFiles");
            directory.mkdirs();
            File file = new File(directory, "textfile.txt");
            FileOutputStream fOut = new FileOutputStream(file);

            /*
            FileOutputStream fOut =
                    openFileOutput("textfile.txt",
                        MODE_WORLD_READABLE);
            */

            OutputStreamWriter osw = new
                    OutputStreamWriter(fOut);

            //---write the string to the file---
            osw.write(str);
            osw.flush();
            osw.close();

            //---display file saved message---
            Toast.makeText(getBaseContext(),
                "File saved successfully!",
                Toast.LENGTH_SHORT).show();

            //---clears the EditText---
            textBox.setText("");
        }
        catch (IOException ioe)
        {
            ioe.printStackTrace();
        }

    }
```

上述代码使用 getExternalStorageDirectory()方法返回外部存储器的完整路径。一般而言，对于真实设备返回/sdcard 路径，对于 Android 模拟器返回/mnt/sdcard 路径。然而，永远不要试图用硬编码的方式指定 SD 卡的路径，因为制造商可能会给 SD 卡指派一个不同的路径名称。因此，应确保使用 getExternalStorageDirectory()方法来返回指向 SD 卡的完整路径。

然后在 SD 卡中创建一个名为 MyFiles 的目录。最终，文件将保存在此目录下。

为了从外部存储器中加载文件，需要为 Load 按钮修改 onClickLoad()方法：

```
        public void onClickLoad(View view) {
            try
            {
                //---SD Storage---
                File sdCard = Environment.getExternalStorageDirectory();
```

```
            File directory = new File (sdCard.getAbsolutePath() +
                "/MyFiles");
            File file = new File(directory, "textfile.txt");
            FileInputStream fIn = new FileInputStream(file);
            InputStreamReader isr = new InputStreamReader(fIn);

            /*
            FileInputStream fIn =
                    openFileInput("textfile.txt");
            InputStreamReader isr = new
                    InputStreamReader(fIn);
            */

            char[] inputBuffer = new char[READ_BLOCK_SIZE];
            String s = "";

            int charRead;
            while ((charRead = isr.read(inputBuffer))>0)
            {
                //---convert the chars to a String---
                String readString =
                    String.copyValueOf(inputBuffer, 0,
                            charRead);
                s += readString;

                inputBuffer = new char[READ_BLOCK_SIZE];
            }
            //---set the EditText to the text that has been
            // read---
            textBox.setText(s);

            Toast.makeText(getBaseContext(),
                    "File loaded successfully!",
                    Toast.LENGTH_SHORT).show();
        }
        catch (IOException ioe) {
            ioe.printStackTrace();
        }
    }
}
```

注意，要写入到外部存储器，需要在 AndroidManifest.xml 文件中添加 WRITE_EXTERNAL_STORAGE 权限：

```
<?xml version="1.0" encoding="utf-8"?>
<manifest xmlns:android="http://schemas.android.com/apk/res/android"
    package="net.learn2develop.Files"
    android:versionCode="1"
    android:versionName="1.0" >
```

```xml
<uses-sdk android:minSdkVersion="14" />
<uses-permission
    android:name="android.permission.WRITE_EXTERNAL_STORAGE" />

<application
    android:icon="@drawable/ic_launcher"
    android:label="@string/app_name" >
    <activity
        android:label="@string/app_name"
        android:name=".FilesActivity" >
        <intent-filter >
            <action android:name="android.intent.action.MAIN" />

            <category android:name="android.intent.category.LAUNCHER" />
        </intent-filter>
    </activity>
</application>

</manifest>
```

如果运行前面修改过的代码，会看到/mnt/sdcard/MyFiles/文件夹下创建了一个文本文件，如图6-12所示。

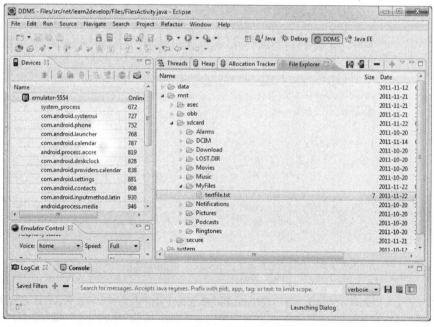

图 6-12

6.2.3 选择最佳存储选项

前面的小节描述了在Android应用程序中保存数据的3种主要方式：SharedPreferences、内部存储器和外部存储器。在应用程序中应该选择哪一种呢？以下是一些建议：

- 如果数据可以用键/值对表示，那么使用 SharedPreferences 对象。例如，如果想要存储用户首选项的数据，如用户姓名、背景色、生日、最后登录日期等，那么使用 SharedPreferences 对象来存储这些数据是一个理想的方法。而且，您也不用亲自操心这些数据是如何存储的；所有需要您做的只是使用 SharedPreferences 对象存储和检索它们。
- 如果需要存储临时数据，使用内部存储器是一个好的选择。例如，应用程序(如一个 RSS 阅读器)可能需要显示从 Web 上下载的图像。在这种情况下，将图像保存在内部存储器上是一个好的解决方法。您也许还希望持久化用户所创建的数据，例如有一个备忘录应用程序可以用来让用户记些笔记并保存它们以备后用。在所有这些情况下，使用内部存储器是一个好的选择。
- 有时需要和其他用户共享应用程序数据。例如，您创建了一个 Android 应用程序来记录用户曾经去过的地点的坐标，并想与其他用户分享这些数据。在这种情况下，可以将文件存储在设备的 SD 卡上，这样用户在后面可以很容易地将数据转移到其他设备(和计算机)上来使用。

6.2.4 使用静态资源

除了在运行时动态地创建和使用文件之外，还可以在设计时将文件添加到包，以便于在运行时使用它。例如，您可能需要与包捆绑一些帮助文件，以便于用户需要时可以显示一些帮助信息。这时，可以在包下面的 res/raw 文件夹(需要自己创建)下添加文件。图 6-13 显示了 res/raw 文件夹包含了一个名为 textfile.txt 的文件。

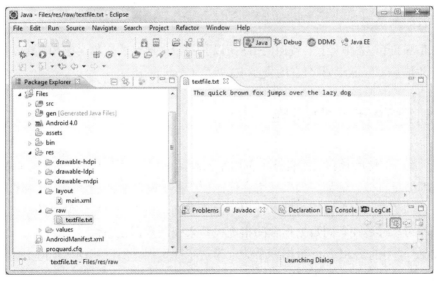

图 6-13

为了在代码中使用此文件，要利用 Activity 类的 getResources()方法返回一个 Resources 对象，然后使用其 openRawResource()方法来打开包含在 res/raw 文件夹下的文件：

```
import java.io.BufferedReader;
```

```java
import java.io.InputStream;

public class FilesActivity extends Activity {
    EditText textBox;
    static final int READ_BLOCK_SIZE = 100;

    /** Called when the activity is first created. */
    @Override
    public void onCreate(Bundle savedInstanceState) {
        super.onCreate(savedInstanceState);
        setContentView(R.layout.main);

        textBox = (EditText) findViewById(R.id.txtText1);

        InputStream is = this.getResources().openRawResource
          (R.raw.textfile);
        BufferedReader br = new BufferedReader(new InputStreamReader(is));
        String str = null;
        try {
            while ((str = br.readLine()) != null) {
                Toast.makeText(getBaseContext(),
                    str, Toast.LENGTH_SHORT).show();
            }
            is.close();
            br.close();
        } catch (IOException e) {
            e.printStackTrace();
        }
    }
}
```

存储在 res/raw 文件夹下的资源的 ID 是以其不带扩展名的文件名来命名的。例如，如果文本文件是 textfile.txt，那么它的资源 ID 就是 R.raw.textfile。

6.3 创建和使用数据库

到目前为止，您所看到的所有技术都用来保存简单的数据集合。对于存储关系型数据，使用数据库会更有效。例如，如果您想存储一所学校所有学生的测验结果，那么使用数据库来表示它们要有效得多，因为可以使用数据库查询来检索一个特定学生的结果。此外，使用数据库可以通过在不同数据集合间指定关系来强制数据完整性。

Android 使用的是 SQLite 数据库系统。为一个应用程序所创建的数据库只能被此应用程序访问；其他应用程序将不能访问它。

本节中，将学习如何以编程方式在 Android 应用程序中创建一个 SQLite 数据库。对于 Android，在一个应用程序中以编程方式创建的 SQLite 数据库总是存储在 /data/data/<package_name>/databases 文件夹下。

6.3.1 创建 DBAdapter 辅助类

处理数据库的一个好的做法是创建一个辅助类来封装访问数据的所有复杂性，使之对于调用它的代码来说是透明的。因此，在本节中将创建一个名为 DBAdapter 的辅助类，用来创建、打开、关闭和使用一个 SQLite 数据库。

本例中，将创建一个名为 MyDB 的数据库，包含一张名为 contacts 的表。这个表有 3 列：_id、name 和 email(如图 6-14 所示)。

图 6-14

试一试 创建数据库辅助类

Databases.zip 代码文件可以在Wrox.com 上下载

(1) 打开 Eclipse，创建一个名为 Databases 的 Android 项目。

(2) 在包中新增一个 Java 类文件，并命名为 DBAdapter (如图 6-15 所示)。

(3) 在 DBAdapter.java 文件中添加下列粗体显示的语句：

图 6-15

```
package net.learn2develop.Databases;

import android.content.ContentValues;
import android.content.Context;
import android.database.Cursor;
import android.database.SQLException;
import android.database.sqlite.SQLiteDatabase;
import android.database.sqlite.SQLiteOpenHelper;
import android.util.Log;

public class DBAdapter {
    static final String KEY_ROWID = "_id";
    static final String KEY_NAME = "name";
    static final String KEY_EMAIL = "email";
    static final String TAG = "DBAdapter";

    static final String DATABASE_NAME = "MyDB";
    static final String DATABASE_TABLE = "contacts";
    static final int DATABASE_VERSION = 1;

    static final String DATABASE_CREATE =
        "create table contacts (_id integer primary key autoincrement, "
        + "name text not null, email text not null);";

    final Context context;

    DatabaseHelper DBHelper;
```

```java
    SQLiteDatabase db;

    public DBAdapter(Context ctx)
    {
        this.context = ctx;
        DBHelper = new DatabaseHelper(context);
    }

    private static class DatabaseHelper extends SQLiteOpenHelper
    {
        DatabaseHelper(Context context)
        {
            super(context, DATABASE_NAME, null, DATABASE_VERSION);
        }

        @Override
        public void onCreate(SQLiteDatabase db)
        {
            try {
                db.execSQL(DATABASE_CREATE);
            } catch (SQLException e) {
                e.printStackTrace();
            }
        }

        @Override
        public void onUpgrade(SQLiteDatabase db, int oldVersion, int
            newVersion)
        {
            Log.w(TAG, "Upgrading database from version " + oldVersion+"to "
                    + newVersion + ", which will destroy all old data");
            db.execSQL("DROP TABLE IF EXISTS contacts");
            onCreate(db);
        }
    }

    //---opens the database---
    public DBAdapter open() throws SQLException
    {
        db = DBHelper.getWritableDatabase();
        return this;
    }

    //---closes the database---
    public void close()
    {
        DBHelper.close();
    }

    //---insert a contact into the database---
    public long insertContact(String name, String email)
    {
```

```
        ContentValues initialValues = new ContentValues();
        initialValues.put(KEY_NAME, name);
        initialValues.put(KEY_EMAIL, email);
        return db.insert(DATABASE_TABLE, null, initialValues);
    }

    //---deletes a particular contact---
    public boolean deleteContact(long rowId)
    {
        return db.delete(DATABASE_TABLE, KEY_ROWID + "=" + rowId, null) > 0;
    }

    //---retrieves all the contacts---
    public Cursor getAllContacts()
    {
        return db.query(DATABASE_TABLE, new String[] {KEY_ROWID, KEY_NAME,
                KEY_EMAIL}, null, null, null, null, null);
    }

    //---retrieves a particular contact---
    public Cursor getContact(long rowId) throws SQLException
    {
        Cursor mCursor =
                db.query(true, DATABASE_TABLE, new String[] {KEY_ROWID,
                KEY_NAME, KEY_EMAIL}, KEY_ROWID + "=" + rowId, null,
                null, null, null, null);
        if (mCursor != null) {
            mCursor.moveToFirst();
        }
        return mCursor;
    }

    //---updates a contact---
    public boolean updateContact(long rowId, String name, String email)
    {
        ContentValues args = new ContentValues();
        args.put(KEY_NAME, name);
        args.put(KEY_EMAIL, email);
        return db.update(DATABASE_TABLE, args, KEY_ROWID + "=" + rowId,
          null) > 0;
    }
}
```

示例说明

首先为在数据库中将要创建的表定义几个常量来包含不同的字段：

```
static final String KEY_ROWID = "_id";
static final String KEY_NAME = "name";
static final String KEY_EMAIL = "email";
static final String TAG = "DBAdapter";

static final String DATABASE_NAME = "MyDB";
```

```
    static final String DATABASE_TABLE = "contacts";
    static final int DATABASE_VERSION = 1;

    static final String DATABASE_CREATE =
        "create table contacts (_id integer primary key autoincrement, "
        + "name text not null, email text not null);";
```

特别是，DATABASE_CREATE 常量包含了用于在 MyDB 数据库中创建 contacts 表的 SQL 语句。

在 DBAdapter 类中，还添加了一个私有类来扩展 SQLiteOpenHelper 类，它是一个在 Android 中用来处理数据库创建和版本管理的辅助类。尤其是，您重写了 onCreate()方法和 onUpgrade()方法：

```
    private static class DatabaseHelper extends SQLiteOpenHelper
    {
        DatabaseHelper(Context context)
        {
            super(context, DATABASE_NAME, null, DATABASE_VERSION);
        }

        @Override
        public void onCreate(SQLiteDatabase db)
        {
            try {
                db.execSQL(DATABASE_CREATE);
            } catch (SQLException e) {
                e.printStackTrace();
            }
        }

        @Override
        public void onUpgrade(SQLiteDatabase db, int oldVersion, int newVersion)
        {
            Log.w(TAG,"Upgrading database from version"+oldVersion+" to "
                    +newVersion + ", which will destroy all old data");
            db.execSQL("DROP TABLE IF EXISTS contacts");
            onCreate(db);
        }
    }
```

onCreate()方法在所需数据库不存在时创建一个新的数据库。onUpgrade()方法在数据库需要升级时调用。这可以通过检查在 DATABASE_VERSION 常量中定义的值来实现。对于 onUpgrade()方法的这一实现，只是删除表并再创建它。

然后，可以定义用于打开和关闭数据库的不同方法，以及用于在表中添加、修改、删除行的方法。

```
        //---opens the database---
```

```java
public DBAdapter open() throws SQLException
{
    db = DBHelper.getWritableDatabase();
    return this;
}

//---closes the database---
public void close()
{
    DBHelper.close();
}

//---insert a contact into the database---
public long insertContact(String name, String email)
{
    ContentValues initialValues = new ContentValues();
    initialValues.put(KEY_NAME, name);
    initialValues.put(KEY_EMAIL, email);
    return db.insert(DATABASE_TABLE, null, initialValues);
}

//---deletes a particular contact---
public boolean deleteContact(long rowId)
{
    return db.delete(DATABASE_TABLE, KEY_ROWID + "=" + rowId, null) > 0;
}

//---retrieves all the contacts---
public Cursor getAllContacts()
{
    return db.query(DATABASE_TABLE, new String[] {KEY_ROWID, KEY_NAME,
            KEY_EMAIL}, null, null, null, null, null);
}

//---retrieves a particular contact---
public Cursor getContact(long rowId) throws SQLException
{
    Cursor mCursor =
            db.query(true, DATABASE_TABLE, new String[] {KEY_ROWID,
                KEY_NAME, KEY_EMAIL}, KEY_ROWID + "=" + rowId, null,
                null, null, null, null);
    if (mCursor != null) {
        mCursor.moveToFirst();
    }
    return mCursor;
}

//---updates a contact---
public boolean updateContact(long rowId, String name, String email)
{
```

```
        ContentValues args = new ContentValues();
        args.put(KEY_NAME, name);
        args.put(KEY_EMAIL, email);
        return db.update(DATABASE_TABLE,args,KEY_ROWID+"="+rowId,null)>0;
    }
```

注意，Android 使用 Cursor 类作为查询的返回值。可以将 Cursor 看成是一个指向数据库查询的结果集的指针。使用 Cursor 可以使 Android 更有效地按需要管理行和列。

使用 ContentValues 对象来存储键/值对。其 put()方法可用于插入具有不同数据类型值的键。

为了使用 DBAdapter 类在应用程序中创建数据库，创建了 DBAdapter 类的一个实例：

```
public DBAdapter(Context ctx)
{
    this.context = ctx;
    DBHelper = new DatabaseHelper(context);
}
```

然而，DBAdapter 类的构造函数创建了 DatabaseHelper 类的一个实例来创建一个新的数据库：

```
DatabaseHelper(Context context)
{
    super(context, DATABASE_NAME, null, DATABASE_VERSION);
}
```

6.3.2 以编程方式使用数据库

创建了 DBAdapter 辅助类后，就可以使用数据库了。在下面的小节中，将学习如何对数据库执行常见的 CRUD(创建、读取、更新和删除)操作。

1. 添加联系人

下面的"试一试"演示了如何将一个联系人添加到表中。

试一试　　向表中添加联系人

Databases.zip 代码文件可以在 Wrox.com 上下载

(1) 打开先前创建的同一个项目，在 DatabasesActivity.java 文件中添加下列粗体显示的语句：

```
package net.learn2develop.Databases;

import android.app.Activity;
import android.os.Bundle;
```

```
public class DatabasesActivity extends Activity {
    /** Called when the activity is first created. */
    @Override
    public void onCreate(Bundle savedInstanceState) {
        super.onCreate(savedInstanceState);
        setContentView(R.layout.main);

        DBAdapter db = new DBAdapter(this);

        //---add a contact---
        db.open();
        long id =
        db.insertContact("Wei-MengLee","weimenglee@learn2develop.net");
        id = db.insertContact("Mary Jackson", "mary@jackson.com");
        db.close();
    }
}
```

(2) 按 F11 键在 Android 模拟器上调试应用程序。

示例说明

在这个例子中，首先创建了一个 DBAdapter 类的实例：

```
DBAdapter db = new DBAdapter(this);
```

insertContact()方法返回所插入行的 ID。如果在操作中发生错误则返回-1。

如果使用 DDMS 检查 Android 设备/模拟器的文件系统，可以看到在 databases 文件夹下创建了 MyDB 数据库(如图 6-16 所示)。

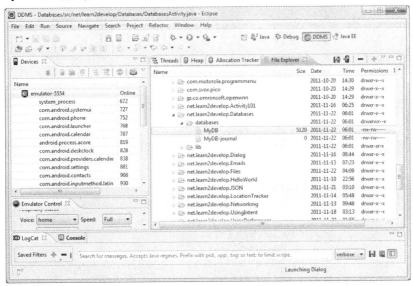

图 6-16

2. 检索所有联系人

为了检索 contacts 表中所有联系人，使用 DBAdapter 类的 getAllContacts()方法，如下面的"试一试"所示。

试一试　　从表中检索所有联系人

Databases.zip 代码文件可以在 Wrox.com 上下载

（1）打开先前创建的同一个项目，在 DatabasesActivity.java 文件中添加下列粗体显示的语句：

```
package net.learn2develop.Databases;

import android.app.Activity;
import android.database.Cursor;
import android.os.Bundle;
import android.widget.Toast;

public class DatabasesActivity extends Activity {
    /** Called when the activity is first created. */
    @Override
    public void onCreate(Bundle savedInstanceState) {
        super.onCreate(savedInstanceState);
        setContentView(R.layout.main);

        DBAdapter db = new DBAdapter(this);

        /*
        //---add a contact---
        db.open();
        ...
        db.close();
        */

        //---get all contacts---
        db.open();
        Cursor c = db.getAllContacts();
        if (c.moveToFirst())
        {
            do {
                DisplayContact(c);
            } while (c.moveToNext());
        }
        db.close();
    }

    public void DisplayContact(Cursor c)
    {
```

```
        Toast.makeText(this,
            "id: " + c.getString(0) + "\n" +
            "Name: " + c.getString(1) + "\n" +
            "Email:  " + c.getString(2),
            Toast.LENGTH_LONG).show();
    }
}
```

(2) 按 F11 键在 Android 模拟器上调试应用程序。图 6-17 展示了 Toast 类所显示的从数据库中检索到的联系人。

示例说明

DBAdapter 类的 getAllContacts()方法检索存储于数据库中的所有联系人。结果以 Cursor 对象的形式返回。为了显示所有联系人，首先需要调用 Cursor 对象的 moveToFirst()方法。如果成功(这意味着至少有一行可用)，则使用 DisplayContact()方法来显示该联系人的详细信息。要移动到下一个联系人，则要调用 Cursor 对象的 moveToNext()方法。

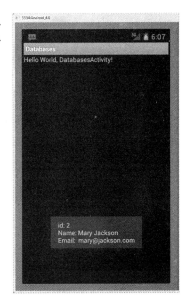

图 6-17

3. 检索单个联系人

要利用联系人的 ID 来检索单个联系人，可按下面的"试一试"所展示的，调用 DBAdapter 类的 getContact()方法。

试一试   从表中检索单个联系人

Databases.zip 代码文件可以在 Wrox.com 上下载

(1) 打开先前创建的同一个项目，在 DatabasesActivity.java 文件中添加下列粗体显示的语句：

```
@Override
public void onCreate(Bundle savedInstanceState) {
    super.onCreate(savedInstanceState);
    setContentView(R.layout.main);

    DBAdapter db = new DBAdapter(this);

    /*
    //---add a contact---
    ...
    //--get all contacts---
    ...
    db.close();
    */
```

```
    //---get a contact---
    db.open();
    Cursor c = db.getContact(2);
    if (c.moveToFirst())
        DisplayContact(c);
    else
        Toast.makeText(this, "No contact found", Toast.LENGTH_LONG).
        show();
    db.close();
}
```

(2) 按 F11 键在 Android 模拟器上调试应用程序。第二个联系人的详细信息将通过 Toast 类显示出来。

示例说明

DBAdapter 类的 getContact()方法使用联系人的 ID 来检索单个联系人。传入联系人的 ID；这里传入的 ID 是 2，表明想要检索第二个联系人：

```
    Cursor c = db.getContact(2);
```

结果以 Cursor 对象的形式返回。如果返回了一行，就使用 DisplayContact()方法来显示联系人的详细信息；否则使用 Toast 类显示一条消息。

4. 更新联系人

通过调用 DBAdapter 类的 updateContact()方法并传递一个想要更新的联系人的 ID，可以实现对某个特定联系人的更新操作，如下面的"试一试"所示。

试一试　　更新表中某个联系人

Databases.zip 代码文件可以在 Wrox.com 上下载

(1) 打开先前创建的同一个项目，在 DatabasesActivity.java 文件中添加下列粗体显示的语句：

```
@Override
public void onCreate(Bundle savedInstanceState) {
    super.onCreate(savedInstanceState);
    setContentView(R.layout.main);
    
    DBAdapter db = new DBAdapter(this);
    
    /*
    //---add a contact---
    ...
    //--get all contacts---
    ...
```

```
            //---get a contact---
            ...
            db.close();
            */

            //---update contact---
            db.open();
            if (db.updateContact(1, "Wei-Meng Lee", "weimenglee@gmail.com"))
                Toast.makeText(this, "Update successful.",
                    Toast.LENGTH_LONG).show();
            else
                Toast.makeText(this, "Update failed.", Toast.LENGTH_LONG).
                    show();
            db.close();
        }
```

(2) 按 F11 键在 Android 模拟器上调试应用程序。如果更新成功,将显示一条消息。

示例说明

DBAdapter 类中的 updateContact()方法利用您打算更新的联系人的 ID 来更新此联系人的详细信息。它返回一个 Boolean 值,表明更新是否成功。

5. 删除一个联系人

使用 DBAdapter 类的 deleteContact()方法并传递一个想要删除的联系人的 ID,可以实现对某个联系人的删除操作,如下面的"试一试"所示。

试一试 删除表中的某个联系人

Databases.zip 代码文件可以在 Wrox.com 上下载

(1) 打开先前创建的同一个项目,在 DatabasesActivity.java 文件中添加下列粗体显示的语句:

```
        @Override
        public void onCreate(Bundle savedInstanceState) {
            super.onCreate(savedInstanceState);
            setContentView(R.layout.main);

            DBAdapter db = new DBAdapter(this);

            /*
            //---add a contact---
            ...
            //--get all contacts---
            ...
            //---get a contact---
            ...
            //---update contact---
```

```
    ...
    db.close();
    */

    //---delete a contact---
    db.open();
    if (db.deleteContact(1))
        Toast.makeText(this,"Delete successful.",Toast.LENGTH_LONG).
            show();
    else
        Toast.makeText(this, "Delete failed.", Toast.LENGTH_LONG).
            show();
    db.close();
}
```

(2) 按 F11 键在 Android 模拟器上调试应用程序。如果删除成功，将显示一条消息。

示例说明

DBAdapter 类中的 deleteContact()方法利用您打算删除的联系人的 ID 来删除此联系人。它返回一个 Boolean 值，表明删除是否成功。

6．升级数据库

有时，在创建和开始使用数据库之后，您可能需要添加另外的表、改变数据库的模式，或者在表中添加一些列。这时，就需要将旧数据库中现有的数据迁移到一个新数据库中。

要升级数据库，需要将 DATABASE_VERSION 常量改为比先前要高的值。例如，如果先前的值是 1，则现在改为 2：

```
public class DBAdapter {
    static final String KEY_ROWID = "_id";
    static final String KEY_NAME = "name";
    static final String KEY_EMAIL = "email";
    static final String TAG = "DBAdapter";

    static final String DATABASE_NAME = "MyDB";
    static final String DATABASE_TABLE = "contacts";
    static final int DATABASE_VERSION = 2;
```

注意：在运行这个示例之前，一定要注释掉前一小节中介绍的删除语句块。否则，因为数据库中的表将被删除，删除操作会失败。

当再一次运行应用程序时，在 Eclipse 的 LogCat 窗口中可以看到以下消息：

```
DBAdapter(8705): Upgrading database from version 1 to 2, which
will destroy all old data
```

在这一例子中，为简单起见，直接删除现有的表并创建一个新表。在实际中，通常需要备份现有的表，然后将其内容复制到新表中。

6.3.3 预创建数据库

在实际的应用中，有时在设计时预创建数据库要比在运行时创建更有效。例如，您可能想创建一个应用程序来记录所有自己去过的地方的坐标。此时，在设计时预创建数据库，然后在运行时使用这个数据库会简单得多。

要预创建一个 SQLite 数据库，可以使用 Internet 上很多可用的免费工具。其中一个这样的工具就是 SQLite Database Browser，其对于不同的平台都是免费可用的(http://sourceforge.net/projects/sqlitebrowser/)。

一旦安装了 SQLite Database Browser，就可以用可视化方式来创建一个数据库。图 6-18 展示了已经创建好一个指明了字段的 contacts 表。

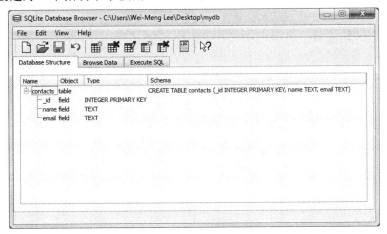

图 6-18

用行填充表也很简单。图 6-19 展示了如何使用 Browse Data 选项卡来为表填充数据。

图 6-19

在设计时创建了数据库后,下一步要做的就是将其与应用程序捆绑,这样就可以在应用程序中使用数据库了。下面的"试一试"展示了如何捆绑。

试一试　　捆绑一个数据库

(1) 使用先前创建的同一个项目,将在前一节中创建的 SQLite 数据库文件拖放到 Eclipse 中 Android 项目的 assets 文件夹下(如图 6-20 所示)。

图 6-20

 注意:添加到 assets 文件夹下的文件名必须采用小写字母格式。因此,像 MyDB 这样的文件名是无效的,而 mydb 是可以的。

(2) 在 DatabasesActivity.java 文件中添加下列粗体显示的语句:

```
package net.learn2develop.Databases;

import java.io.File;
import java.io.FileNotFoundException;
import java.io.FileOutputStream;
import java.io.IOException;
import java.io.InputStream;
import java.io.OutputStream;

import android.app.Activity;
import android.database.Cursor;
import android.os.Bundle;
import android.widget.Toast;

public class DatabasesActivity extends Activity {
    /** Called when the activity is first created. */
    @Override
    public void onCreate(Bundle savedInstanceState) {
        super.onCreate(savedInstanceState);
        setContentView(R.layout.main);

        DBAdapter db = new DBAdapter(this);
        try {
```

```
            String destPath = "/data/data/" + getPackageName() +
                "/databases";
            File f = new File(destPath);
            if (!f.exists()) {
                f.mkdirs();
                f.createNewFile();

                //---copy the db from the assets folder into
                // the databases folder---
                CopyDB(getBaseContext().getAssets().open("mydb"),
                    new FileOutputStream(destPath + "/MyDB"));
            }
        } catch (FileNotFoundException e) {
            e.printStackTrace();
        } catch (IOException e) {
            e.printStackTrace();
        }

        //---get all contacts---
        db.open();
        Cursor c = db.getAllContacts();
        if (c.moveToFirst())
        {
            do {
                DisplayContact(c);
            } while (c.moveToNext());
        }
        db.close();
    }

    public void CopyDB(InputStream inputStream,
    OutputStream outputStream) throws IOException {
        //---copy 1K bytes at a time---
        byte[] buffer = new byte[1024];
        int length;
        while ((length = inputStream.read(buffer)) > 0) {
            outputStream.write(buffer, 0, length);
        }
        inputStream.close();
        outputStream.close();
    }

    public void DisplayContact(Cursor c)
    {
        Toast.makeText(this,
            "id: " + c.getString(0) + "\n" +
            "Name: " + c.getString(1) + "\n" +
            "Email: " + c.getString(2),
            Toast.LENGTH_LONG).show();
    }
}
```

(3) 按 F11 键在 Android 模拟器上调试应用程序。当应用程序运行时，它将把 mydb 数据库文件以 MyDB 为名复制到/data/data/net.learn2develop.Databases/databases/文件夹下。

示例说明

首先将 CopyDB()方法定义为将数据库文件从一处复制到另一处：

```
public void CopyDB(InputStream inputStream,
OutputStream outputStream) throws IOException {
    //---copy 1K bytes at a time---
    byte[] buffer = new byte[1024];
    int length;
    while ((length = inputStream.read(buffer)) > 0) {
        outputStream.write(buffer, 0, length);
    }
    inputStream.close();
    outputStream.close();
}
```

注意，在这种情况下，使用 InputStream 对象来从源文件中读取数据，然后使用 OutputStream 对象将其写入到目标文件中。

当活动被创建时，将位于 assets 文件夹下的数据库文件复制到 Android 设备(或模拟器)上的/data/data/net.learn2develop.Databases/databases/文件夹下：

```
try {
    String destPath = "/data/data/" + getPackageName() +
        "/databases";
    File f = new File(destPath);
    if (!f.exists()) {
        f.mkdirs();
        f.createNewFile();

        //---copy the db from the assets folder into
        // the databases folder---
        CopyDB(getBaseContext().getAssets().open("mydb"),
            new FileOutputStream(destPath + "/MyDB"));
    }
} catch (FileNotFoundException e) {
    e.printStackTrace();
} catch (IOException e) {
    e.printStackTrace();
}
```

只有在目标文件夹下不存在数据库文件时才执行复制操作。如果不进行这一检查，每一次创建活动时，数据库文件将被 assets 文件夹下的文件所覆盖。这也许是您所不希望的，因为应用程序在运行时很可能会对数据库文件做修改，这将消除到目前为止您对数据库所做出的一切改变。

为了确保数据库文件的确被复制了,在测试应用程序之前要确保在模拟器中删除了数据库文件(如果还存在的话)。可以使用 DDMS 来删除数据库(如图 6-21 所示)。

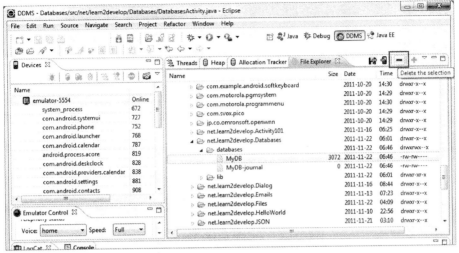

图 6-21

6.4 本章小结

在本章中,我们学习了将持久性数据保存在 Android 设备中的不同方法。对于简单的非结构化数据,使用 SharedPreferences 对象是一个理想的方案。如果需要存储批量数据,可以考虑使用传统的文件系统。最后,对于结构化数据,在关系数据库管理系统中进行存储会更有效率。为此,Android 提供了可以利用公开的 API 轻松访问的 SQLite 数据库。

注意,对于 SharedPreferences 对象和 SQLite 数据库来说,数据只能被创建它的应用程序访问。换句话说,它是不可共享的。如果需要在不同应用程序之间共享数据,则要创建一个内容提供者(content provider)。第 7 章将详细讨论内容提供者。

练 习

1. 如何使用活动显示应用程序的首选项?
2. 说出能够获取 Android 设备的外部存储器的路径的方法。
3. 当向外部存储器写入文件时需要声明什么权限?

练习答案参见附录 C。

本章主要内容

主 题	关 键 概 念
保存简单的用户数据	使用 SharedPreferences 对象
在同一应用程序的活动之间共享数据	使用 getSharedPreferences()方法
保存到文件	使用 FileOutputStream 和 OutputStreamReader 类

(续表)

主　　题	关　键　概　念
读文件	使用 FileInputStream 和 InputStreamReader 类
保存到外部存储器	使用 getExternalStorageDirectory()方法返回指向外部存储器的路径
访问位于 res/raw 文件夹下的文件	使用 Resources 对象(通过 getResources()方法获得)的 openRawResource()方法
创建数据库辅助类	扩展 SQLiteOpenHelper 类

第 7 章

内容提供者

本章将介绍以下内容:

- 内容提供者简介
- 如何在 Android 中使用内容提供者
- 如何创建和使用自己的内容提供者

在第 6 章中,我们学习了持久化数据的不同方法——使用共享首选项、文件系统以及 SQLite 数据库。虽然使用数据库方法来保存结构化的复杂数据是值得推荐的方式,但数据共享是一个挑战,因为数据库只能被创建它的包访问。

我们将在本章学习 Android 通过使用内容提供者来共享数据的方法。您将学习到如何使用内置的内容提供者以及通过实现自己的内容提供者来跨包共享数据。

7.1 在 Android 中共享数据

在 Android 中,推荐使用内容提供者来实现跨包的数据共享。可以将内容提供者视为一种数据存储。它存储数据的方式和使用它的应用程序无关,重要的是包如何以一致的编程接口来访问存储其中的数据。内容提供者的行为方式与数据库很像——您可以查询它,编辑以及增加或删除其内容。然而,与数据库不同的是,内容提供者可以使用不同的方式来存储数据。数据可以存储于数据库、文件中甚至网络上。

Android 附带了许多有用的内容提供者,包括:

- Browser——存储诸如浏览器书签、浏览器历史记录等数据
- CallLog——存储诸如未接电话、通话详细信息等数据
- Contacts——存储联系人详细信息
- MediaStore——存储媒体文件,如音频、视频和图像
- Settings——存储设备的设置和首选项

除了一些内置的内容提供者以外,还可以创建自己的内容提供者。

为了查询内容提供者，可为特定行使用一个可选的说明符，以 URI 形式指定查询字符串。查询 URI 的格式如下所示：

```
<standard_prefix>://<authority>/<data_path>/<id>
```

URI 的不同部分如下所示：

- 内容提供者的 standard prefix 始终是 content://。
- authority 指定了内容提供者的名称，如内置的 Contacts 内容提供者的名称为 contacts。对于第三方内容提供者，将采用完全限定的名称，如 com.wrox.provider 或者 net.learn2develop.provider。
- data path 指定了请求数据的类型。例如，如果您是从 Contacts 内容提供者获取所有联系人，那么 data path 就是 people，而 URI 为 content://contacts/people。
- id 指定了请求的特定记录。例如，如果您从 Contacts 内容提供者中查找 2 号联系人，那么 URI 为 content://contacts/people/2。

表 7-1 列出了一些查询字符串的示例。

表 7-1 查询字符串的示例

查询字符串	描　　述
content://media/internal/images	返回一个存储在设备上的所有内部图像的列表
content://media/external/images	返回一个存储在设备的外部存储器(如 SD 卡)上的所有图像的列表
content://call_log/calls	返回一个在 CallLog 中记录的所有通话的列表
content://browser/bookmarks	返回一个存储在浏览器中的书签列表

7.2 使用内容提供者

理解内容提供者的最佳方法就是实际地运用它。下面的"试一试"展示了在 Android 应用程序中如何使用内容提供者。

试一试　使用 Contacts 内容提供者

Provider.zip 代码文件可以在 Wrox.com 上下载

(1) 打开 Eclipse，创建一个新的 Android 项目，命名为 Provider。
(2) 在 main.xml 文件中添加下列粗体显示的语句：

```xml
<?xml version="1.0" encoding="utf-8"?>
<LinearLayout xmlns:android="http://schemas.android.com/apk/res/android"
    android:layout_width="fill_parent"
    android:layout_height="fill_parent"
    android:orientation="vertical" >
```

```xml
<ListView
    android:id="@+id/android:list"
    android:layout_width="fill_parent"
    android:layout_height="wrap_content"
    android:layout_weight="1"
    android:stackFromBottom="false"
    android:transcriptMode="normal" />

<TextView
    android:id="@+id/contactName"
    android:textStyle="bold"
    android:layout_width="wrap_content"
    android:layout_height="wrap_content" />

<TextView
    android:id="@+id/contactID"
    android:layout_width="fill_parent"
    android:layout_height="wrap_content" />

</LinearLayout>
```

(3) 在 ProviderActivity.java 类中编写如下代码：

```java
package net.learn2develop.Provider;

import android.app.ListActivity;
import android.content.CursorLoader;
import android.database.Cursor;
import android.net.Uri;
import android.os.Bundle;
import android.provider.ContactsContract;
import android.widget.CursorAdapter;
import android.widget.SimpleCursorAdapter;

public class ProviderActivity extends ListActivity {
    /** Called when the activity is first created. */
    @Override
    public void onCreate(Bundle savedInstanceState) {
        super.onCreate(savedInstanceState);
        setContentView(R.layout.main);

        Uri allContacts = Uri.parse("content://contacts/people");

        Cursor c;
        if (android.os.Build.VERSION.SDK_INT <11) {
            //---before Honeycomb---
            c = managedQuery(allContacts, null, null, null, null);
        } else {
            //---Honeycomb and later---
            CursorLoader cursorLoader = new CursorLoader(
```

```
                    this,
                    allContacts,
                    null,
                    null,
                    null ,
                    null);
            c = cursorLoader.loadInBackground();
        }

        String[] columns = new String[] {
            ContactsContract.Contacts.DISPLAY_NAME,
            ContactsContract.Contacts._ID};

        int[] views = new int[] {R.id.contactName, R.id.contactID};

        SimpleCursorAdapter adapter;

        if (android.os.Build.VERSION.SDK_INT <11) {
            //---before Honeycomb---
            adapter = new SimpleCursorAdapter(
                    this, R.layout.main, c, columns, views);
        } else {
            //---Honeycomb and later---
            adapter = new SimpleCursorAdapter(
                    this, R.layout.main, c, columns, views,
                    CursorAdapter.FLAG_REGISTER_CONTENT_OBSERVER);
        }

        this.setListAdapter(adapter);   }
}
```

(4) 在 AndroidManifest.xml 文件中添加下列粗体显示的语句：

```
<?xml version="1.0" encoding="utf-8"?>
<manifest xmlns:android="http://schemas.android.com/apk/res/android"
    package="net.learn2develop.Provider"
    android:versionCode="1"
    android:versionName="1.0" >

    <uses-sdk android:minSdkVersion="14" />
    <uses-permission android:name="android.permission.READ_CONTACTS"/>

    <application
        android:icon="@drawable/ic_launcher"
        android:label="@string/app_name" >
        <activity
            android:label="@string/app_name"
            android:name=".ProviderActivity" >
            <intent-filter >
                <action android:name="android.intent.action.MAIN" />

                <category android:name="android.intent.category.LAUNCHER"/>
            </intent-filter>
```

```
        </activity>
    </application>

</manifest>
```

(5) 启动一个 AVD 并在 Android 模拟器中创建一些联系人。要添加联系人，进入 Phone 应用程序，单击顶部的星型图标，如图 7-1 所示。单击模拟器的 MENU 按钮，然后单击 New contact 菜单项。模拟器将提醒备份联系人。单击 Keep Local 按钮，然后输入几个人的姓名、电话号码和电子邮件地址。

(6) 按 F11 键在 Android 模拟器上调试应用程序。图 7-2 展示了活动显示刚刚创建的联系人列表。

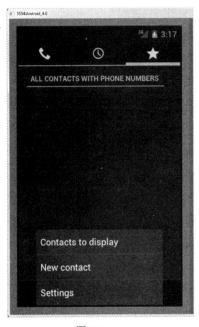

图 7-1   图 7-2

示例说明

在这一示例中，检索所有存储在 Contacts 应用程序中的联系人并在 ListView 中进行显示。首先，指定了访问 Contacts 应用程序的 URI：

```
Uri allContacts = Uri.parse("content://contacts/people");
```

然后，用一个条件检查来检测当前运行应用程序的设备的版本：

```
Cursor c;
if (android.os.Build.VERSION.SDK_INT <11) {
    //---before Honeycomb---
    c = managedQuery(allContacts, null, null, null, null);
} else {
    //---Honeycomb and later---
    CursorLoader cursorLoader = new CursorLoader(
```

```
        this,
        allContacts,
        null,
        null,
        null ,
        null);
    c = cursorLoader.loadInBackground();
}
```

如果应用程序运行在早于 Honeycomb 的设备上(android.os.Build.VERSION.SDK_INT 变量的值小于 11),那么可以使用 Activity 类的 managedQuery()方法来检索一个托管游标。托管游标处理在应用程序暂停时卸载其自身和在应用程序重启时重新查询自身的所有工作。语句:

```
Cursor c = managedQuery(allContacts, null, null, null, null);
```

等价于:

```
Cursor c=getContentResolver().query(allContacts,null,null,null,null);
//---allows the activity to manage the Cursor's
// lifecyle based on the activity's lifecycle---
startManagingCursor(c);
```

getContentResolver()方法返回一个 ContentResolver 对象,它可以使用合适的内容提供者来帮助解析内容 URI。

但是,从 Android API 级别 11 开始(Honeycomb 及更高版本),manageQuery()方法被弃用了(仍然可以使用,但是不建议这么做)。对于运行 Honeycomb 或更高版本的设备,应该使用 CursorLoader 类:

```
CursorLoader cursorLoader = new CursorLoader(
        this,
        allContacts,
        null,
        null,
        null ,
        null);
c = cursorLoader.loadInBackground();
```

CursorLoader 类(从 Android API 级别 11 才开始可用)在后台线程中执行游标查询,因此不会阻塞应用程序的用户界面。

SimpleCursorAdapter 对象将一个游标映射到在 XML 文件(main.xml)中定义的 TextView(或者 ImageView),并将数据(由 columns 表示)映射到视图(由 views 表示)上:

```
String[] columns = new String[] {
    ContactsContract.Contacts.DISPLAY_NAME,
    ContactsContract.Contacts._ID};

  int[] views = new int[] {R.id.contactName, R.id.contactID};
```

```
    SimpleCursorAdapter adapter;

    if (android.os.Build.VERSION.SDK_INT <11) {
        //---before Honeycomb---
        adapter = new SimpleCursorAdapter(
                this, R.layout.main, c, columns, views);
    } else {
        //---Honeycomb and later---
        adapter = new SimpleCursorAdapter(
                this, R.layout.main, c, columns, views,
                CursorAdapter.FLAG_REGISTER_CONTENT_OBSERVER);
    }

    this.setListAdapter(adapter);
```

与 managedQuery()方法类似，SimpleCursorAdapter 类的一个构造函数已被弃用。对于 Honeycomb 及更高版本的设备，需要使用 SimpleCursorAdapter 类的一个新构造函数，它带有一个额外的参数：

```
    //---Honeycomb and later---
    adapter = new SimpleCursorAdapter(
            this, R.layout.main, c, columns, views,
            CursorAdapter.FLAG_REGISTER_CONTENT_OBSERVER);
```

这个标志注册了当内容提供者中发生更改时所通知的适配器。

注意，为了使应用程序可以访问 Contacts 程序，需要在 AndroidManifest.xml 文件中有 READ_CONTACTS 权限。

7.2.1 预定义查询字符串常量

除了使用查询 URI，还可以利用 Android 中的一个预定义查询字符串常量的列表来为不同数据类型指定 URI。例如，除了使用查询 content://contacts/people，还可以使用 Android 中的一个预定义常量将以下语句：

```
    Uri allContacts = Uri.parse("content://contacts/people");
```

重写为：

```
    Uri allContacts = ContactsContract.Contacts.CONTENT_URI;
```

注意：对于 Android 2.0 或更高版本，需要使用 ContactsContract.Contacts. CONTENT_URI 这个 URI 来查询基本的 Contacts 记录。

以下是一些预定义查询字符串常量的示例：
- Browser.BOOKMARKS_URI

- Browser.SEARCHES_URI
- CallLog.CONTENT_URI
- MediaStore.Images.Media.INTERNAL_CONTENT_URI
- MediaStore.Images.Media.EXTERNAL_CONTENT_URI
- Settings.CONTENT_URI

如果想要检索第一个联系人，则可按如下方式指定此联系人的 ID：

```
Uri allContacts = Uri.parse("content://contacts/people/1");
```

或者，可以结合 ContentUris 类的 withAppendedId()方法来使用预定义常量：

```
import android.content.ContentUris;
...
        Uri allContacts = ContentUris.withAppendedId(
            ContactsContract.Contacts.CONTENT_URI, 1);
```

除了绑定到 ListView，还可以使用 Cursor 对象将结果输出来，如下所示：

```
package net.learn2develop.Provider;

import android.app.ListActivity;
import android.content.CursorLoader;
import android.database.Cursor;
import android.net.Uri;
import android.os.Bundle;
import android.provider.ContactsContract;
import android.widget.CursorAdapter;
import android.widget.SimpleCursorAdapter;
import android.util.Log;
public class ProviderActivity extends ListActivity {
    /** Called when the activity is first created. */
    @Override
    public void onCreate(Bundle savedInstanceState) {
        super.onCreate(savedInstanceState);
        setContentView(R.layout.main);

        Uri allContacts = ContactsContract.Contacts.CONTENT_URI;
        ...
        ...
        if (android.os.Build.VERSION.SDK_INT <11) {
            //---before Honeycomb---
            adapter = new SimpleCursorAdapter(
                    this, R.layout.main, c, columns, views);
        } else {
            //---Honeycomb and later---
            adapter = new SimpleCursorAdapter(
                    this, R.layout.main, c, columns, views,
                    CursorAdapter.FLAG_REGISTER_CONTENT_OBSERVER);
        }
```

```
        this.setListAdapter(adapter);
        PrintContacts(c);
    }

    private void PrintContacts(Cursor c)
    {
        if (c.moveToFirst()) {
            do{
            String contactID = c.getString(c.getColumnIndex(
                    ContactsContract.Contacts._ID));
            String contactDisplayName =
                    c.getString(c.getColumnIndex(
                        ContactsContract.Contacts.DISPLAY_NAME));
                Log.v("Content Providers", contactID + ", " +
                    contactDisplayName);
            } while (c.moveToNext());
        }
    }
}
```

 注意：如果对如何查看 LogCat 窗口不太熟悉，可参阅附录 A 快速了解一下 Eclipse IDE。

PrintContacts()方法将在 LogCat 窗口中输出如下内容：

```
12-13 08:32:50.471: V/Content Providers(12346): 1, Wei-Meng Lee
12-13 08:32:50.471: V/Content Providers(12346): 2, Linda Chen
12-13 08:32:50.471: V/Content Providers(12346): 3, Joanna Yip
```

它输出了存储在 Contacts 应用程序中的每一个联系人的 ID 和姓名。这时，通过访问 ContactsContract.Contacts._ID 字段来获取联系人的 ID，访问 ContactsContract.Contacts.DISPLAY_NAME 来得到联系人的姓名。如果想显示联系人的电话号码，由于这个信息存储于另外一张表中，因此需要再次查询内容提供者：

```
    private void PrintContacts(Cursor c)
    {
        if (c.moveToFirst()) {
            do{
                String contactID = c.getString(c.getColumnIndex(
                        ContactsContract.Contacts._ID));
                String contactDisplayName =
                    c.getString(c.getColumnIndex(
                            ContactsContract.Contacts.DISPLAY
                                _NAME));
                Log.v("Content Providers", contactID + ", " +
                    contactDisplayName);
```

```
            //---get phone number---
            int hasPhone =
                c.getInt(c.getColumnIndex(
                    ContactsContract.Contacts.HAS_PHONE_
                    NUMBER));
        if (hasPhone == 1) {
            Cursor phoneCursor =
                getContentResolver().query(
                    ContactsContract.CommonDataKinds.Phone.
                     CONTENT_URI, null,
                    ContactsContract.CommonDataKinds.Phone.CONTACT
                     _ID + " = " +
                    contactID, null, null);
            while (phoneCursor.moveToNext()) {
                Log.v("Content Providers",
                    phoneCursor.getString(
                        phoneCursor.getColumnIndex(
                            ContactsContract.CommonDataKinds.
                            Phone.NUMBER)));
            }
            phoneCursor.close();
        }

    } while (c.moveToNext());
  }
}
```

注意：为了访问一个联系人的电话号码,需要对存储于 ContactsContract.CommonDataKinds.Phone.CONTENT_URI 中的 URI 进行查询。

在上面的代码片段中,首先使用 ContactsContract.Contacts.HAS_PHONE_NUMBER 字段检查一个联系人是否有电话号码。如果其至少有一个电话号码,就可以基于此联系人的 ID 对内容提供者再次查询。一旦检索到这一(些)号码,就可以迭代遍历并把它们输出来。应有如下显示内容：

```
12-13 08:59:31.881: V/Content Providers(13351): 1, Wei-Meng Lee
12-13 08:59:32.311: V/Content Providers(13351): +651234567
12-13 08:59:32.321: V/Content Providers(13351): 2, Linda Chen
12-13 08:59:32.511: V/Content Providers(13351): +1 876-543-21
12-13 08:59:32.545: V/Content Providers(13351): 3, Joanna Yip
12-13 08:59:32.641: V/Content Providers(13351): +239 846 5522
```

7.2.2 投影

managedQuery()方法的第 2 个参数(CursorLoader 类的第 3 个参数)控制查询返回的列数。这个参数被称为投影(projection),之前将其指定为 null：

```
Cursor c;
if (android.os.Build.VERSION.SDK_INT <11) {
    //---before Honeycomb---
    c = managedQuery(allContacts, null, null, null, null);
} else {
    //---Honeycomb and later---
    CursorLoader cursorLoader = new CursorLoader(
            this,
            allContacts,
            null,
            null,
            null ,
            null);
    c = cursorLoader.loadInBackground();
}
```

可以通过创建一个包含需要返回的列名的数组来指定要返回的确切列，如下所示：

```
String[] projection = new String[]
        {ContactsContract.Contacts._ID,
         ContactsContract.Contacts.DISPLAY_NAME,
         ContactsContract.Contacts.HAS_PHONE_NUMBER};

Cursor c;
if (android.os.Build.VERSION.SDK_INT <11) {
    //---before Honeycomb---
    c = managedQuery(allContacts, projection, null, null, null);
} else {
    //---Honeycomb and later---
    CursorLoader cursorLoader = new CursorLoader(
            this,
            allContacts,
            projection,
            null,
            null ,
            null);
    c = cursorLoader.loadInBackground();
}
```

在上述情况下，_ID、DISPLAY_NAME 以及 HAS_PHONE_NUMBER 字段都将被检索。

7.2.3 筛选

managedQuery()方法的第 3 个和第 4 个参数(CursorLoader 类的第 4 个和第 5 个参数)可以用来指定一个 SQL 的 WHERE 子句来对查询结果进行筛选。例如，下列语句只检索名字以 Lee 结尾的人：

```
Cursor c;
if (android.os.Build.VERSION.SDK_INT <11) {
```

```
            //---before Honeycomb---
            c = managedQuery(allContacts, projection,
                    ContactsContract.Contacts.DISPLAY_NAME + " LIKE '%Lee'",
                    null, null);
        } else {
            //---Honeycomb and later---
            CursorLoader cursorLoader = new CursorLoader(
                    this,
                    allContacts,
                    projection,
                    ContactsContract.Contacts.DISPLAY_NAME + " LIKE '%Lee'",
                    null ,
                    null);
            c = cursorLoader.loadInBackground();
        }
```

这里，managedQuery()方法的第 3 个参数(CursorLoader 构造函数的第 4 个参数)包含了一个 SQL 语句，其中含有要搜索的名字(Lee)。还可以将搜索字符串放在方法/构造函数的下一个参数中，如下所示：

```
    Cursor c;
    if (android.os.Build.VERSION.SDK_INT <11) {
        //---before Honeycomb---
        c = managedQuery(allContacts, projection,
                ContactsContract.Contacts.DISPLAY_NAME + " LIKE ?",
                new String[] {"%Lee"}, null);
    } else {
        //---Honeycomb and later---
        CursorLoader cursorLoader = new CursorLoader(
                this,
                allContacts,
                projection,
                ContactsContract.Contacts.DISPLAY_NAME + " LIKE ?",
                new String[] {"%Lee"},
                null);
        c = cursorLoader.loadInBackground();
    }
```

7.2.4 排序

managedQuery()方法的最后一个参数以及 Cursorloader 类的构造函数可以用来指定一个 SQL 的 ORDER BY 子句来对查询结果排序。例如，下列语句将联系人名字按升序进行排序：

```
    Cursor c;
    if (android.os.Build.VERSION.SDK_INT <11) {
        //---before Honeycomb---
        c = managedQuery(allContacts, projection,
                ContactsContract.Contacts.DISPLAY_NAME + " LIKE ?",
                new String[] {"%Lee"},
                ContactsContract.Contacts.DISPLAY_NAME + " ASC");
```

```
    } else {
        //---Honeycomb and later---
        CursorLoader cursorLoader = new CursorLoader(
            this,
            allContacts,
            projection,
            ContactsContract.Contacts.DISPLAY_NAME + " LIKE ?",
            new String[] {"%Lee"},
            ContactsContract.Contacts.DISPLAY_NAME + " ASC");
        c = cursorLoader.loadInBackground();
    }
```

7.3 创建自己的内容提供者

在 Android 中创建自己的内容提供者非常简单。所有您需要做的是扩展抽象类 ContentProvider，并重写其中定义的各种方法。

本节将学习如何创建一个简单的内容提供者来存储一个图书列表。为了便于说明，内容提供者将图书存储在一个包含 3 个字段的数据库表中，如图 7-3 所示。

下面的"试一试"展示了具体的操作步骤。

图 7-3

试一试 创建自己的内容提供者

ContentProviders.zip 代码文件可以在 Wrox.com 上下载

(1) 打开 Eclipse，创建一个新的 Android 项目并命名为 ContentProviders。
(2) 在项目的 src 文件夹下增加一个新的 Java 类文件，并命名为 BooksProvider.java。
(3) 按如下所示填充 BooksProvider.java 文件：

```
package net.learn2develop.ContentProviders;

import android.content.ContentProvider;
import android.content.ContentUris;
import android.content.ContentValues;
import android.content.Context;
import android.content.UriMatcher;
import android.database.Cursor;
import android.database.SQLException;
import android.database.sqlite.SQLiteDatabase;
import android.database.sqlite.SQLiteOpenHelper;
import android.database.sqlite.SQLiteQueryBuilder;
import android.net.Uri;
import android.text.TextUtils;
import android.util.Log;
```

```java
public class BooksProvider extends ContentProvider {
    static final String PROVIDER_NAME =
        "net.learn2develop.provider.Books";

    static final Uri CONTENT_URI =
        Uri.parse("content://"+ PROVIDER_NAME + "/books");

    static final String _ID = "_id";
    static final String TITLE = "title";
    static final String ISBN = "isbn";

    static final int BOOKS = 1;
    static final int BOOK_ID = 2;

    private static final UriMatcher uriMatcher;
    static{
        uriMatcher = new UriMatcher(UriMatcher.NO_MATCH);
        uriMatcher.addURI(PROVIDER_NAME, "books", BOOKS);
        uriMatcher.addURI(PROVIDER_NAME, "books/#", BOOK_ID);
    }

    //---for database use---
    SQLiteDatabase booksDB;
    static final String DATABASE_NAME = "Books";
    static final String DATABASE_TABLE = "titles";
    static final int DATABASE_VERSION = 1;
    static final String DATABASE_CREATE =
        "create table " + DATABASE_TABLE +
        " (_id integer primary key autoincrement, "
        + "title text not null, isbn text not null);";

    private static class DatabaseHelper extends SQLiteOpenHelper
    {
        DatabaseHelper(Context context) {
            super(context, DATABASE_NAME, null, DATABASE_VERSION);
        }

        @Override
        public void onCreate(SQLiteDatabase db)
        {
            db.execSQL(DATABASE_CREATE);
        }

        @Override
        public void onUpgrade(SQLiteDatabase db, int oldVersion,
                int newVersion) {
            Log.w("Content provider database",
                    "Upgrading database from version " +
                            oldVersion + " to " + newVersion +
                    ", which will destroy all old data");
```

```java
            db.execSQL("DROP TABLE IF EXISTS titles");
            onCreate(db);
        }
    }

    @Override
    public int delete(Uri arg0, String arg1, String[] arg2) {
        // arg0 = uri
        // arg1 = selection
        // arg2 = selectionArgs
        int count=0;
        switch (uriMatcher.match(arg0)){
        case BOOKS:
            count = booksDB.delete(
                    DATABASE_TABLE,
                    arg1,
                    arg2);
            break;
        case BOOK_ID:
            String id = arg0.getPathSegments().get(1);
            count = booksDB.delete(
                    DATABASE_TABLE,
                    _ID + " = " + id +
                    (!TextUtils.isEmpty(arg1) ? " AND (" +
                            arg1 + ')' : ""),
                            arg2);
            break;
        default: throw new IllegalArgumentException("Unknown URI " + arg0);
        }
        getContext().getContentResolver().notifyChange(arg0, null);
        return count;
    }

    @Override
    public String getType(Uri uri) {
        switch (uriMatcher.match(uri)){
        //---get all books---
        case BOOKS:
          return "vnd.android.cursor.dir/vnd.learn2develop.books ";

        //---get a particular book---
        case BOOK_ID:
            return "vnd.android.cursor.item/vnd.learn2develop.books ";

        default:
            throw new IllegalArgumentException("Unsupported URI: " + uri);
        }
    }

    @Override
    public Uri insert(Uri uri, ContentValues values) {
```

```java
            //---add a new book---
            long rowID = booksDB.insert(
                    DATABASE_TABLE,
                    "",
                    values);

            //---if added successfully---
            if (rowID>0)
            {
                Uri _uri = ContentUris.withAppendedId(CONTENT_URI, rowID);
                getContext().getContentResolver().notifyChange(_uri, null);
                returnuri;
            }
            throw new SQLException("Failed to insert row into " + uri);
    }

    @Override
    public boolean onCreate() {
        Context context = getContext();
        DatabaseHelper dbHelper = new DatabaseHelper(context);
        booksDB = dbHelper.getWritableDatabase();
        return (booksDB == null)? false:true;
    }

    @Override
    public Cursor query(Uri uri, String[] projection, String selection,
            String[] selectionArgs, String sortOrder) {
        SQLiteQueryBuilder sqlBuilder = new SQLiteQueryBuilder();
        sqlBuilder.setTables(DATABASE_TABLE);

        if (uriMatcher.match(uri) == BOOK_ID)
            //---if getting a particular book---
            sqlBuilder.appendWhere(
                    _ID + " = " + uri.getPathSegments().get(1));

        if (sortOrder==null || sortOrder=="")
            sortOrder = TITLE;

        Cursor c = sqlBuilder.query(
                booksDB,
                projection,
                selection,
                selectionArgs,
                null ,
                null,
                sortOrder);

        //---register to watch a content URI for changes---
        c.setNotificationUri(getContext().getContentResolver(), uri);
        return c;
```

```java
    }

    @Override
    public int update(Uri uri, ContentValues values, String selection,
            String[] selectionArgs) {
        int count = 0;
        switch (uriMatcher.match(uri)){
        case BOOKS:
            count = booksDB.update(
                    DATABASE_TABLE,
                    values,
                    selection,
                    selectionArgs);
            break;
        case BOOK_ID:
            count = booksDB.update(
                    DATABASE_TABLE,
                    values,
                    _ID + " = " + uri.getPathSegments().get(1) +
                    (!TextUtils.isEmpty(selection) ? " AND (" +
                            selection + ')' : ""),
                    selectionArgs);
            break;
        default: throw new IllegalArgumentException("Unknown URI " + uri);
        }
        getContext().getContentResolver().notifyChange(uri, null);
        return count;
    }
}
```

(4) 在 AndroidManifest.xml 文件中添加下列粗体显示的语句：

```xml
<?xml version="1.0" encoding="utf-8"?>
<manifest xmlns:android="http://schemas.android.com/apk/res/android"
    package="net.learn2develop.ContentProviders"
    android:versionCode="1"
    android:versionName="1.0" >

    <uses-sdk android:minSdkVersion="14" />

    <application
        android:icon="@drawable/ic_launcher"
        android:label="@string/app_name" >
        <activity
            android:label="@string/app_name"
            android:name=".ContentProvidersActivity" >
            <intent-filter >
                <action android:name="android.intent.action.MAIN" />

                <category android:name="android.intent.category.LAUNCHER" />
            </intent-filter>
```

```xml
    </activity>
    <provider android:name="BooksProvider"
        android:authorities="net.learn2develop.provider.Books">
    </provider>
</application>
</manifest>
```

示例说明

在本例中,首先创建一个名为 BooksProvider 的类,它扩展了 ContentProvider 基类。这个类中重写的各个方法如下所示:

- getType()——返回给定 URI 上的数据的 MIME 类型
- onCreate()——当启动提供者时调用
- query()——接收客户端请求,结果以 Cursor 对象形式返回
- insert()——向内容提供者中插入一条新记录
- delete()——从内容提供者中删除一条现有记录
- update()——在内容提供者中更新一条现有记录

在内容提供者中,可以自由选择如何存储数据——传统的文件系统、XML、数据库或者通过 Web 服务。本例中,使用前一章讨论的 SQLite 数据库方法。

接着在 BooksProvider 类中定义以下常量:

```java
static final String PROVIDER_NAME =
    "net.learn2develop.provider.Books";

static final Uri CONTENT_URI =
    Uri.parse("content://"+ PROVIDER_NAME + "/books");

static final String _ID = "_id";
static final String TITLE = "title";
static final String ISBN = "isbn";

static final int BOOKS = 1;
static final int BOOK_ID = 2;

private static final UriMatcher uriMatcher;
static{
    uriMatcher = new UriMatcher(UriMatcher.NO_MATCH);
    uriMatcher.addURI(PROVIDER_NAME, "books", BOOKS);
    uriMatcher.addURI(PROVIDER_NAME, "books/#", BOOK_ID);
}

//---for database use---
SQLiteDatabase booksDB;
static final String DATABASE_NAME = "Books";
static final String DATABASE_TABLE = "titles";
static final int DATABASE_VERSION = 1;
```

```
static final String DATABASE_CREATE = 
    "create table " + DATABASE_TABLE + 
    " (_id integer primary key autoincrement, "
    + "title text not null, isbn text not null);";
```

在上面的代码中可看到,使用一个 UriMatcher 对象来解析通过一个 ContentResolver 传递给内容提供者的内容 URI。例如,下面的内容 URI 代表了一个对内容提供者中的所有图书的请求:

```
content://net.learn2develop.provider.Books/books
```

下面的内容 URI 代表了一个对 _id 为 5 的特定图书的请求:

```
content://net.learn2develop.provider.Books/books/5
```

内容提供者使用 SQLite 数据库存储图书。注意,使用 SQLiteOpenHelper 辅助类来协助管理数据库:

```
private static class DatabaseHelper extends SQLiteOpenHelper
{
    DatabaseHelper(Context context) {
        super(context, DATABASE_NAME, null, DATABASE_VERSION);
    }

    @Override
    public void onCreate(SQLiteDatabase db)
    {
        db.execSQL(DATABASE_CREATE);
    }

    @Override
    public void onUpgrade(SQLiteDatabase db, int oldVersion,
            int newVersion) {
        Log.w("Content provider database",
                "Upgrading database from version " +
                    oldVersion + " to " + newVersion +
                ", which will destroy all old data");
        db.execSQL("DROP TABLE IF EXISTS titles");
        onCreate(db);
    }
}
```

接下来,通过重写 getType() 方法来唯一地描述内容提供者的数据类型。使用 UriMatcher 对象,对于单本图书返回 vnd.android.cursor.item/vnd.learn2develop.books,对于多本图书返回 vnd.android.cursor.dir/vnd.learn2develop.books:

```
@Override
public String getType(Uri uri) {
    switch (uriMatcher.match(uri)){
        //---get all books---
```

```java
        case BOOKS:
            return "vnd.android.cursor.dir/vnd.learn2develop.books ";

        //---get a particular book---
        case BOOK_ID:
            return "vnd.android.cursor.item/vnd.learn2develop.books ";

        default:
            throw new IllegalArgumentException("Unsupported URI: " + uri);
    }
}
```

下一步,重写onCreate()方法以在内容提供者启动时打开一个到数据库的连接:

```java
@Override
public boolean onCreate() {
    Context context = getContext();
    DatabaseHelper dbHelper = new DatabaseHelper(context);
    booksDB = dbHelper.getWritableDatabase();
    return (booksDB == null)? false:true;
}
```

重写query()方法,以允许客户端查询图书:

```java
@Override
public Cursor query(Uri uri, String[] projection, String selection,
        String[] selectionArgs, String sortOrder) {
    SQLiteQueryBuilder sqlBuilder = new SQLiteQueryBuilder();
    sqlBuilder.setTables(DATABASE_TABLE);

    if (uriMatcher.match(uri) == BOOK_ID)
        //---if getting a particular book---
        sqlBuilder.appendWhere(
            _ID + " = " + uri.getPathSegments().get(1));

    if (sortOrder==null || sortOrder=="")
        sortOrder = TITLE;

    Cursor c = sqlBuilder.query(
        booksDB,
        projection,
        selection,
        selectionArgs,
        null,
        null,
        sortOrder);

    //---register to watch a content URI for changes---
    c.setNotificationUri(getContext().getContentResolver(), uri);
    return c;
}
```

默认情况下，查询的结果按 title 字段进行排序。查询结果以 Cursor 对象形式返回。
为了在内容提供者中插入一本新图书，需要重写 insert()方法：

```
@Override
public Uri insert(Uri uri, ContentValues values) {
    //---add a new book---
    long rowID = booksDB.insert(
            DATABASE_TABLE,
            "",
            values);

    //---if added successfully---
    if (rowID>0)
    {
        Uri _uri = ContentUris.withAppendedId(CONTENT_URI, rowID);
        getContext().getContentResolver().notifyChange(_uri, null);
        return _uri;
    }
    throw new SQLException("Failed to insert row into " + uri);
}
```

一旦记录被成功插入，则调用 ContentResolver 的 notifyChange()方法。这将通知已注册的观察者更新了一行。

若要删除一本图书，则重写 delete()方法如下：

```
@Override
public int delete(Uri arg0, String arg1, String[] arg2) {
    // arg0 = uri
    // arg1 = selection
    // arg2 = selectionArgs
    int count=0;
    switch (uriMatcher.match(arg0)){
    case BOOKS:
        count = booksDB.delete(
                DATABASE_TABLE,
                arg1,
                arg2);
        break;
    case BOOK_ID:
        String id = arg0.getPathSegments().get(1);
        count = booksDB.delete(
                DATABASE_TABLE,
                _ID + " = " + id +
                (!TextUtils.isEmpty(arg1) ? " AND (" +
                        arg1 + ')' : ""),
                arg2);
        break;
    default: throw new IllegalArgumentException("Unknown URI " + arg0);
    }
```

```
            getContext().getContentResolver().notifyChange(arg0, null);
            return count;
        }
```

同样，在删除操作后要调用 ContentResolver 的 notifyChange()方法。这将通知已注册的观察者删除了一行。

最后，若要更新一本图书，则重写 update()方法如下：

```
        @Override
        public int update(Uri uri, ContentValues values, String selection,
                String[] selectionArgs) {
            int count = 0;
            switch (uriMatcher.match(uri)){
            case BOOKS:
                count = booksDB.update(
                        DATABASE_TABLE,
                        values,
                        selection,
                        selectionArgs);
                break;
            case BOOK_ID:
                count = booksDB.update(
                        DATABASE_TABLE,
                        values,
                        _ID + " = " + uri.getPathSegments().get(1) +
                        (!TextUtils.isEmpty(selection) ? " AND (" +
                                selection + ')' : ""),
                                selectionArgs);
                break;
            default: throw new IllegalArgumentException("Unknown URI " + uri);
            }
            getContext().getContentResolver().notifyChange(uri, null);
            return count;
        }
```

与 insert()和 delete()方法一样，在更新后调用 ContentResolver 的 notifyChange()方法。这将通知已注册的观察者更新了一行。

最后，为了将内容提供者注册到 Android，可以修改 AndroidManifest.xml 文件，添加 <provider>元素。

7.4 使用内容提供者

既然已经构建了自己的新的内容提供者，那就可以在 Android 应用程序中测试它。下面的"试一试"展示了是如何做到这一点的。

试一试　　使用新建的内容提供者

(1) 使用在前一节中所创建的同一个项目，在 main.xml 文件中添加下列粗体显示的语句：

```xml
<?xml version="1.0" encoding="utf-8"?>
<LinearLayout xmlns:android="http://schemas.android.com/apk/res/android"
    android:layout_width="fill_parent"
    android:layout_height="fill_parent"
    android:orientation="vertical" >

<TextView
    android:layout_width="fill_parent"
    android:layout_height="wrap_content"
    android:text="ISBN" />

<EditText
    android:id="@+id/txtISBN"
    android:layout_height="wrap_content"
    android:layout_width="fill_parent" />

<TextView
    android:layout_width="fill_parent"
    android:layout_height="wrap_content"
    android:text="Title" />

<EditText
    android:id="@+id/txtTitle"
    android:layout_height="wrap_content"
    android:layout_width="fill_parent" />

<Button
    android:text="Add title"
    android:id="@+id/btnAdd"
    android:layout_width="fill_parent"
    android:layout_height="wrap_content"
    android:onClick="onClickAddTitle" />

<Button
    android:text="Retrieve titles"
    android:id="@+id/btnRetrieve"
    android:layout_width="fill_parent"
    android:layout_height="wrap_content"
    android:onClick="onClickRetrieveTitles" />

</LinearLayout>
```

(2) 在 ContentProvidersActivity.java 文件中添加下列粗体显示的语句：

```java
package net.learn2develop.ContentProviders;

import android.app.Activity;
import android.content.ContentValues;
import android.content.CursorLoader;
import android.database.Cursor;
import android.net.Uri;
```

```java
import android.os.Bundle;
import android.view.View;
import android.widget.EditText;
import android.widget.Toast;

public class ContentProvidersActivity extends Activity {
    /** Called when the activity is first created. */
    @Override
    public void onCreate(Bundle savedInstanceState) {
        super.onCreate(savedInstanceState);
        setContentView(R.layout.main);
    }

    public void onClickAddTitle(View view) {
        //---add a book---
        ContentValues values = new ContentValues();
        values.put(BooksProvider.TITLE, ((EditText)
                findViewById(R.id.txtTitle)).getText().toString());
        values.put(BooksProvider.ISBN, ((EditText)
                findViewById(R.id.txtISBN)).getText().toString());
        Uri uri = getContentResolver().insert(
                BooksProvider.CONTENT_URI, values);
        Toast.makeText(getBaseContext(),uri.toString(),
                Toast.LENGTH_LONG).show();
    }

    public void onClickRetrieveTitles(View view) {
        //---retrieve the titles---
        Uri allTitles = Uri.parse(
                "content://net.learn2develop.provider.Books/books");
        Cursor c;
        if (android.os.Build.VERSION.SDK_INT <11) {
            //---before Honeycomb---
            c = managedQuery(allTitles, null, null, null,
                    "title desc");
        } else {
            //---Honeycomb and later---
            CursorLoader cursorLoader = new CursorLoader(
                    this,
                    allTitles, null, null, null,
                    "title desc");
            c = cursorLoader.loadInBackground();
        }
        if (c.moveToFirst()) {
            do{
                Toast.makeText(this,
                    c.getString(c.getColumnIndex(
                        BooksProvider._ID)) + ", " +
                    c.getString(c.getColumnIndex(
                        BooksProvider.TITLE)) + ", " +
```

```
            c.getString(c.getColumnIndex(
                BooksProvider.ISBN)),
            Toast.LENGTH_SHORT).show();
        } while (c.moveToNext());
    }
  }
}
```

(3) 按 F11 键在 Android 模拟器上调试应用程序。

(4) 输入一本图书的 ISBN 和书名并单击 Add title 按钮。图 7-4 展示了由 Toast 类显示添加到内容提供者的图书的 URI。为了检索存储在内容提供者中的所有书名，单击 Retrieve titles 按钮并观察使用 Toast 类显示的值。

示例说明

首先，修改活动，这样用户就可以输入一本图书的 ISBN 和书名来添加到刚刚创建的内容提供者中。

要添加一本图书到内容提供者中，可以创建一个新的 ContentValues 对象，然后用与此图书有关的各种信息来填充这一对象：

图 7-4

```
//---add a book---
ContentValues values = new ContentValues();
values.put(BooksProvider.TITLE, ((EditText)
        findViewById(R.id.txtTitle)).getText().toString());
values.put(BooksProvider.ISBN, ((EditText)
        findViewById(R.id.txtISBN)).getText().toString());
Uri uri = getContentResolver().insert(
        BooksProvider.CONTENT_URI, values);
```

注意，因为内容提供者是在同一个包中，所以可以分别使用 BooksProvider.TITLE 和 BooksProvider.ISBN 常量来表示 title 和 isbn 字段。如果是从另一个包来访问内容提供者的，那么将不能够使用这些常量。在那种情况下，需要直接指定字段名称，如下所示：

```
ContentValues values = new ContentValues();
values.put("title", ((EditText)
    findViewById(R.id.txtTitle)).getText().toString());
values.put("isbn", ((EditText)
    findViewById(R.id.txtISBN)).getText().toString());
Uri uri = getContentResolver().insert(
        Uri.parse(
            "content://net.learn2develop.provider.Books/books"),
            values);
```

另外还要注意，对于外部包，需要使用完全限定的名称来引用内容 URI：

```
    Uri.parse(
        "content://net.learn2develop.provider.Books/books"),
```

要检索内容提供者中的所有书名,可使用下面的代码片段:

```
//---retrieve the titles---
Uri allTitles = Uri.parse(
    "content://net.learn2develop.provider.Books/books");
Cursor c;
if (android.os.Build.VERSION.SDK_INT <11) {
    //---before Honeycomb---
    c = managedQuery(allTitles, null, null, null,
        "title desc");
} else {
    //---Honeycomb and later---
    CursorLoader cursorLoader = new CursorLoader(
        this,
        allTitles, null, null, null,
        "title desc");
    c = cursorLoader.loadInBackground();
}
if (c.moveToFirst()) {
    do{
        Toast.makeText(this,
            c.getString(c.getColumnIndex(
                BooksProvider._ID)) + ", " +
            c.getString(c.getColumnIndex(
                BooksProvider.TITLE)) + ", " +
            c.getString(c.getColumnIndex(
                BooksProvider.ISBN)),
            Toast.LENGTH_SHORT).show();
    } while (c.moveToNext());
}
```

前面的查询将返回按 title 字段降序排列的结果。

如果想更新一本图书的详细信息,调用 update() 方法,使用内容 URI 来指示图书的 ID:

```
ContentValues editedValues = new ContentValues();
editedValues.put(BooksProvider.TITLE, "Android Tips and Tricks");
getContentResolver().update(
    Uri.parse(
        "content://net.learn2develop.provider.Books/books/2"),
    editedValues,
    null,
    null);
```

要删除一本图书,调用 delete() 方法,使用内容 URI 来指示图书的 ID:

```
//---delete a title---
getContentResolver().delete(
    Uri.parse("content://net.learn2develop.provider.
```

```
            Books/books/2"),
        null, null);
```

要删除所有图书，只要在内容 URI 中省略图书的 ID：
```
//---delete all titles---
getContentResolver().delete(
        Uri.parse("content://net.learn2develop.provider.
        Books/books"),
    null, null);
```

7.5 本章小结

在这一章中，我们了解了什么是内容提供者，以及如何使用一些内置在 Android 中的内容提供者。特别是，我们学习了如何使用 Contacts 内容提供者。Google 做出的提供内容提供者的决定使得应用程序可以通过一套标准的编程接口进行数据共享。除了内置的内容提供者以外，还可以由自己创建自定义的内容提供者来与其他包实现数据共享。

练 习

1. 写一个查询，实现从 Contacts 应用程序中检索所有包含单词 jack 的联系人。
2. 说出在实现自己的内容提供者时必须重写的方法。
3. 如何在 AndroidManifest.xml 文件中注册一个内容提供者？

练习答案参见附录 C。

本章主要内容

主　题	关　键　概　念
检索一个托管游标	使用 managedQuery()方法(对于早于 Honeycomb 的设备)或使用 CursorLoader 类(对于 Honeycomb 或之后的设备)
为内容提供者指定查询的两种方法	使用查询 URI 或者预定义查询字符串常量
在一个内容提供者中检索一列的值	使用 getColumnIndex()方法
访问联系人姓名所用的查询 URI	ContactsContract.Contacts.CONTENT_URI
访问联系人电话号码所用的查询 URI	ContactsContract.CommonDataKinds.Phone.CONTENT_URI
创建自己的内容提供者	创建一个扩展 ContentProvider 类的类

第 8 章

消息传递

本章将介绍以下内容:
- 如何以编程方式通过应用程序发送 SMS 消息
- 如何使用内置的 Messaging 应用程序发送 SMS 消息
- 如何接收传入的 SMS 消息
- 如何通过应用程序发送电子邮件消息

一旦启动并运行基本的 Android 应用程序,下一个有趣的事情就是为其添加与外界通信的能力。您可能希望在一件事情发生(如您到达了一个特定的地理位置)时应用程序可以给另一部手机发送 SMS 消息,或者可能希望访问一个提供特定服务(如汇率、天气等)的 Web 服务。

在本章中,将学习如何以编程方式从 Android 应用程序中发送和接收 SMS 消息。还将学习如何在 Android 应用程序中调用 Mail 应用程序来向其他用户发送电子邮件。

8.1 SMS 消息传递

SMS 消息传递是当今手机上的一个主要的杀手级应用——对于一些用户来说,这跟手机本身一样必不可少。当前您购买的任何手机都至少应该具有 SMS 消息传递的功能,几乎所有年龄段的用户都知道如何发送和接收这类消息。Android 带有一个内置的 SMS 应用程序,可以接收和发送 SMS 消息。不过,在某些情况下,您可能想要将 SMS 功能集成到您自己的应用程序中。举个例子,您也许打算写一个能够按固定时间间隔自动发送 SMS 消息的应用程序。例如,您想追踪孩子的位置时这就非常有用——只要给他们一个 Android 设备,可以每 30 分钟发出一条包含地理位置信息的 SMS 消息就行了。这下,您就对他们放了学是否真的去了图书馆了解得一清二楚(当然,这也意味着您不得不为发送这类短信而花钱)。

本节介绍如何在 Android 应用程序中以编程方式发送和接收 SMS 消息。对 Android 开发人员来说，好消息是不需要用一个真正的设备来对 SMS 消息传递进行测试：免费的 Android 模拟器提供了这一功能。

8.1.1 以编程方式发送 SMS 消息

首先，您将学习如何以编程方式通过应用程序发送 SMS 消息。使用这种方法，应用程序可以自动发送 SMS 消息给收件人，而无须用户干预。下面的"试一试"将告诉您这是如何做到的。

试一试 发送 SMS 消息

SMS.zip 代码文件可以在Wrox.com 上下载

(1) 打开 Eclipse，创建一个新的 Android 项目，命名为 SMS。
(2) 在 main.xml 文件中添加下列粗体显示的语句来替换 TextView：

```xml
<?xml version="1.0" encoding="utf-8"?>
<LinearLayout xmlns:android="http://schemas.android.com/apk/res/android"
    android:layout_width="fill_parent"
    android:layout_height="fill_parent"
    android:orientation="vertical" >

<Button
    android:id="@+id/btnSendSMS"
    android:layout_width="fill_parent"
    android:layout_height="wrap_content"
    android:text="Send SMS"
    android:onClick="onClick" />

</LinearLayout>
```

(3) 在 AndroidManifest.xml 文件中添加下列粗体显示的语句：

```xml
<?xml version="1.0" encoding="utf-8"?>
<manifest xmlns:android="http://schemas.android.com/apk/res/android"
    package="net.learn2develop.SMS"
    android:versionCode="1"
    android:versionName="1.0" >

    <uses-sdk android:minSdkVersion="14" />
    <uses-permission android:name="android.permission.SEND_SMS"/>

    <application
        android:icon="@drawable/ic_launcher"
        android:label="@string/app_name" >
        <activity
            android:label="@string/app_name"
            android:name=".SMSActivity" >
            <intent-filter >
```

```xml
            <action android:name="android.intent.action.MAIN" />
            <category android:name="android.intent.category.LAUNCHER" />
        </intent-filter>
    </activity>
</application>
</manifest>
```

(4) 在 SMSActivity.java 文件中添加下列粗体显示的语句：

```java
package net.learn2develop.SMS;

import android.app.Activity;
import android.os.Bundle;

import android.telephony.SmsManager;
import android.view.View;

public class SMSActivity extends Activity {
    /** Called when the activity is first created. */
    @Override
    public void onCreate(Bundle savedInstanceState) {
        super.onCreate(savedInstanceState);
        setContentView(R.layout.main);
    }

    public void onClick(View v) {
        sendSMS("5556", "Hello my friends!");
    }

    //---sends an SMS message to another device---
    private void sendSMS(String phoneNumber, String message)
    {
        SmsManager sms = SmsManager.getDefault();
        sms.sendTextMessage(phoneNumber, null, message, null, null);
    }
}
```

(5) 按 F11 键在 Android 模拟器上调试应用程序。使用 Android SDK 和 AVD Manager 启动另一个 AVD。

(6) 在第 1 个 Android 模拟器(5544)上，单击 Send SMS 按钮来发送 SMS 消息到第 2 个模拟器(5556)。图 8-1 展示了由第 2 个模拟器收到的 SMS 消息(注意在第 2 个模拟器顶端的通知栏)。

示例说明

Android 使用基于权限的策略，应用程序所需的所有权限都必须在 AndroidManifest.xml 文件中指定。这可以确保在应用程序安装时，用户确切地知道它需要哪些访问权限。

图 8-1

由于发送 SMS 消息会导致用户支付额外的费用,因此在 AndroidManifest.xml 文件中指明 SMS 权限可以使用户决定是否允许安装应用程序。

使用 SmsManager 类,可以以编程方式来发送 SMS 消息。与其他类不同,不能直接实例化这个类,而是要调用 getDefault()静态方法获得一个 SmsManager 对象。然后,使用 sendTextMessage()方法来发送 SMS 消息:

```
//---sends an SMS message to another device---
private void sendSMS(String phoneNumber, String message)
{
    SmsManager sms = SmsManager.getDefault();
    sms.sendTextMessage(phoneNumber, null, message, null, null);
}
```

以下是 sendTextMessage()方法用到的 5 个参数:

- destinationAddress——收件人的电话号码
- scAddress——服务中心地址,null 代表默认的 SMSC
- text——SMS 消息的内容
- sentIntent——发送消息后调用的挂起的意图(下一节将详细讨论)
- deliveryIntent——消息递送后调用的挂起的意图(下一节详细讨论)

 注意:如果使用 SmsManager 类以编程方式发送 SMS 消息,发送的消息不会出现在发送者的内置的 Messaging 应用程序中。

8.1.2 发送消息后获取反馈

前一节学习了如何使用 SmsManager 类以编程方式发送 SMS 消息，但如何知道消息已被正确发送了呢？要达到此目的，可以创建两个 PendingIntent 对象来监视 SMS 消息发送过程中的状态。这两个 PendingIntent 对象传递给 sendTextMessage()方法的最后两个参数。下面的代码片段展示了如何监视被发送的 SMS 消息的状态：

```java
package net.learn2develop.SMS;

import android.app.Activity;
import android.app.PendingIntent;
import android.content.BroadcastReceiver;
import android.content.Context;
import android.content.Intent;
import android.content.IntentFilter;
import android.os.Bundle;

import android.telephony.SmsManager;
import android.view.View;
import android.widget.Toast;

public class SMSActivity extends Activity {
    String SENT = "SMS_SENT";
    String DELIVERED = "SMS_DELIVERED";
    PendingIntent sentPI, deliveredPI;
    BroadcastReceiver smsSentReceiver, smsDeliveredReceiver;

    /** Called when the activity is first created. */
    @Override
    public void onCreate(Bundle savedInstanceState) {
        super.onCreate(savedInstanceState);
        setContentView(R.layout.main);

        sentPI = PendingIntent.getBroadcast(this, 0,
                new Intent(SENT), 0);

        deliveredPI = PendingIntent.getBroadcast(this, 0,
                new Intent(DELIVERED), 0);
    }

    @Override
    public void onResume() {
        super.onResume();

        //---create the BroadcastReceiver when the SMS is sent---
        smsSentReceiver = new BroadcastReceiver(){
            @Override
            public void onReceive(Context arg0, Intent arg1) {
                switch (getResultCode())
```

```java
            {
            case Activity.RESULT_OK:
                Toast.makeText(getBaseContext(), "SMS sent",
                        Toast.LENGTH_SHORT).show();
                break;
            case SmsManager.RESULT_ERROR_GENERIC_FAILURE:
                Toast.makeText(getBaseContext(), "Generic failure",
                        Toast.LENGTH_SHORT).show();
                break;
            case SmsManager.RESULT_ERROR_NO_SERVICE:
                Toast.makeText(getBaseContext(), "No service",
                        Toast.LENGTH_SHORT).show();
                break;
            case SmsManager.RESULT_ERROR_NULL_PDU:
                Toast.makeText(getBaseContext(), "Null PDU",
                        Toast.LENGTH_SHORT).show();
                break;
            case SmsManager.RESULT_ERROR_RADIO_OFF:
                Toast.makeText(getBaseContext(), "Radio off",
                        Toast.LENGTH_SHORT).show();
                break;
            }
        }
    };

    //---create the BroadcastReceiver when the SMS is delivered---
    smsDeliveredReceiver = new BroadcastReceiver(){
        @Override
        public void onReceive(Context arg0, Intent arg1) {
            switch (getResultCode())
            {
            case Activity.RESULT_OK:
                Toast.makeText(getBaseContext(), "SMS delivered",
                        Toast.LENGTH_SHORT).show();
                break;
            case Activity.RESULT_CANCELED:
                Toast.makeText(getBaseContext(), "SMS not delivered",
                        Toast.LENGTH_SHORT).show();
                break;
            }
        }
    };

    //---register the two BroadcastReceivers---
    registerReceiver(smsDeliveredReceiver, new IntentFilter(DELIVERED));
    registerReceiver(smsSentReceiver, new IntentFilter(SENT));
}

@Override
public void onPause() {
```

```
        super.onPause();
        //---unregister the two BroadcastReceivers---
        unregisterReceiver(smsSentReceiver);
        unregisterReceiver(smsDeliveredReceiver);
    }

    public void onClick(View v) {
        sendSMS("5556", "Hello my friends!");
    }

    //---sends an SMS message to another device---
    private void sendSMS(String phoneNumber, String message)
    {
        SmsManager sms = SmsManager.getDefault();
        sms.sendTextMessage(phoneNumber, null, message, sentPI, deliveredPI);
    }
}
```

上述示例在 onCreate()方法中创建了两个 PendingIntent 对象：

```
sentPI = PendingIntent.getBroadcast(this, 0,
        new Intent(SENT), 0);

deliveredPI = PendingIntent.getBroadcast(this, 0,
        new Intent(DELIVERED), 0);
```

这两个 PendingIntent 对象用于后面在 SMS 消息被发送(SMS_SENT)和递送(SMS_DELIVERED)后发送广播。

然后，在 onResume()方法中创建并注册了两个 BroadcastReceiver。这两个 BroadcastReceiver 侦听与 SMS_SENT 和 SMS_DELIVERED 匹配的意图(分别在发送和递送消息后由 SmsManager 触发)：

```
//---register the two BroadcastReceivers---
registerReceiver(smsDeliveredReceiver, new IntentFilter(DELIVERED));
registerReceiver(smsSentReceiver, new IntentFilter(SENT));
```

在每一个 BroadcastReceiver 中，重写 onReceive()方法并得到当前的结果码。

两个 PendingIntent 对象被传递给 sendTextMessage()方法的最后两个参数：

```
SmsManager sms = SmsManager.getDefault();
sms.sendTextMessage(phoneNumber, null, message, sentPI, deliveredPI);
```

在这种情况下，无论消息是被正确发送还是递送失败，都将通过这两个 PendingIntent 对象来通知其状态。

最后，在 onPause()方法中，注销这两个 BroadcastReceiver 对象。

注意：如果在 Android 模拟器上测试应用程序，只有 sentPI PendingIntent 对象会被触发，deliveredPI PendingIntent 对象却不会。在实际设备上，两个 PendingIntent 对象都会触发。

8.1.3 使用意图发送 SMS 消息

使用 SmsManager 类，可以通过应用程序发送 SMS 消息，而不需要涉及内置的 Messaging 应用程序。但有时候，如果可以直接调用内置的 Messaging 应用程序并让它做发送消息的所有工作，发送 SMS 消息会变得更容易些。

要在应用程序中激活内置的 Messaging 应用程序，可以使用一个 Intent 对象和 MIME 类型 vnd.android-dir/mms-sms，如下面的代码片段所示：

```
Intent i = new
        Intent(android.content.Intent.ACTION_VIEW);
i.putExtra("address", "5556; 5558; 5560");

i.putExtra("sms_body", "Hello my friends!");
i.setType("vnd.android-dir/mms-sms");
startActivity(i);
```

这将调用 Messaging 应用程序，如图 8-2 所示。注意，可以发送 SMS 给多个收件人，只需要用(在 putExtra()方法中)分号分隔每个电话号码就行了。这些号码在 Messaging 应用程序中将用逗号分隔开。

图 8-2

注意:如果使用这个方法调用 Messaging 应用程序,就没有必要在 AndroidManifest.xml 文件中指定 SEND_SMS 权限,因为您的应用程序并非是最终发送消息的那个。

8.1.4 接收 SMS 消息

除了从 Android 应用程序发送 SMS 消息外,还可以在应用程序中使用 BroadcastReceiver 对象接收传入的 SMS 消息。如果希望应用程序在收到一条特定的 SMS 消息时执行一个动作,这就很有用了。例如,您可能想追踪您的手机位置以防丢失或被盗。在这种情况下,可以编写一个应用程序,用来自动侦听包含一些秘密代码的 SMS 消息。一旦收到此类信息,就可以给发送者发回一条包含位置坐标的 SMS 消息。

下面的"试一试"展示了如何以编程方式侦听传入的 SMS 消息。

试一试 接收 SMS 消息

(1) 使用在前一节所创建的同一个项目,在 AndroidManifest.xml 文件中添加下列粗体显示的语句:

```xml
<?xml version="1.0" encoding="utf-8"?>
<manifest xmlns:android="http://schemas.android.com/apk/res/android"
    package="net.learn2develop.SMS"
    android:versionCode="1"
    android:versionName="1.0" >

    <uses-sdk android:minSdkVersion="10" />
    <uses-permission android:name="android.permission.SEND_SMS"/>
    <uses-permission android:name="android.permission.RECEIVE_SMS"/>

    <application
        android:icon="@drawable/ic_launcher"
        android:label="@string/app_name" >
        <activity
            android:label="@string/app_name"
            android:name=".SMSActivity" >
            <intent-filter >
                <action android:name="android.intent.action.MAIN" />
                <category android:name="android.intent.category.LAUNCHER" />
            </intent-filter>
        </activity>
        <receiver android:name=".SMSReceiver">
            <intent-filter>
                <action android:name=
                    "android.provider.Telephony.SMS_RECEIVED" />
            </intent-filter>
        </receiver>
```

```
</application>

</manifest>
```

(2) 在项目的 src 文件夹中,在包名下增加一个新的类文件,并命名为 SMSReceiver(如图 8-3 所示)。

图 8-3

(3) 按如下所示编写 SMSReceiver.java 文件:

```
package net.learn2develop.SMS;

import android.content.BroadcastReceiver;
import android.content.Context;
import android.content.Intent;
import android.os.Bundle;
import android.telephony.SmsMessage;
import android.util.Log;
import android.widget.Toast;

public class SMSReceiver extends BroadcastReceiver
{
    @Override
    public void onReceive(Context context, Intent intent)
    {
        //---get the SMS message passed in---
        Bundle bundle = intent.getExtras();
        SmsMessage[] msgs = null;
```

```
    String str = "SMS from ";
    if (bundle != null)
    {
        //---retrieve the SMS message received---
        Object[] pdus = (Object[]) bundle.get("pdus");
        msgs = new SmsMessage[pdus.length];
        for (int i=0; i<msgs.length; i++){
            msgs[i] = SmsMessage.createFromPdu((byte[])pdus[i]);
            if (i==0) {
                //---get the sender address/phone number---
                str += msgs[i].getOriginatingAddress();
                str += ": ";
            }
            //---get the message body---
            str += msgs[i].getMessageBody().toString();
        }
        //---display the new SMS message---
        Toast.makeText(context, str, Toast.LENGTH_SHORT).show();
        Log.d("SMSReceiver", str);
    }
}
}
```

(4) 按 F11 键在 Android 模拟器上调试应用程序。

(5) 使用 DDMS，给模拟器发送一条消息。应用程序将能够接收到这条消息并用 Toast 类进行显示(如图 8-4 所示)。

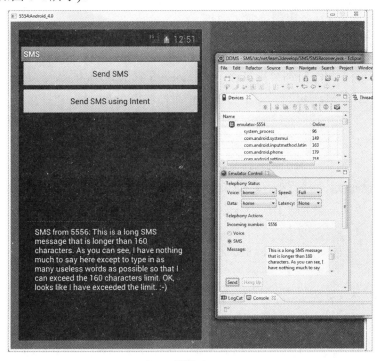

图 8-4

示例说明

要侦听传入的 SMS 消息，需要创建一个 BroadcastReceiver 类。BroadcastReceiver 类使应用程序接收其他应用程序使用 sendBroadcast()方法发送的意图。从本质上讲，它使您的应用程序可以处理由其他程序应用所引发的事件。当接收到一个意图对象时，调用 onReceive()方法。因此，需要重写这一方法。

onReceive()方法在收到一个传入的 SMS 消息时被触发。SMS 消息通过一个 Bundle 对象包含在 Intent 对象(即 intent；onReceive()方法的第二个参数)中。注意收到的每条 SMS 消息都会调用 onReceive()方法。如果设备收到了 5 条 SMS 消息，onReceive()方法就会被调用 5 次。

每条 SMS 消息以 PDU 格式存储在一个 Object 数组中。如果 SMS 消息少于 160 个字符，那么数组中只包含一个元素。否则，该消息将被分割成多条更小的消息，作为数组中的多个元素存储。

要提取每条消息的内容，可以使用 SmsMessage 类的静态方法 createFromPdu()。发件人的电话号码通过 getOriginatingAddress()方法来获得，因此如果需要给发送者发一个自动回复，就可以使用该方法获得发送者的电话号码。要提取消息的正文，需要使用 getMessageBody()方法。

BroadcastReceiver 有一个有趣的特性：即使应用程序不在运行，您也可以继续侦听传入的 SMS 消息；只要应用程序已经安装在设备上，任何传入的 SMS 消息都将被该应用程序所接收。

1. 阻止 Messaging 应用程序接收消息

在前一节您可能注意到了，每次向模拟器(或设备)发送 SMS 消息时，您的应用程序和内置的 Messaging 应用程序都会接收它。这是因为在收到 SMS 消息时，Android 设备上的所有应用程序(包括 Messaging 应用程序)会轮流处理传入的消息。但是，有时候不希望看到这种行为——例如，您可能希望自己的应用程序收到消息，然后阻止将消息发送到其他应用程序。这种功能十分有用，特别是在构建某种跟踪应用程序时。

解决的办法很简单。为了阻止内置的 Messaging 应用程序处理传入的消息。应用程序只需要在 Messaging 应用程序之前处理该消息。为此，在<intent-filter>元素中添加 android:priority 属性，如下所示：

```
<receiver android:name=".SMSReceiver">
    <intent-filter android:priority="100">
        <action android:name=
            "android.provider.Telephony.SMS_RECEIVED" />
    </intent-filter>
</receiver>
```

将这个属性设为一个比较大的数字，例如 100。数字越大，Android 就会越早执行您的应用程序。当收到传入的消息时，您的应用程序将首先执行，此时您可以决定怎么处理消

第 8 章 消息传递

息。为了防止其他应用程序看到此消息，只需要在 BroadReceiver 类中调用 abortBroadcast() 方法：

```
@Override
public void onReceive(Context context, Intent intent)
{
    //---get the SMS message passed in---
    Bundle bundle = intent.getExtras();
    SmsMessage[] msgs = null;
    String str = "SMS from ";
    if (bundle != null)
    {
        //---retrieve the SMS message received---
        Object[] pdus = (Object[]) bundle.get("pdus");
        msgs = new SmsMessage[pdus.length];
        for (int i=0; i<msgs.length; i++){
            msgs[i] = SmsMessage.createFromPdu((byte[])pdus[i]);
            if (i==0) {
                //---get the sender address/phone number---
                str += msgs[i].getOriginatingAddress();
                str += ": ";
            }
            //---get the message body---
            str += msgs[i].getMessageBody().toString();
        }

        //---display the new SMS message---
        Toast.makeText(context, str, Toast.LENGTH_SHORT).show();
        Log.d("SMSReceiver", str);

        //---stop the SMS message from being broadcasted---
        this.abortBroadcast();
    }
}
```

这样，其他任何应用程序都不会收到您的 SMS 消息。

注意：在设备上安装了前面这个应用程序后，所有传入的 SMS 消息都将被这个应用程序拦截，再也不会出现在 Messaging 应用程序中。

2. 通过 BroadcastReceiver 更新一个活动

上一节介绍了如何使用一个 BroadcastReceiver 类侦听传入的 SMS 消息，然后使用 Toast 类来显示接收到的 SMS 消息。通常情况下，您想要将 SMS 消息发回给应用程序的主活动。例如，您可能希望消息在一个 TextView 中显示。下面的"试一试"将告诉您如何做到这一点。

试一试　　创建一个基于视图的应用程序项目

(1) 使用在前一节所创建的同一个项目，在 main.xml 文件中添加下列粗体显示的行：

```xml
<?xml version="1.0" encoding="utf-8"?>
<LinearLayout xmlns:android="http://schemas.android.com/apk/res/android"
    android:layout_width="fill_parent"
    android:layout_height="fill_parent"
    android:orientation="vertical" >

<Button
    android:id="@+id/btnSendSMS"
    android:layout_width="fill_parent"
    android:layout_height="wrap_content"
    android:text="Send SMS"
    android:onClick="onClick" />

<TextView
    android:id="@+id/textView1"
    android:layout_width="wrap_content"
    android:layout_height="wrap_content" />

</LinearLayout>
```

(2) 在 SMSReceiver.java 文件中添加下列粗体显示的语句：

```java
package net.learn2develop.SMS;

import android.content.BroadcastReceiver;
import android.content.Context;
import android.content.Intent;
import android.os.Bundle;
import android.telephony.SmsMessage;
import android.util.Log;
import android.widget.Toast;

public class SMSReceiver extends BroadcastReceiver
{
    @Override
    public void onReceive(Context context, Intent intent)
    {
        //---get the SMS message passed in---
        Bundle bundle = intent.getExtras();
        SmsMessage[] msgs = null;
        String str = "SMS from ";
        if (bundle != null)
        {
            //---retrieve the SMS message received---
            Object[] pdus = (Object[]) bundle.get("pdus");
            msgs = new SmsMessage[pdus.length];
            for (int i=0; i<msgs.length; i++){
```

```
                    msgs[i]=SmsMessage.createFromPdu((byte[])pdus[i]);
                    if (i==0) {
                        //---get the sender address/phone number---
                        str += msgs[i].getOriginatingAddress();
                        str += ": ";
                    }
                    //---get the message body---
                    str += msgs[i].getMessageBody().toString();
                }
                //---display the new SMS message---
                Toast.makeText(context, str, Toast.LENGTH_SHORT).show();
                Log.d("SMSReceiver", str);

                //---send a broadcast intent to update the SMS received in the
                activity---
                Intent broadcastIntent = new Intent();
                broadcastIntent.setAction("SMS_RECEIVED_ACTION");
                broadcastIntent.putExtra("sms", str);
                context.sendBroadcast(broadcastIntent);
            }
        }
}
```

(3) 在 SMSActivity.java 文件中添加下列粗体显示的语句：

```
package net.learn2develop.SMS;

import android.app.Activity;
import android.app.PendingIntent;
import android.content.BroadcastReceiver;
import android.content.Context;
import android.content.Intent;
import android.content.IntentFilter;
import android.os.Bundle;

import android.telephony.SmsManager;
import android.view.View;
import android.widget.TextView;
import android.widget.Toast;

public class SMSActivity extends Activity {
    String SENT = "SMS_SENT";
    String DELIVERED = "SMS_DELIVERED";
    PendingIntent sentPI, deliveredPI;
    BroadcastReceiver smsSentReceiver, smsDeliveredReceiver;
    IntentFilter intentFilter;

    private BroadcastReceiver intentReceiver = new BroadcastReceiver() {
        @Override
        public void onReceive(Context context, Intent intent) {
            //---display the SMS received in the TextView---
```

```java
        TextView SMSes = (TextView) findViewById(R.id.textView1);
        SMSes.setText(intent.getExtras().getString("sms"));
    }
};

/** Called when the activity is first created. */
@Override
public void onCreate(Bundle savedInstanceState) {
    super.onCreate(savedInstanceState);
    setContentView(R.layout.main);

    sentPI = PendingIntent.getBroadcast(this, 0,
            new Intent(SENT), 0);

    deliveredPI = PendingIntent.getBroadcast(this, 0,
            new Intent(DELIVERED), 0);

    //---intent to filter for SMS messages received---
    intentFilter = new IntentFilter();
    intentFilter.addAction("SMS_RECEIVED_ACTION");
}

@Override
public void onResume() {
    super.onResume();

    //---register the receiver---
    registerReceiver(intentReceiver, intentFilter);

    //---create the BroadcastReceiver when the SMS is sent---
    smsSentReceiver = new BroadcastReceiver(){
        @Override
        public void onReceive(Context arg0, Intent arg1) {
            switch (getResultCode())
            {
            case Activity.RESULT_OK:
                Toast.makeText(getBaseContext(), "SMS sent",
                        Toast.LENGTH_SHORT).show();
                break;
            case SmsManager.RESULT_ERROR_GENERIC_FAILURE:
                Toast.makeText(getBaseContext(), "Generic failure",
                        Toast.LENGTH_SHORT).show();
                break;
            case SmsManager.RESULT_ERROR_NO_SERVICE:
                Toast.makeText(getBaseContext(), "No service",
                        Toast.LENGTH_SHORT).show();
                break;
            case SmsManager.RESULT_ERROR_NULL_PDU:
                Toast.makeText(getBaseContext(), "Null PDU",
                        Toast.LENGTH_SHORT).show();
```

```java
                    break;
                case SmsManager.RESULT_ERROR_RADIO_OFF:
                    Toast.makeText(getBaseContext(), "Radio off",
                            Toast.LENGTH_SHORT).show();
                    break;
                }
            }
        };

        //---create the BroadcastReceiver when the SMS is delivered---
        smsDeliveredReceiver = new BroadcastReceiver(){
            @Override
            public void onReceive(Context arg0, Intent arg1) {
                switch (getResultCode())
                {
                case Activity.RESULT_OK:
                    Toast.makeText(getBaseContext(), "SMS delivered",
                            Toast.LENGTH_SHORT).show();
                    break;
                case Activity.RESULT_CANCELED:
                    Toast.makeText(getBaseContext(), "SMS not delivered",
                            Toast.LENGTH_SHORT).show();
                    break;
                }
            }
        };

        //---register the two BroadcastReceivers---
        registerReceiver(smsDeliveredReceiver, new IntentFilter
           (DELIVERED));
        registerReceiver(smsSentReceiver, new IntentFilter(SENT));
    }

    @Override
    public void onPause() {
        super.onPause();

        //---unregister the receiver---
        unregisterReceiver(intentReceiver);

        //---unregister the two BroadcastReceivers---
        unregisterReceiver(smsSentReceiver);
        unregisterReceiver(smsDeliveredReceiver);
    }

    public void onClick(View v) {
        sendSMS("5556", "Hello my friends!");
    }

    public void onSMSIntentClick (View v) {
```

```
            Intent i = new
                    Intent(android.content.Intent.ACTION_VIEW);
            i.putExtra("address", "5556; 5558; 5560");

            i.putExtra("sms_body", "Hello my friends!");
            i.setType("vnd.android-dir/mms-sms");
            startActivity(i);
        }

        //--sends an SMS message to another device¡ª-
        private void sendSMS(String phoneNumber, String message)
        {
            SmsManager sms = SmsManager.getDefault();
            sms.sendTextMessage(phoneNumber, null, message, sentPI,
              deliveredPI);
        }
    }
```

(4) 按 F11 键在 Android 模拟器上调试应用程序。使用 DDMS 向模拟器发送一条 SMS 消息。图 8-5 展示了分别由 Toast 类和 TextView 所显示的收到的消息。

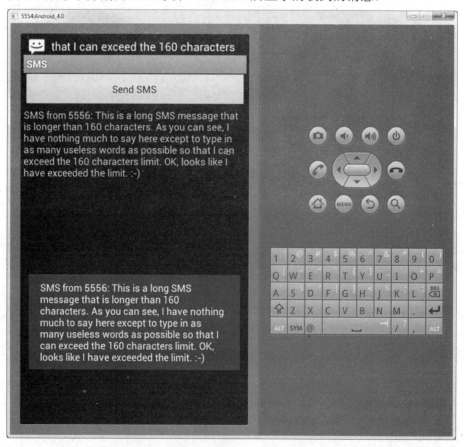

图 8-5

示例说明

首先在活动中添加一个 TextView,可以用它来显示接收到的 SMS 消息。

接下来修改 SMSReceiver 类。这样当它接收到一条 SMS 消息时,将广播另一个 Intent 对象,使得侦听这一意图的任何应用程序可以得到通知(我们接下来将在活动中实现)。收到的 SMS 也将通过这个意图发送出去。

```
//---send a broadcast intent to update the SMS received in the activity---
Intent broadcastIntent = new Intent();
broadcastIntent.setAction("SMS_RECEIVED_ACTION");
broadcastIntent.putExtra("sms", str);
context.sendBroadcast(broadcastIntent);
```

下一步,在活动中创建一个 BroadcastReceiver 对象来侦听广播意图:

```
private BroadcastReceiver intentReceiver = new BroadcastReceiver() {
    @Override
    public void onReceive(Context context, Intent intent) {
        //---display the SMS received in the TextView---
        TextView SMSes = (TextView) findViewById(R.id.textView1);
        SMSes.setText(intent.getExtras().getString("sms"));
    }
};
```

当收到一个广播意图时,更新 TextView 中的 SMS 消息。

您需要创建一个 IntentFilter 对象以便侦听一个特定的意图。这里,意图是 SMS_RECEIVED_ACTION:

```
@Override
public void onCreate(Bundle savedInstanceState) {
    super.onCreate(savedInstanceState);
    setContentView(R.layout.main);

    sentPI = PendingIntent.getBroadcast(this, 0,
        new Intent(SENT), 0);

    deliveredPI = PendingIntent.getBroadcast(this, 0,
        new Intent(DELIVERED), 0);

    //---intent to filter for SMS messages received---
    intentFilter = new IntentFilter();
    intentFilter.addAction("SMS_RECEIVED_ACTION");
}
```

最后,在活动的 onResume()事件中注册 BroadcastReceiver,在 onPause()事件中进行注销:

```
@Override
protected void onResume() {
    //--register the receiver--
    registerReceiver(intentReceiver, intentFilter);
```

```java
        super.onResume();
    }

    @Override
    protected void onPause() {
        //--unregister the receiver--
        unregisterReceiver(intentReceiver);
        super.onPause();
    }

    @Override
    public void onResume() {
        super.onResume();

        //---register the receiver---
        registerReceiver(intentReceiver, intentFilter);

        //---create the BroadcastReceiver when the SMS is sent---
        //...
    }

    @Override
    public void onPause() {
        super.onPause();

        //---unregister the receiver---
        unregisterReceiver(intentReceiver);

        //---unregister the two BroadcastReceivers---
        //...
    }
```

这意味着，只有在收到 SMS 消息且活动在屏幕上可见时，TextView 才会显示该消息。如果接收到 SMS 消息时活动不在前台，TextView 将不会被更新。

3. 通过 BroadcastReceiver 调用一个活动

前面的示例说明了如何传递接收到的 SMS 消息来在活动中显示。然而，在许多情况下，当接收 SMS 消息时活动可能在后台。在这种情况下，当接收一条消息时，如果能将活动推到前台将是很有用的。下面的"试一试"展示了该如何做到这一点。

试一试　　调用一个活动

(1) 使用在前一节所创建的同一个项目，在 SMSActivity.java 文件中添加下列粗体显示的行：

```java
/** Called when the activity is first created. */
@Override
public void onCreate(Bundle savedInstanceState) {
```

```java
        super.onCreate(savedInstanceState);
        setContentView(R.layout.main);
        
        sentPI = PendingIntent.getBroadcast(this, 0,
                new Intent(SENT), 0);
        
        deliveredPI = PendingIntent.getBroadcast(this, 0,
                new Intent(DELIVERED), 0);
        
        //---intent to filter for SMS messages received---
        intentFilter = new IntentFilter();
        intentFilter.addAction("SMS_RECEIVED_ACTION");
        
        //---register the receiver---
        registerReceiver(intentReceiver, intentFilter);
    }
    
    @Override
    public void onResume() {
        super.onResume();
        
        //---register the receiver---
        //registerReceiver(intentReceiver, intentFilter);
        
        //---create the BroadcastReceiver when the SMS is sent---
        smsSentReceiver = new BroadcastReceiver(){
            @Override
            public void onReceive(Context arg0, Intent arg1) {
                switch (getResultCode())
                {
                case Activity.RESULT_OK:
                    Toast.makeText(getBaseContext(), "SMS sent",
                            Toast.LENGTH_SHORT).show();
                    break;
                case SmsManager.RESULT_ERROR_GENERIC_FAILURE:
                    Toast.makeText(getBaseContext(), "Generic failure",
                            Toast.LENGTH_SHORT).show();
                    break;
                case SmsManager.RESULT_ERROR_NO_SERVICE:
                    Toast.makeText(getBaseContext(), "No service",
                            Toast.LENGTH_SHORT).show();
                    break;
                case SmsManager.RESULT_ERROR_NULL_PDU:
                    Toast.makeText(getBaseContext(), "Null PDU",
                            Toast.LENGTH_SHORT).show();
                    break;
                case SmsManager.RESULT_ERROR_RADIO_OFF:
                    Toast.makeText(getBaseContext(), "Radio off",
                            Toast.LENGTH_SHORT).show();
                    break;
```

```
            }
        }
    };

    //---create the BroadcastReceiver when the SMS is delivered---
    smsDeliveredReceiver = new BroadcastReceiver(){
        @Override
        public void onReceive(Context arg0, Intent arg1) {
            switch (getResultCode())
            {
            case Activity.RESULT_OK:
                Toast.makeText(getBaseContext(), "SMS delivered",
                        Toast.LENGTH_SHORT).show();
                break;
            case Activity.RESULT_CANCELED:
                Toast.makeText(getBaseContext(), "SMS not delivered",
                        Toast.LENGTH_SHORT).show();
                break;
            }
        }
    };

    //---register the two BroadcastReceivers---
    registerReceiver(smsDeliveredReceiver, new IntentFilter
        (DELIVERED));
    registerReceiver(smsSentReceiver, new IntentFilter(SENT));
}

@Override
public void onPause() {
    super.onPause();

    //---unregister the receiver---
    //unregisterReceiver(intentReceiver);

    //---unregister the two BroadcastReceivers---
    unregisterReceiver(smsSentReceiver);
    unregisterReceiver(smsDeliveredReceiver);
}

@Override
protected void onDestroy() {
    super.onDestroy();

    //---unregister the receiver---
    unregisterReceiver(intentReceiver);
}
```

(2) 在 SMSReceiver.java 文件中添加下列粗体显示的语句：
```
@Override
```

```java
public void onReceive(Context context, Intent intent)
{
    //---get the SMS message passed in---
    Bundle bundle = intent.getExtras();
    SmsMessage[] msgs = null;
    String str = "SMS from ";
    if (bundle != null)
    {
        //---retrieve the SMS message received---
        Object[] pdus = (Object[]) bundle.get("pdus");
        msgs = new SmsMessage[pdus.length];
        for (int i=0; i<msgs.length; i++){
            msgs[i] = SmsMessage.createFromPdu((byte[])pdus[i]);
            if (i==0) {
                //---get the sender address/phone number---
                str += msgs[i].getOriginatingAddress();
                str += ": ";
            }
            //---get the message body---
            str += msgs[i].getMessageBody().toString();
        }
        //---display the new SMS message---
        Toast.makeText(context, str, Toast.LENGTH_SHORT).show();
        Log.d("SMSReceiver", str);

        //---launch the SMSActivity---
        Intent mainActivityIntent = new Intent(context, SMSActivity.
           class);
        mainActivityIntent.setFlags(Intent.FLAG_ACTIVITY_NEW_TASK);
        context.startActivity(mainActivityIntent);

        //---send a broadcast intent to update the SMS received
        in the activity---
        Intent broadcastIntent = new Intent();
        broadcastIntent.setAction("SMS_RECEIVED_ACTION");
        broadcastIntent.putExtra("sms", str);
        context.sendBroadcast(broadcastIntent);
    }
}
```

(3) 按如下所示修改 AndroidManifest.xml 文件：

```xml
<activity
    android:label="@string/app_name"
    android:name=".SMSActivity"
    android:launchMode="singleTask" >
    <intent-filter >
        <action android:name="android.intent.action.MAIN" />
        <category android:name="android.intent.category.LAUNCHER" />
    </intent-filter>
</activity>
```

(4) 按 F11 键在 Android 模拟器上调试应用程序。当 SMSActivity 显示时，单击 Home 按钮将活动发送到后台。

(5) 使用 DDMS 再次向模拟器发送一条 SMS 消息。这一次，注意活动将被推到前台，显示接收的 SMS 消息。

示例说明

在 SMSActivity 类中，首先在活动的 onCreate( )事件而不是 onResume( )事件中注册 BroadcastReceiver，并在 onDestroy()事件而不是 onPause()事件中进行注销。这确保了活动即使是在后台，它仍能够侦听广播意图。

接下来，修改 SMSReceiver 类中的 onReceive()事件，在广播另一个意图之前使用一个意图将活动推到前台：

```
//---launch the SMSActivity---
Intent mainActivityIntent = new Intent(context, SMSActivity.class);
mainActivityIntent.setFlags(Intent.FLAG_ACTIVITY_NEW_TASK);
context.startActivity(mainActivityIntent);

//---send a broadcast intent to update the SMS received in the activity---
Intent broadcastIntent = new Intent();
broadcastIntent.setAction("SMS_RECEIVED_ACTION");
broadcastIntent.putExtra("sms", str);
context.sendBroadcast(broadcastIntent);
```

startActivity()方法启动活动，并将它推到前台。注意，需要设置 Intent.FLAG_ACTIVITY_NEW_TASK 标志，因为从一个活动上下文的外部调用 startActivity()需要 FLAG_ACTIVITY_NEW_TASK 标志。

还需要将 AndroidManifest.xml 文件中<activity>元素的 launchMode 属性设置为 singleTask：

```
<activity
    android:label="@string/app_name"
    android:name=".SMSActivity"
    android:launchMode="singleTask" >
```

如果不进行设置，在应用程序收到 SMS 消息时，将启动活动的多个实例。

注意在这个示例中，当活动在后台时(如单击 Home 按钮来显示主屏幕)，随着 SMS 消息的接收，活动将被推到前台并且 TextView 得到更新。但是，如果活动被终止(如单击 Back 按钮来销毁它)，活动会再次启动，但 TextView 不会被更新。

8.1.5 说明和警告

虽然发送和接收 SMS 消息的能力使 Android 成为开发复杂应用程序的一个非常引人注目的平台，但这种灵活性是要付出代价的。貌似无害的应用程序可能在幕后发送 SMS 消息而用户对此一无所知。正如最近一个基于 SMS 的 Android 木马程序的例子(http://forum.vodafone.

co.nz/topic/5719-android-sms-trojan-warning/)，该应用程序自称是一个媒体播放器，但是一旦安装，它将向一个收费昂贵的号码发送 SMS 消息，导致用户产生巨额话费。

尽管用户需要显式地将权限(例如访问 Internet、发送和接收 SMS 消息等)授予应用程序，但也仅仅是在安装时才显示对权限的请求。如果用户单击 Install 按钮，他或她将被认为授予了权限，允许应用程序发送和接收 SMS 消息。这是很危险的，因为在应用程序安装后，它可以发送和接收 SMS 消息，而不再给用户任何提示。

此外，应用程序还可以"嗅探"传入的 SMS 消息。例如，在前一节所学到的技术基础上，可以很容易地编写出检查 SMS 消息中某些关键字的应用程序。当 SMS 消息包含正在查找的关键字时，可以使用 Location Manager(将在第 9 章中讨论)来获取您的地理位置，然后给 SMS 消息的发送者发回坐标。这样，这个发送者就可以很容易地追踪您的位置。所有这些任务都可以很容易地在您一无所知的情况下完成！也就是说，用户应该尽量避免安装来历不明(例如来自未知的网站、陌生人等)的 Android 应用程序。

8.2 发送电子邮件

与 SMS 消息传递类似，Android 还支持电子邮件。Android 上的 Gmail/Email 应用程序可以使您使用 POP3 或 IMAP 来配置电子邮件账户。除了使用 Gmail/Email 应用程序发送和接收电子邮件外，还可以通过编程方式从 Android 应用程序中发送电子邮件。下面的"试一试"将告诉您如何做。

试一试 以编程方式发送电子邮件

Emails.zip 代码文件可以在 Wrox.com 上下载

(1) 打开 Eclipse，创建一个名为 Emails 的新的 Android 项目。
(2) 在 main.xml 文件中添加下列粗体显示的语句来替换 TextView：

```xml
<?xml version="1.0" encoding="utf-8"?>
<LinearLayout xmlns:android="http://schemas.android.com/apk/res/android"
    android:layout_width="fill_parent"
    android:layout_height="fill_parent"
    android:orientation="vertical" >

<Button
    android:id="@+id/btnSendEmail"
    android:layout_width="fill_parent"
    android:layout_height="wrap_content"
    android:text="Send Email"
    android:onClick="onClick" />

</LinearLayout>
```

(3) 在 SMSActivity.java 文件中添加下列粗体显示的语句:

```java
package net.learn2develop.Emails;

import android.app.Activity;
import android.content.Intent;
import android.net.Uri;
import android.os.Bundle;
import android.view.View;

public class EmailsActivity extends Activity {
    /** Called when the activity is first created. */
    @Override
    public void onCreate(Bundle savedInstanceState) {
        super.onCreate(savedInstanceState);
        setContentView(R.layout.main);
    }

    public void onClick(View v) {
        //---replace the following email addresses with real ones---
        String[] to =
            {"someguy@example.com",
             "anotherguy@example.com"};
        String[] cc = {"busybody@example.com"};
        sendEmail(to, cc, "Hello", "Hello my friends!");
    }

    //---sends an SMS message to another device---
    private void sendEmail(String[] emailAddresses, String[] carbonCopies,
    String subject, String message)
    {
        Intent emailIntent = new Intent(Intent.ACTION_SEND);
        emailIntent.setData(Uri.parse("mailto:"));
        String[] to = emailAddresses;
        String[] cc = carbonCopies;
        emailIntent.putExtra(Intent.EXTRA_EMAIL, to);
        emailIntent.putExtra(Intent.EXTRA_CC, cc);
        emailIntent.putExtra(Intent.EXTRA_SUBJECT, subject);
        emailIntent.putExtra(Intent.EXTRA_TEXT, message);
        emailIntent.setType("message/rfc822");
        startActivity(Intent.createChooser(emailIntent, "Email"));
    }
}
```

(4) 按 F11 键在 Android 模拟器/设备上调试应用程序(确保在尝试这个示例之前配置了电子邮件)。单击 Send Email 按钮可以看到在模拟器/设备上启动了 Email 应用程序,如图 8-6 所示。

第 8 章 消息传递

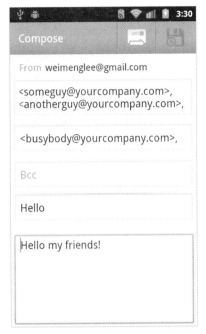

图 8-6

关于测试电子邮件的一点提示

如果您是一名 Gmail 的用户，那么在进行测试时，可以利用这一事实：可以在账户的用户名后使用加号(+)和任何字符串。此时，仍然可以接收电子邮件，但好处是可以进行筛选。例如，本书作者的 Gmail 账户是 weimenglee@gmail.com。在进行测试时，作者会把收件人地址配置为 weimenglee+android_book@gmail.com。这样就很容易筛选出这些消息，方便以后删除。另外，还可以在 Gmail 地址中加入点号。例如，如果想要测试发送一封电子邮件给多个人，可以发送到 weimeng.lee@gmail.com、wei.meng.lee@gmail.com 和 wei.menglee@gmail.com，发给这些地址的电子邮件最终都会作为一个单独的消息发送到作者的账户 weimenglee@gmail.com 中。这对于测试很有帮助。

最后，看一看 http://smtp4dev.codeplex.com，上面包含了很多虚拟的 SMTP 服务器，可用于调试电子邮件消息。

示例说明

在这个示例中，启动内置的 Email 应用程序来发送一封电子邮件消息。要做到这一点，可以使用一个 Intent 对象并使用 setData()、putExtra()和 setType()方法来设置各种参数：

```
Intent emailIntent = new Intent(Intent.ACTION_SEND);
emailIntent.setData(Uri.parse("mailto:"));
String[] to = emailAddresses;
String[] cc = carbonCopies;
emailIntent.putExtra(Intent.EXTRA_EMAIL, to);
emailIntent.putExtra(Intent.EXTRA_CC, cc);
emailIntent.putExtra(Intent.EXTRA_SUBJECT, subject);
emailIntent.putExtra(Intent.EXTRA_TEXT, message);
```

```
emailIntent.setType("message/rfc822");
startActivity(Intent.createChooser(emailIntent, "Email"));
```

8.3 本章小结

本章介绍了应用程序与外界通信的两种关键的方式。首先学习了如何发送和接收 SMS 消息。使用 SMS 可以构建多种多样的依赖于移动运营商提供的服务的应用程序。第 9 章将展示一个极佳的例子,说明如何使用 SMS 消息传递构建一个位置跟踪应用程序。

还学习了如何从 Android 应用程序中发送电子邮件。这是通过使用 Intent 对象调用内置的 Email 应用程序实现的。

练 习

1. 说出在 Android 应用程序中可以用来发送 SMS 消息的两种方式。
2. 说出为发送和接收 SMS 消息需要在 AndroidManifest.xml 文件中声明的权限。
3. 如何通过 BroadcastReceiver 通知一个活动?

练习答案参见附录 C。

本章主要内容

主　题	关 键 概 念
以编程方式发送 SMS 消息	使用 SmsManager 类
发送消息后获取反馈	在 sendTextMessage()方法中使用两个 PendingIntent 对象
使用意图发送 SMS 消息	将意图类型设置为 vnd.android-dir/mms-sms
接收 SMS 消息	实现一个 BroadcastReceiver 并在 AndroidManifest.xml 文件中设置它
使用意图发送电子邮件	将意图类型设置为 message/rfc822

第 9 章

基于位置的服务

本章将介绍以下内容：
- 如何在 Android 应用程序中显示 Google Maps
- 如何在地图上显示缩放控件
- 如何在不同的地图视图间切换
- 如何在地图上添加标记
- 如何获取在地图上触摸的位置的地址
- 如何进行地理编码和反向地理编码
- 如何使用 GPS、Cell-ID 和 Wi-Fi 三角测量法来获取地理数据
- 如何监控一个位置
- 如何构建一个位置跟踪应用程序

近些年来，我们都看到了移动应用程序的爆炸性增长。其中一类非常流行的应用程序是基于位置的服务，即 LBS。LBS 应用程序跟踪您的位置，并可提供额外服务，如定位附近的便利设施以及提供路线规划建议等。当然，LBS 应用程序中的一个关键因素是地图，它可以对您的位置进行可视化表示。

本章中将学习如何在 Android 应用程序中使用 Google Maps，以及如何以编程方式操作它。此外，还将学习如何利用 Android SDK 中提供的 LocationManager 类获得您的地理位置。本章最后将建立一个位置跟踪应用程序，可以把它安装在一个 Android 设备上，用来使用 SMS 消息跟踪朋友和亲人的位置。

9.1 显示地图

Google Maps 是与 Android 平台捆绑在一起的众多应用程序之一。除了直接使用 Maps 应用程序外，还可以将它嵌入到您自己的应用程序中来做一些非常酷的事情。本节介绍如何在 Android 应用程序中使用 Google Maps 以及用编程方式完成以下功能：

- 改变 Google Maps 的视图。
- 在 Google Maps 中获取位置的经度和纬度。
- 进行地理编码和反向地理编码(将一个地址转换为经纬度或反之)。
- 在 Google Maps 上添加标记。

9.1.1 创建项目

开始时,需要首先创建一个 Android 项目以便在活动中显示 Google Maps。

试一试　　创建 Google APIS 项目

LBS.zip 代码文件可以在 Wrox.com 上下载

(1) 使用 Eclipse,创建一个 Android 项目,命名为 LBS。

注意:为了在 Android 应用程序中使用 Google Maps,需要确保选择 Google APIs 作为构建目标。Google Maps 并不是标准 Android SDK 的一部分,因此需要在 Google APIs 附加组件中找到它。

(2) 创建项目后,可以看到在 Google APIs 文件夹下多出来一个 JAR 文件(maps.jar),如图 9-1 所示。

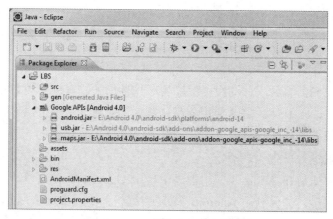

图 9-1

示例说明

这个简单的动作创建了一个使用 Google APIs 附加组件的 Android 项目。Google APIs 附加组件包括了标准的 Android 库和 USB 库,以及打包在 maps.jar 文件中的 Maps 库。

9.1.2 获取 Maps API 密钥

从 Android SDK 发行版 1.0 开始,在 Android 应用程序中集成 Google Maps 前需要先申请一个免费的 Google Maps API 密钥。当申请这个密钥时,必须同意 Google 的使用条款,所以一定要仔细阅读。

要申请一个密钥，可按照下面列出的一系列步骤进行。

注意：Google 提供的关于申请 Maps API 密钥的详细文档参见 http://code.google.com/android/add-ons/google-apis/mapkey.html。

首先，如果您正在直接连接到您的开发机的 Android 模拟器或 Android 设备上进行应用程序测试，则需要在默认文件夹(对于 Windows 7 的用户来说是 C:\User\<username>\.android)下找到 SDK 调试证书。您可以进入 Eclipse 并选择 Window | Preferences 来验证调试证书是否存在。展开 Android 项，并选择 Build(如图 9-2 所示)。在窗口的右侧，可以看到调试证书所在的位置。

注意：对于 Windows XP 用户来说，默认的 Android 文件夹是 C:\Documents and Settings\<username>\Local Settings\Application Data\Android。

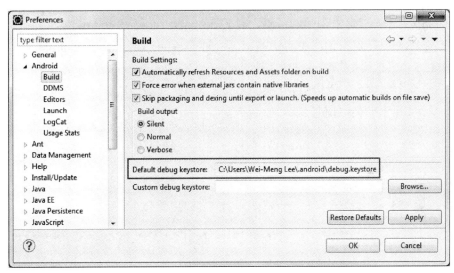

图 9-2

调试密钥库的文件名为 debug.keystore。这是 Eclipse 用来为应用程序进行签名的证书，以便应用程序可以在 Android 模拟器或设备上运行。

使用调试密钥库，需要使用安装 JDK 时包含的 keytool.exe 应用程序来提取其 MD5 指纹。这个指纹是用来申请免费 Google Maps 密钥的。通常可以在 C:\Program Files\Java\<JDK_version_number>\bin 文件夹下找到 keytool.exe。

发出以下命令(如图 9-3 所示)来提取 MD5 指纹：

```
keytool.exe -list -alias androiddebugkey -keystore
"C:\Users\<username>\.android\debug.keystore" -storepass android
-keypass android -v
```

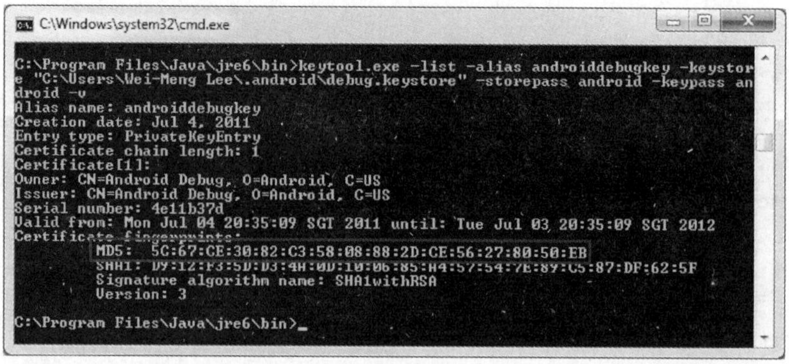

图 9-3

在这个例子中，我的 MD5 指纹是 5C:67:CE:30:82:C3:58:08:88:2D:CE:56:27:80:50:EB。
在前面的命令中，使用了如下参数：

- -list——显示关于指定密钥库的详细信息。
- -alias——密钥库的别名，对于 debug.keystore 是 androiddebugkey。
- -keystore——指定密钥库的位置。
- -storepass——指定密钥库的密码，对于 debug.keystore 是 android。
- -keypass——指定密钥库中密钥的密码，对于 debug.keystore 是 android。

复制 MD5 证书指纹并将 Web 浏览器转到 http://code.google.com/android/maps-api-signup.html。按照页面上的指令完成申请并获取 Google Maps 密钥。当这一切完成后，就应该可以看到如图 9-4 所示的类似信息。

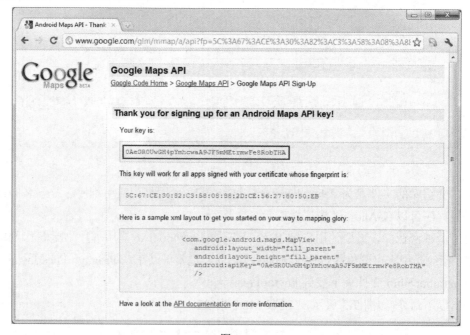

图 9-4

第9章 基于位置的服务

 注意：虽然可以使用调试密钥库的 MD5 指纹来获取用于在 Android 模拟器或设备上调试应用程序的 Maps API 密钥，但如果试图将 Android 应用程序作为一个 APK 文件部署，密钥将不再有效。一旦准备将应用程序部署到 Android Market(或者使用其他发布方法)，就需要使用可以为您的应用程序进行签名的证书来重新申请一个 Maps API 密钥。第 12 章将对此进行详细讨论。

9.1.3 显示地图

现在就要准备在 Android 应用程序中显示 Google Maps 了。这将包含两个主要任务：
- 修改 AndroidManifest.xml 文件，添加<uses-library>元素和 INTERNET 权限。
- 在您的用户界面中添加 MapView 元素。

下面的"试一试"将告诉您应该如何做。

试一试　显示 Google Maps

(1) 使用在前一节中所创建的项目，在 main.xml 文件中用下列粗体显示的行替换 TextView(一定要用您先前获取的 API 密钥替换 android:apiKey 属性的值)：

```xml
<?xml version="1.0" encoding="utf-8"?>
<LinearLayout xmlns:android="http://schemas.android.com/apk/res/android"
    android:layout_width="fill_parent"
    android:layout_height="fill_parent"
    android:orientation="vertical" >

    <com.google.android.maps.MapView
        android:id="@+id/mapView"
        android:layout_width="fill_parent"
        android:layout_height="fill_parent"
        android:enabled="true"
        android:clickable="true"
        android:apiKey="0AeGR0UwGH4pYmhcwaA9JF5mMEtrmwFe8RobTHA" />

</LinearLayout>
```

(2) 在 AndroidManifest.xml 文件中添加下列粗体显示的行：

```xml
<?xml version="1.0" encoding="utf-8"?>
<manifest xmlns:android="http://schemas.android.com/apk/res/android"
    package="net.learn2develop.LBS"
    android:versionCode="1"
    android:versionName="1.0" >

    <uses-sdk android:minSdkVersion="14" />
    <uses-permission android:name="android.permission.INTERNET"/>

    <application
        android:icon="@drawable/ic_launcher"
```

343

```xml
        android:label="@string/app_name" >
    <uses-library android:name="com.google.android.maps" />
    <activity
        android:label="@string/app_name"
        android:name=".LBSActivity" >
        <intent-filter >
            <action android:name="android.intent.action.MAIN" />
            <category android:name="android.intent.category.
               LAUNCHER" />
        </intent-filter>
    </activity>
</application>

</manifest>
```

(3) 在 LBSActivity.java 文件中添加下列粗体显示的语句。注意，LBSActivity 现在扩展 MapActivity 基类。

```java
package net.learn2develop.LBS;

import com.google.android.maps.MapActivity;
import android.os.Bundle;

public class LBSActivity extends MapActivity {
    /** Called when the activity is first created. */
    @Override
    public void onCreate(Bundle savedInstanceState) {
        super.onCreate(savedInstanceState);
        setContentView(R.layout.main);
    }

    @Override
    protected boolean isRouteDisplayed() {
        // TODO Auto-generated method stub
        return false;
    }
}
```

(4) 按 F11 键在 Android 模拟器上调试应用程序。图 9-5 展示了在应用程序的活动中显示的 Google Maps。

示例说明

为了在应用程序中显示 Google Maps，首先需要在清单文件中拥有 INTERNET 权限。然后在 UI 文件中添加<com.google.android.maps.MapView>元素来把地图嵌入到活动中。特别重要的是，活动现在必须扩展 MapActivity 基类，后者本身是 Activity 类的扩展。对于 MapActivity 类，需要实现一个方法：isRouteDisplayed()。这一方法用于 Google 的计算目的，如果在地图上显示了路径信息，则此方法应返回 true。对于一些最简单的情况，可以只是返回 false。

第 9 章 基于位置的服务

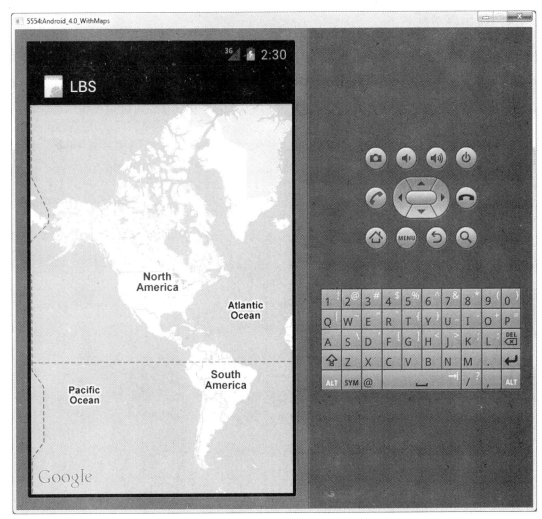

图 9-5

为了在 Android 模拟器上测试应用程序，一定要在创建 AVD 时选择 Google Maps API 作为目标。

如果看不到地图

如果您看到的只是一个带有网格的空白屏幕而没有显示 Google Maps，那最有可能的是在 main.xml 文件中使用了错误的 API 密钥。还有可能是在 AndroidManifest.xml 文件中缺少 INTERNET 权限。最后，确保在您的模拟器/设备上具有 Internet 访问权限。

如果程序没有运行(也即程序崩溃)，那可能是您忘记在 AndroidManifest.xml 文件中添加以下语句：

`<uses-library android:name="com.google.android.maps" />`

注意这条语句在 AndroidManifest.xml 文件中的位置：它应该位于<Application>元素内。

9.1.4 显示缩放控件

前一节展示了如何在 Android 应用程序中显示 Google Maps。可以将地图平移到任何想要的位置上,地图能够随即进行动态更新。然而,在模拟器上是没有办法对地图上一个特定的位置直接进行放大或缩小的(在一个真正的 Android 设备上,可以用手指捏放地图进行缩放)。因此,在本节中将学习如何利用内置的缩放控件,使用户可以对地图进行放大或缩小的操作。

试一试　　**显示内置的缩放控件**

(1) 使用前一节创建的项目,添加下列粗体显示的语句:

```java
package net.learn2develop.LBS;

import com.google.android.maps.MapActivity;
import com.google.android.maps.MapView;

import android.os.Bundle;

public class LBSActivity extends MapActivity {
    MapView mapView;
    /** Called when the activity is first created. */
    @Override
    public void onCreate(Bundle savedInstanceState) {
        super.onCreate(savedInstanceState);
        setContentView(R.layout.main);

        mapView = (MapView) findViewById(R.id.mapView);
        mapView.setBuiltInZoomControls(true);
    }

    @Override
    protected boolean isRouteDisplayed() {
        // TODO Auto-generated method stub
        return false;
    }
}
```

(2) 按 F11 键在 Android 模拟器上调试应用程序。当单击并拖拽地图时,可以注意到在地图底部显示的内置缩放控件(如图 9-6 所示)。可以通过单击减号(-)图标缩小地图,单击加号(+)图标放大地图。

第 9 章 基于位置的服务

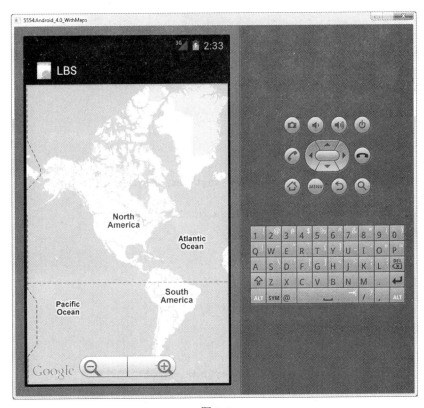

图 9-6

示例说明

要显示内置的缩放控件,首先要获取对 MapView 的引用并调用 setBuiltInZoomControls()方法:

```
mapView = (MapView) findViewById(R.id.mapView);
mapView.setBuiltInZoomControls(true);
```

除了显示缩放控件,还可以使用 MapController 类的 zoomIn()或 zoomOut()方法以编程方式来实现对地图的缩放。下面的"试一试"将告诉您如何做到这一点。

试一试 以编程方式缩放地图

(1) 使用前一节创建的项目,在 LBSActivity.java 文件中添加下列粗体显示的语句:

```
package net.learn2develop.LBS;

import com.google.android.maps.MapActivity;
import com.google.android.maps.MapController;
import com.google.android.maps.MapView;

import android.os.Bundle;
import android.view.KeyEvent;

public class LBSActivity extends MapActivity {
    MapView mapView;
```

347

```java
/** Called when the activity is first created. */
@Override
public void onCreate(Bundle savedInstanceState) {
    super.onCreate(savedInstanceState);
    setContentView(R.layout.main);

    mapView = (MapView) findViewById(R.id.mapView);
    mapView.setBuiltInZoomControls(true);
}

public boolean onKeyDown(int keyCode, KeyEvent event)
{
    MapController mc = mapView.getController();
    switch (keyCode)
    {
        case KeyEvent.KEYCODE_3:
            mc.zoomIn();
            break;
        case KeyEvent.KEYCODE_1:
            mc.zoomOut();
            break;
    }
    return super.onKeyDown(keyCode, event);
}

@Override
protected boolean isRouteDisplayed() {
    // TODO Auto-generated method stub
    return false;
}
}
```

(2) 按 F11 键在 Android 模拟器上调试应用程序。现在，可以通过按下模拟器上的数字键 3 来放大地图，按数字键 1 来缩小地图。

示例说明

为了处理在活动上的按键操作，需要处理 onKeyDown 事件：

```java
public boolean onKeyDown(int keyCode, KeyEvent event)
{
    MapController mc = mapView.getController();
    switch (keyCode)
    {
        case KeyEvent.KEYCODE_3:
            mc.zoomIn();
            break;
        case KeyEvent.KEYCODE_1:
            mc.zoomOut();
            break;
    }
```

```
        return super.onKeyDown(keyCode, event);
    }
```

为了管理对地图的平移和缩放，需要从 MapView 对象中获得一个 MapController 类的实例。MapController 类包含了 zoomIn()和 zoomOut()方法(再加上其他一些用来控制地图的方法)，可以使用户对地图进行放大或缩小。

注意，如果将应用程序部署到真实设备上，可能无法测试这个应用程序的缩放功能，因为现在大多数 Android 设备并没有物理键盘。

9.1.5 改变视图

默认情况下，Google Maps 是以地图视图来显示的，它基本上描绘了感兴趣的街道和地方。还可以使用 MapView 类的 setSatellite()方法将 Google Maps 设置为以卫星视图显示。

```
    @Override
    public void onCreate(Bundle savedInstanceState) {
        super.onCreate(savedInstanceState);
        setContentView(R.layout.main);

        mapView = (MapView) findViewById(R.id.mapView);
        mapView.setBuiltInZoomControls(true);
        mapView.setSatellite(true);
    }
```

图 9-7 展示了以卫星视图显示的 Google Maps。

图 9-7

如果想要在地图上显示交通状况,可以使用setTraffic()方法:

```
@Override
public void onCreate(Bundle savedInstanceState) {
    super.onCreate(savedInstanceState);
    setContentView(R.layout.main);

    mapView = (MapView) findViewById(R.id.mapView);
    mapView.setBuiltInZoomControls(true);
    mapView.setSatellite(true);
    mapView.setTraffic(true);
}
```

图 9-8 所示的地图上显示了当前的交通状况(需要缩放地图来查看道路)。不同的颜色反映不同的交通状况。一般来说,绿色相当于约每小时 50 英里的车速,交通比较顺畅;黄色相当于约每小时 25～50 英里的车速,交通状况中等;红色相当于低于约每小时 25 英里的车速,交通比较拥堵。

图 9-8

注意,只有美国、法国、英国、澳大利亚以及加拿大的主要城市提供交通状况信息,不过新的国家和城市在不断加入。

9.1.6 导航到特定位置

默认情况下,当 Google Maps 首次加载时,显示的是美国地图。然而,也可以将 Google

Maps 设置为显示一个特定的位置。在这种情况下，可以使用 MapController 类的 animateTo() 方法。

下面的"试一试"向您展示了如何以编程方式将 Google Maps 动画显示到一个特定位置。

试一试　　将地图导航到一个特定位置并进行显示

(1) 使用前一节创建的项目，在 LBSActivity.java 文件中添加下列粗体显示的语句：

```java
package net.learn2develop.LBS;

import com.google.android.maps.GeoPoint;
import com.google.android.maps.MapActivity;
import com.google.android.maps.MapController;
import com.google.android.maps.MapView;

import android.os.Bundle;
import android.view.KeyEvent;

public class LBSActivity extends MapActivity {
    MapView mapView;
    MapController mc;
    GeoPoint p;

    /** Called when the activity is first created. */
    @Override
    public void onCreate(Bundle savedInstanceState) {
        super.onCreate(savedInstanceState);
        setContentView(R.layout.main);

        mapView = (MapView) findViewById(R.id.mapView);
        mapView.setBuiltInZoomControls(true);
        mapView.setSatellite(true);
        mapView.setTraffic(true);

        mc = mapView.getController();
        String coordinates[] = {"1.352566007", "103.78921587"};
        double lat = Double.parseDouble(coordinates[0]);
        double lng = Double.parseDouble(coordinates[1]);

        p = new GeoPoint(
            (int) (lat * 1E6),
            (int) (lng * 1E6));

        mc.animateTo(p);
        mc.setZoom(13);
        mapView.invalidate();
    }

    public boolean onKeyDown(int keyCode, KeyEvent event)
    {
```

```
        //...
    }

    @Override
    protected boolean isRouteDisplayed() {
        //...
    }
}
```

(2) 按 F11 键在 Android 模拟器上调试应用程序。当地图被加载后，可注意到它动画显示到新加坡的一个特定位置(如图 9-9 所示)。

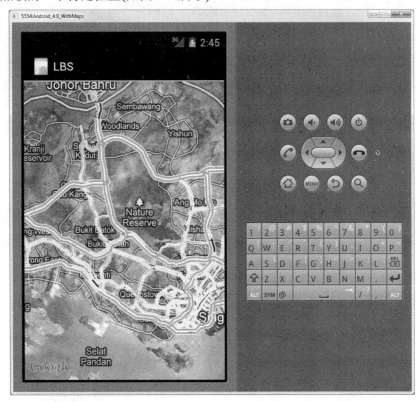

图 9-9

示例说明

在前面的代码中，首先从 MapView 实例中获取一个地图控制器并将其赋给一个 MapController 对象(mc)。然后使用一个 GeoPoint 对象来代表一个地理位置。注意，对于 GeoPoint 类，一个位置的经度和纬度是用微度来表示的。这意味着它们是以整数值存储。例如，对于一个纬度值 40.747778，需要乘以 1e6(10 的 6 次方，一百万)得到 40747778。

要将地图导航到特定位置，可以使用 MapController 类的 animateTo()方法。setZoom()方法可以用来指定显示地图所采用的缩放级别(数字越大，地图上就能显示更多细节)。invalidate()方法强制重绘 MapView。

9.1.7 添加标记

在地图上添加标记来指明感兴趣的位置,这样可以使用户很容易地定位他们所要查找的地方。下面的"试一试"展示了如何在 Google Maps 上添加标记。

试一试　在地图上添加标记

(1) 创建一个包含一个图钉的 GIF 图像(如图 9-10 所示),并将其复制到项目的 res/drawable-mdpi 文件夹下。为达到最好效果,将图像背景变为透明,以防止添加其后会遮挡住部分地图。

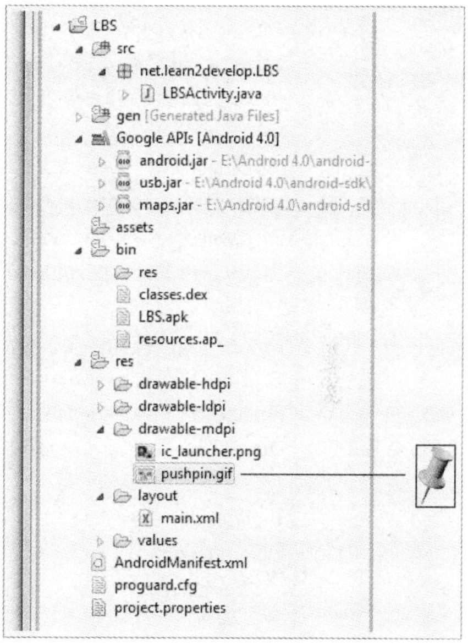

(2) 使用先前的活动中创建的项目,在 LBSActivity.java 文件中添加下列粗体显示的语句:

```
package net.learn2develop.LBS;

import java.util.List;

import om.google.android.maps.GeoPoint;
import com.google.android.maps.MapActivity;
import com.google.android.maps.MapController;
import com.google.android.maps.MapView;
import com.google.android.maps.Overlay;

import android.graphics.Bitmap;
import android.graphics.BitmapFactory;
import android.graphics.Canvas;
import android.graphics.Point;
import android.os.Bundle;
import android.view.KeyEvent;

public class LBSActivity extends MapActivity {
    MapView mapView;
    MapController mc;
    GeoPoint p;

    private class MapOverlay extends com.google.android.maps.Overlay
    {
        @Override
        public boolean draw(Canvas canvas, MapView mapView,
        boolean shadow, long when)
        {
            super.draw(canvas, mapView, shadow);

            //---translate the GeoPoint to screen pixels---
            Point screenPts = new Point();
            mapView.getProjection().toPixels(p, screenPts);
```

图 9-10

```java
            //---add the marker---
            Bitmap bmp = BitmapFactory.decodeResource(
                getResources(), R.drawable.pushpin);
            canvas.drawBitmap(bmp, screenPts.x, screenPts.y-50, null);
            return true;
        }
    }

    /** Called when the activity is first created. */
    @Override
    public void onCreate(Bundle savedInstanceState) {
        super.onCreate(savedInstanceState);
        setContentView(R.layout.main);

        mapView = (MapView) findViewById(R.id.mapView);
        mapView.setBuiltInZoomControls(true);
        mapView.setSatellite(true);
        mapView.setTraffic(true);

        mc = mapView.getController();
        String coordinates[] = {"1.352566007", "103.78921587"};
        double lat = Double.parseDouble(coordinates[0]);
        double lng = Double.parseDouble(coordinates[1]);

        p = new GeoPoint(
            (int) (lat * 1E6),
            (int) (lng * 1E6));

        mc.animateTo(p);
        mc.setZoom(13);

        //---Add a location marker---
        MapOverlay mapOverlay = new MapOverlay();
        List<Overlay> listOfOverlays = mapView.getOverlays();
        listOfOverlays.clear();
        listOfOverlays.add(mapOverlay);

        mapView.invalidate();
    }

    public boolean onKeyDown(int keyCode, KeyEvent event)
    {
        //...
    }

    @Override
    protected boolean isRouteDisplayed() {
        //...
    }
}
```

(3) 按 F11 键在 Android 模拟器上调试应用程序。图 9-11 显示了添加到地图上的标记。

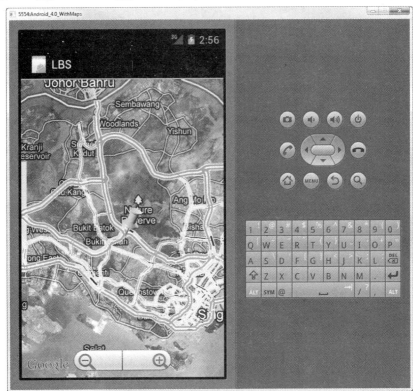

图 9-11

示例说明

为了在地图上添加标记,首先需要定义一个扩展 Overlay 类的类:

```
private class MapOverlay extends com.google.android.maps.Overlay
    {
        @Override
        public boolean draw(Canvas canvas, MapView mapView,
        boolean shadow, long when)
        {
            //...
        }
    }
```

覆盖代表可以在地图上绘制的单独的一项。可以添加任意多个覆盖。在 MapOverlay 类中,重写 draw()方法,这样就可以在地图上绘制出图钉图像。特别要注意的是,需要将地理位置(由 GeoPoint 对象 p 表示)转换成屏幕坐标:

```
//---translate the GeoPoint to screen pixels---
Point screenPts = new Point();
mapView.getProjection().toPixels(p, screenPts);
```

因为想让图钉的钉尖来指示地点的具体位置,所以需要从这个点(如图 9-12 所示)的 y

坐标扣除图像的高度(50 像素)，并在该位置绘制图像：

```
//---add the marker---
Bitmap bmp = BitmapFactory.decodeResource(
    getResources(), R.drawable.pushpin);
canvas.drawBitmap(bmp, screenPts.x, screenPts.y-50, null);
```

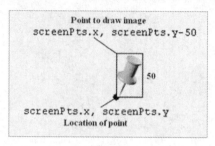

图 9-12

要添加标记，创建 MapOverlay 类的一个实例并把它添加到 MapView 对象的可用覆盖列表中：

```
//---Add a location marker---
MapOverlay mapOverlay = new MapOverlay();
List<Overlay> listOfOverlays = mapView.getOverlays();
listOfOverlays.clear();
listOfOverlays.add(mapOverlay);
```

9.1.8 获取触摸的位置

使用 Google Maps 一段时间之后，您可能想知道刚刚在屏幕上触碰到的位置对应地点的经纬度。知道这个信息非常有用，因为这样就可以确定一个位置的地址。这一过程被称为反向地理编码(我们将在下一节学习如何做到这一点)。

如果已经在地图上添加了一个覆盖,可以在 MapOverlay 类中重写 onTouchEvent()方法。每次用户触摸地图时都会触发这一方法。此方法有两个参数：MotionEvent 和 MapView。使用 MotionEvent 参数，可以利用 getAction()方法来判断用户是否已经从屏幕上抬起了他/她的手指。在下面的代码片段中，如果用户触碰了屏幕，随即又抬起了手指，那么将显示出所触碰位置对应的经度和纬度。

```
package net.learn2develop.LBS;

import java.util.List;

import com.google.android.maps.GeoPoint;
import com.google.android.maps.MapActivity;
import com.google.android.maps.MapController;
import com.google.android.maps.MapView;
import com.google.android.maps.Overlay;

import android.graphics.Bitmap;
```

```java
import android.graphics.BitmapFactory;
import android.graphics.Canvas;
import android.graphics.Point;
import android.os.Bundle;
import android.view.KeyEvent;
import android.view.MotionEvent;
import android.widget.Toast;

public class LBSActivity extends MapActivity {
    MapView mapView;
    MapController mc;
    GeoPoint p;

    private class MapOverlay extends com.google.android.maps.Overlay
    {
        @Override
        public boolean draw(Canvas canvas, MapView mapView,
        boolean shadow, long when)
        {
            //...
        }

        @Override
        public boolean onTouchEvent(MotionEvent event, MapView mapView)
        {
            //---when user lifts his finger---
            if (event.getAction() == 1) {
                GeoPoint p = mapView.getProjection().fromPixels(
                    (int) event.getX(),
                    (int) event.getY());
                    Toast.makeText(getBaseContext(),
                        "Location: "+
                        p.getLatitudeE6() / 1E6 + "," +
                        p.getLongitudeE6() /1E6 ,
                        Toast.LENGTH_SHORT).show();
            }
            return false;
        }
    }
    //...
}
```

getProjection()方法返回一个用于在屏幕像素坐标和经纬度坐标之间进行转换的投影。然后，fromPixels()方法将屏幕坐标转换成 GeoPoint 对象。

图 9-13 展示了当用户单击地图上一个位置时所显示的一组坐标。

图 9-13

9.1.9 地理编码和反向地理编码

正如前一节所述，如果知道某个位置的纬度和经度，就可以使用一个称为反向地理编码的过程来找到它的地址。Android 中的 Google Maps 是通过 Geocoder 类来支持这一点的。下面的代码片段展示了如何利用 getFromLocation()方法来获取您刚刚所触碰位置的地址：

```java
package net.learn2develop.LBS;

import java.io.IOException;
import java.util.List;
import java.util.Locale;

import com.google.android.maps.GeoPoint;
import com.google.android.maps.MapActivity;
import com.google.android.maps.MapController;
import com.google.android.maps.MapView;
import com.google.android.maps.Overlay;

import android.graphics.Bitmap;
import android.graphics.BitmapFactory;
import android.graphics.Canvas;
import android.graphics.Point;
import android.location.Address;
```

```java
import android.location.Geocoder;
import android.os.Bundle;
import android.view.KeyEvent;
import android.view.MotionEvent;
import android.widget.Toast;

public class LBSActivity extends MapActivity {
    MapView mapView;
    MapController mc;
    GeoPoint p;

    private class MapOverlay extends com.google.android.maps.Overlay
    {
        @Override
        public boolean draw(Canvas canvas, MapView mapView,
        boolean shadow, long when)
        {
            //...
        }

        @Override
        public boolean onTouchEvent(MotionEvent event, MapView mapView)
        {
            //---when user lifts his finger---
            if (event.getAction() == 1) {
                GeoPoint p = mapView.getProjection().fromPixels(
                        (int) event.getX(),
                        (int) event.getY());

                /*
                    Toast.makeText(getBaseContext(),
                        "Location: "+
                        p.getLatitudeE6() / 1E6 + "," +
                        p.getLongitudeE6() /1E6 ,
                        Toast.LENGTH_SHORT).show();
                 */

                Geocoder geoCoder = new Geocoder(
                        getBaseContext(), Locale.getDefault());
                try {
                    List<Address> addresses = geoCoder.getFromLocation(
                            p.getLatitudeE6() / 1E6,
                            p.getLongitudeE6() / 1E6, 1);

                    String add = "";
                    if (addresses.size() > 0)
                    {
                        for (int i=0;i<addresses.get(0).getMaxAddress
                        LineIndex();
                                i++)
```

```
                    add += addresses.get(0).getAddressLine(i) + "\n";
                }
                Toast.makeText(getBaseContext(),add,Toast.LENGTH_SHORT).
                show();
            }
            catch (IOException e) {
                e.printStackTrace();
            }
            return true;
        }
        return false;
    }
    //...
}
```

Geocoder 对象使用 getFromLocation()方法将经度和纬度转换成一个地址。一旦获得地址之后，使用 Toast 类来显示它。图 9-14 展示了应用程序显示在地图上所触碰位置的地址。

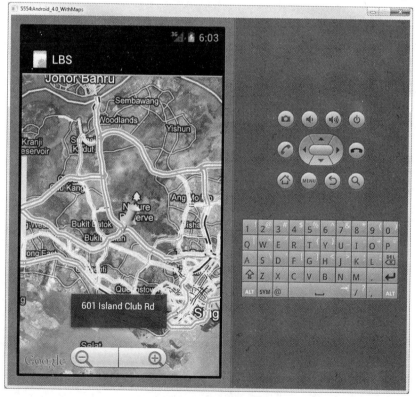

图 9-14

如果知道一个位置的地址，但想要知道它的经度和纬度，那么可以通过地理编码做到这一点。同样，要达到此目的，可以使用 Geocoder 类。下面的代码演示了如何利用 getFromLocationName()方法来获取帝国大厦的准确位置：

```
@Override
```

```java
public void onCreate(Bundle savedInstanceState) {
    super.onCreate(savedInstanceState);
    setContentView(R.layout.main);
    
    mapView = (MapView) findViewById(R.id.mapView);
    mapView.setBuiltInZoomControls(true);
    mapView.setSatellite(true);
    mapView.setTraffic(true);
    
    mc = mapView.getController();
    
    /*
    String coordinates[] = {"1.352566007", "103.78921587"};
    double lat = Double.parseDouble(coordinates[0]);
    double lng = Double.parseDouble(coordinates[1]);
    
    p = new GeoPoint(
        (int) (lat * 1E6),
        (int) (lng * 1E6));
    
    mc.animateTo(p);
    mc.setZoom(13);
    */
    
    //---geo-coding---
    Geocoder geoCoder = new Geocoder(this, Locale.getDefault());
    try {
        List<Address> addresses = geoCoder.getFromLocationName(
            "empire state building", 5);
        
        if (addresses.size() > 0) {
            p = new GeoPoint(
                    (int) (addresses.get(0).getLatitude() * 1E6),
                    (int) (addresses.get(0).getLongitude() * 1E6));
            mc.animateTo(p);
            mc.setZoom(20);
        }
    } catch (IOException e) {
        e.printStackTrace();
    }
    
    //---Add a location marker---
    MapOverlay mapOverlay = new MapOverlay();
    List<Overlay> listOfOverlays = mapView.getOverlays();
    listOfOverlays.clear();
    listOfOverlays.add(mapOverlay);
    
    mapView.invalidate();
}
```

图 9-15 显示了地图被导航到帝国大厦的位置。

图 9-15

9.2 获取位置数据

如今，移动设备普遍配备了 GPS 接收器。由于有许多卫星绕地球运行，因此使用 GPS 接收器可以很容易找到您所在的位置。然而，GPS 工作时要求在空旷的地方，因此在室内或卫星无法穿透的地方(如穿山隧道)，GPS 常常是无效的。

另一种用于定位的有效方式是通过发射塔三角测量法。当移动电话处于开机状态时，它会不断地与其周围的基站联系。在知晓发射塔的标识后，通过使用含有发射塔的标识和它们所处的确切地理位置的各种数据库，可以将这一信息转换为物理位置。发射塔三角测量法的优点是在室内也起作用，无须获得来自卫星的信息。然而，由于该方法的准确性取决于重叠信号的覆盖范围，其变化相当多，因此它不如 GPS 来得精确。发射塔三角测量法在人口稠密地区最为有效，因为那里的发射塔离得很近。

第三种定位方法是依靠 Wi-Fi 三角测量法。设备连接到 Wi-Fi 网络而不是连接到发射塔，并对照数据库来确定服务提供商所服务的位置。这里所描述的 3 种方法中，Wi-Fi 三角测量法是最不准确的。

Android SDK 提供了 LocationManager 类，可帮助您的设备确定用户的物理位置。下面的"试一试"展示了如何用代码做到这一点。

试一试   将地图导航到特定位置

(1) 使用前一节创建的同一个项目,在 LBSActivity.java 文件中添加下列粗体显示的语句:

```
package net.learn2develop.LBS;

import java.io.IOException;
import java.util.List;
import java.util.Locale;

import com.google.android.maps.GeoPoint;
import com.google.android.maps.MapActivity;
import com.google.android.maps.MapController;
import com.google.android.maps.MapView;
import com.google.android.maps.Overlay;

import android.content.Context;

import android.graphics.Bitmap;
import android.graphics.BitmapFactory;
import android.graphics.Canvas;
import android.graphics.Point;
import android.location.Address;
import android.location.Geocoder;

import android.location.Location;
import android.location.LocationListener;
import android.location.LocationManager;

import android.os.Bundle;
import android.view.KeyEvent;
import android.view.MotionEvent;
import android.widget.Toast;

public class LBSActivity extends MapActivity {
    MapView mapView;
    MapController mc;
    GeoPoint p;

    LocationManager lm;
    LocationListener locationListener;

    private class MapOverlay extends com.google.android.maps.Overlay
    {
        //...
    }

    /** Called when the activity is first created. */
    @Override
    public void onCreate(Bundle savedInstanceState) {
```

```java
        super.onCreate(savedInstanceState);
        setContentView(R.layout.main);
        
        mapView = (MapView) findViewById(R.id.mapView);
        mapView.setBuiltInZoomControls(true);
        mapView.setSatellite(true);
        mapView.setTraffic(true);
        
        mc = mapView.getController();
        
        /*
        String coordinates[] = {"1.352566007", "103.78921587"};
        double lat = Double.parseDouble(coordinates[0]);
        double lng = Double.parseDouble(coordinates[1]);
        
        p = new GeoPoint(
            (int) (lat * 1E6),
            (int) (lng * 1E6));
        
        mc.animateTo(p);
        mc.setZoom(13);
        */
        
        //---geo-coding---
        Geocoder geoCoder = new Geocoder(this, Locale.getDefault());
        try {
            //...
        }
        
        //---Add a location marker---
        MapOverlay mapOverlay = new MapOverlay();
        List<Overlay> listOfOverlays = mapView.getOverlays();
        listOfOverlays.clear();
        listOfOverlays.add(mapOverlay);
        
        mapView.invalidate();
        
        //---use the LocationManager class to obtain locations data---
        lm = (LocationManager)
            getSystemService(Context.LOCATION_SERVICE);
        
        locationListener = new MyLocationListener();
    }
    
    @Override
    public void onResume() {
        super.onResume();
        
        //---request for location updates---
        lm.requestLocationUpdates(
```

```java
                    LocationManager.GPS_PROVIDER,
                    0,
                    0,
                    locationListener);
    }

    @Override
    public void onPause() {
        super.onPause();

        //---remove the location listener---
        lm.removeUpdates(locationListener);
    }

    private class MyLocationListener implements LocationListener
    {
        public void onLocationChanged(Location loc) {
            if (loc != null) {
                Toast.makeText(getBaseContext(),
                    "Location changed : Lat: " + loc.getLatitude() +
                    " Lng: " + loc.getLongitude(),
                    Toast.LENGTH_SHORT).show();

                p = new GeoPoint(
                    (int) (loc.getLatitude() * 1E6),
                    (int) (loc.getLongitude() * 1E6));

                mc.animateTo(p);
                mc.setZoom(18);
            }
        }

        public void onProviderDisabled(String provider) {
        }

        public void onProviderEnabled(String provider) {
        }

        public void onStatusChanged(String provider, int status,
            Bundle extras) {
        }
    }

    public boolean onKeyDown(int keyCode, KeyEvent event)
    {
        //...
    }

    @Override
    protected boolean isRouteDisplayed() {
```

```
        //...
    }
}
```

(2) 在 AndroidManifest.xml 文件中添加下列粗体显示的行：

```
<?xml version="1.0" encoding="utf-8"?>
<manifest xmlns:android="http://schemas.android.com/apk/res/android"
    package="net.learn2develop.LBS"
    android:versionCode="1"
    android:versionName="1.0" >

    <uses-sdk android:minSdkVersion="14" />
    <uses-permission android:name="android.permission.INTERNET"/>
    <uses-permission android:name="android.permission.ACCESS_FINE_
    LOCATION"/>

    <application
        android:icon="@drawable/ic_launcher"
        android:label="@string/app_name" >
        <uses-library android:name="com.google.android.maps" />
        <activity
            android:label="@string/app_name"
            android:name=".LBSActivity" >
            <intent-filter>
                <action android:name="android.intent.action.MAIN" />
                <category android:name="android.intent.category.LAUNCHER" />
            </intent-filter>
        </activity>
    </application>

</manifest>
```

(3) 按 F11 键在 Android 模拟器上调试应用程序。

(4) 为了模拟 Android 模拟器收到的 GPS 数据，可以使用 Eclipse 的 DDMS 透视图中的 Location Controls 工具（如图 9-16 所示）。

(5) 首先确保已经在 Devices 选项卡中选中了模拟器，然后在 Emulator Control 选项卡中找到 Location Controls 工具，选择 Manual 选项卡。输入经度和纬度，然后单击 Send 按钮。

(6) 现在可观察到模拟器上的地图动画显示到另一个位置(如图 9-17 所示)。这证明应用程序已经收到了 GPS 数据。

图 9-16

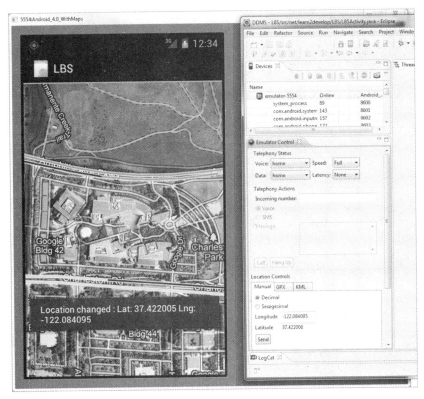

图 9-17

示例说明

在 Android 中，基于位置的服务是由 LocationManager 类提供的，位于 android.location 包中。使用 LocationManager 类，应用程序可以定期获取设备的地理位置的更新，并在它进入某个位置附近时触发一个意图。

在 LBSActivity.java 文件中，首先使用 getSystemService()方法获取一个指向 LocationManager 类的引用。这是在 LBSActivity 的 onCreate()方法中完成的：

```
//---use the LocationManager class to obtain locations data---
lm = (LocationManager)
    getSystemService(Context.LOCATION_SERVICE);

locationListener = new MyLocationListener();
```

接下来，创建了 MyLocationListener 类的一个实例并进行了定义。
MyLocationListener 类实现了 LocationListener 抽象类。在其实现中需要重写 4 个方法：

- onLocationChanged(Location location)——当位置发生变化时调用
- onProviderDisabled(String provider)——当用户禁用了提供者时调用
- onProviderEnabled(String provider)——当用户启动了提供者时调用
- onStatusChanged(String provider, int status, Bundle extras)——当提供者的状态改变时调用

在本例中，你对位置改变时发生了什么更感兴趣，因此在 onLocationChanged()方法中编写了一些代码。具体来说，当位置发生变化时，在屏幕上显示一个小对话框，给出新的位置信息：纬度和经度。这个对话框是使用 Toast 类显示的：

```
public void onLocationChanged(Location loc) {
    if (loc != null) {
        Toast.makeText(getBaseContext(),
            "Location changed : Lat: " + loc.getLatitude() +
            " Lng: " + loc.getLongitude(),
            Toast.LENGTH_SHORT).show();

        p = new GeoPoint(
            (int) (loc.getLatitude() * 1E6),
            (int) (loc.getLongitude() * 1E6));

        mc.animateTo(p);
        mc.setZoom(18);
    }
}
```

在前面的方法中，还将地图定位到接收到的位置。

为了在位置发生变化时得到通知，需要注册一个位置变化的请求，这样才会定期地通知您的程序。这是通过在活动的 onResume()方法中使用 requestLocationUpdates()方法来完成的：

```
@Override
public void onResume() {
    super.onResume();

    //---request for location updates---
    lm.requestLocationUpdates(
        LocationManager.GPS_PROVIDER,
        0,
        0,
        locationListener);
}
```

requestLocationUpdates()方法接受 4 个参数：

- provider——您所注册的服务提供商的名称。在本例中，使用 GPS 来获取地理位置数据。
- minTime——进行通知的最短时间间隔，用毫秒作为单位。0 表示想要持续得到位置变化的通知。
- minDistance——进行通知的最短距离，用米作为单位。0 表示想要持续得到位置变化的通知。
- listener——一个对象，在每一次位置更新时将调用其 onLocationChanged()方法。

最后，在 onPause()方法中，当活动被销毁或者进入后台时将删除侦听器(这样应用程序就不再侦听位置变化，从而节约了设备的电量)。这是使用 removeUpdates()方法完成的：

```
@Override
public void onPause() {
    super.onPause();

    //---remove the location listener---
    lm.removeUpdates(locationListener);
}
```

如果想使用 Cell-ID 和 Wi-Fi 三角测量法(对于室内使用很重要)获得您的位置数据,那么可以使用网络位置服务提供商,如下所示:

```
@Override
public void onResume() {
    super.onResume();

    //---request for location updates---
    lm.requestLocationUpdates(
            LocationManager.GPS_PROVIDER,
            0,
            0,
            locationListener);
}
```

为使用网络位置服务提供商,需要在 AndroidManifest.xml 文件中添加 ACCESS_COARSE_LOCATION 权限:

```
<?xml version="1.0" encoding="utf-8"?>
<manifest xmlns:android="http://schemas.android.com/apk/res/android"
    package="net.learn2develop.LBS"
    android:versionCode="1"
    android:versionName="1.0" >

    <uses-sdk android:minSdkVersion="14" />
    <uses-permission android:name="android.permission.INTERNET"/>
    <uses-permission
        android:name="android.permission.ACCESS_FINE_LOCATION"/>
    <uses-permission
        android:name="android.permission.ACCESS_COARSE_LOCATION"/>

    <application
        android:icon="@drawable/ic_launcher"
        android:label="@string/app_name" >
        <uses-library android:name="com.google.android.maps" />
        <activity
            android:label="@string/app_name"
            android:name=".LBSActivity" >
            <intent-filter >
                <action android:name="android.intent.action.MAIN" />

                <category android:name="android.intent.category.LAUNCHER"
```

```
            />
                </intent-filter>
        </activity>
    </application>

</manifest>
```

注意：在 Android 模拟器上不能使用网络提供商。如果在模拟器上测试前面的代码，会导致一个非法参数异常。所以，需要在一个真实的设备上测试代码。

在应用程序中，可以把 GPS 位置服务提供商和网络位置服务提供商结合起来。

```
@Override
public void onResume() {
    super.onResume();

    //---request for location updates---
    lm.requestLocationUpdates(
            LocationManager.GPS_PROVIDER,
            0,
            0,
            locationListener);

    //---request for location updates---
    lm.requestLocationUpdates(
            LocationManager.NETWORK_PROVIDER,
            0,
            0,
            locationListener);
}
```

然而要注意，因为 GPS 位置服务提供商和网络位置服务提供商都会尝试用自己的方法获得位置信息(GPS 与 Wi-Fi 和 Cell ID 三角测量法)，所以这样做会使应用程序收到两组不同的坐标。因此，在设备中监视两个位置服务提供商的状态并使用合适的那个十分重要。通过实现 MyLocationListener 类的如下三个方法(粗体显示)，可以检查这两个位置提供商的状态：

```
private class MyLocationListener implements LocationListener
{
    @Override
    public void onLocationChanged(Location loc) {
        if (loc != null) {
            Toast.makeText(getBaseContext(),
                    "Location changed : Lat: " + loc.getLatitude() +
                    " Lng: " + loc.getLongitude(),
                    Toast.LENGTH_SHORT).show();
```

```java
            p = new GeoPoint(
                    (int) (loc.getLatitude() * 1E6),
                    (int) (loc.getLongitude() * 1E6));

            mc.animateTo(p);
            mc.setZoom(18);
        }
    }

    //---called when the provider is disabled---
    public void onProviderDisabled(String provider) {
        Toast.makeText(getBaseContext(),
                provider + " disabled",
                Toast.LENGTH_SHORT).show();
    }

    //---called when the provider is enabled---
    public void onProviderEnabled(String provider) {
        Toast.makeText(getBaseContext(),
                provider + " enabled",
                Toast.LENGTH_SHORT).show();
    }

    //---called when there is a change in the provider status---
    public void onStatusChanged(String provider, int status,
        Bundle extras) {
        String statusString = "";
        switch (status) {
            case android.location.LocationProvider.AVAILABLE:
                statusString = "available";
            case android.location.LocationProvider.OUT_OF_SERVICE:
                statusString = "out of service";
            case android.location.LocationProvider.TEMPORARILY_
                UNAVAILABLE:
                statusString = "temporarily unavailable";
        }

        Toast.makeText(getBaseContext(),
                provider + " " + statusString,
                Toast.LENGTH_SHORT).show();
    }
}
```

9.3 监控一个位置

LocationManager 类的一个非常酷的功能是它能够监视一个特定的位置。这是使用 addProximityAlert()方法来实现的。

下面的代码片段显示了如何监控一个特定位置。如果用户位于从该位置起 5 米的半径内，应用程序将触发一个意图来启动 Web 浏览器：

```
import android.app.PendingIntent;
import android.content.Intent;
import android.net.Uri;

        //---use the LocationManager class to obtain locations data---
        lm = (LocationManager)
            getSystemService(Context.LOCATION_SERVICE);

        //---PendingIntent to launch activity if the user is within
        // some locations---
        PendingIntent pendingIntent = PendingIntent.getActivity(
            this, 0, new
            Intent(android.content.Intent.ACTION_VIEW,
              Uri.parse("http://www.amazon.com")), 0);

        lm.addProximityAlert(37.422006,-122.084095,5,-1,pendingIntent);
```

addProximityAlert()方法接受 5 个参数：纬度、经度、半径(以米为单位)、有效期限(接近警报的有效时间，时间过后将删除警报；-1 表示不会过期)，以及挂起的意图。

注意，如果 Android 设备的屏幕进入睡眠状态，也是每 4 分钟检查一次接近状态，这样可以以延长设备的电池寿命。

9.4 项目——创建一个位置跟踪应用程序

现在您已经知道了如何创建一个基于位置的 Android 应用程序，可以实际运用这些知识了。接下来将把本章中介绍的技术和第 8 章介绍的技术结合起来，创建一个很酷、很实用的应用程序。所构建的这个位置跟踪应用程序可以安装到用户的 Android 设备上。向用户的设备发送包含特殊代码的 SMS 消息时，用户的设备将自动发回一条 SMS 消息，其中包含了设备的位置。这类位置跟踪应用程序可以用来跟踪孩子的位置或者独自生活的一位年长的亲属(并且被跟踪人可能不知道这一点)。

警告：在把这款应用程序发布给用户之前，注意在一些国家中，没有当事人的同意就跟踪他/她的位置是非法行为。如果在某个用户的手机上安装了位置跟踪应用程序，那么每当有人向该手机发送一条 Where are you?的 SMS 消息时，该手机就会向发送者发回其位置信息。因此，如果想要在现实生活中使用此项目，必须告知潜在用户该应用程序的功能，这样他们可以选择不暴露自己的位置。

试一试　　调用一个活动

(1) 使用 Eclipse 创建一个新的 Android 项目，命名为 LocationTracker。

(2) 在 AndroidManifest.xml 文件中添加下列粗体显示的代码：

```xml
<?xml version="1.0" encoding="utf-8"?>
<manifest xmlns:android="http://schemas.android.com/apk/res/android"
    package="net.learn2develop.LocationTracker"
    android:versionCode="1"
    android:versionName="1.0">
    <uses-sdk android:minSdkVersion="14" />

    <uses-permission android:name="android.permission.RECEIVE_SMS" />
    <uses-permission android:name="android.permission.SEND_SMS" />
    <uses-permission android:name="android.permission.ACCESS_COARSE_
        LOCATION" />

    <application android:icon="@drawable/icon" android:label="@string/
        app_name">
        <activity android:name=".LocationTrackerActivity"
                android:label="@string/app_name">
            <intent-filter>
                <action android:name="android.intent.action.MAIN" />
                <category android:name="android.intent.category.
                    LAUNCHER" />
            </intent-filter>
        </activity>

        <!-- put this here so that even if the app is not running,
        your app can be woken up when there is an incoming SMS message -->
        <receiver android:name=".SMSReceiver">
            <intent-filter android:priority="100">
                <action
                    android:name="android.provider.Telephony.SMS_RECEIVED"/>
            </intent-filter>
        </receiver>
    </application>
</manifest>
```

(3) 在项目的包名中添加一个新的 Java 类，命名为 SMSReceiver。现在，项目的包名下应该有一个名为 SMSReceiver.java 的 Java 文件，如图 9-18 所示。

图 9-18

(4) 使用下列粗体显示的代码填充 SMSReceiver.java 文件：

```java
package net.learn2develop.LocationTracker;
```

```java
import android.content.BroadcastReceiver;
import android.content.Context;
import android.content.Intent;
import android.location.Location;
import android.location.LocationListener;
import android.location.LocationManager;
import android.os.Bundle;
import android.telephony.SmsManager;
import android.telephony.SmsMessage;

public class SMSReceiver extends BroadcastReceiver
{
    LocationManager lm;
    LocationListener locationListener;
    String senderTel;

    @Override
    public void onReceive(Context context, Intent intent)
    {
        //---get the SMS message that was received---
        Bundle bundle = intent.getExtras();
        SmsMessage[] msgs = null;
        String str="";
        if (bundle != null)
        {
            senderTel = "";
            //---retrieve the SMS message received---
            Object[] pdus = (Object[]) bundle.get("pdus");
            msgs = new SmsMessage[pdus.length];
            for (int i=0; i<msgs.length; i++){
                msgs[i] = SmsMessage.createFromPdu((byte[])pdus[i]);
                if(i==0) {
                    //---get the sender address/phone number---
                    senderTel = msgs[i].getOriginatingAddress();
                }
                //---get the message body---
                str += msgs[i].getMessageBody().toString();
            }

            if (str.startsWith("Where are you?")) {
                //---use the LocationManager class to obtain locations
                data---
            lm = (LocationManager)
                    context.getSystemService(Context.LOCATION_SERVICE);

                //---request location updates---
                locationListener = new MyLocationListener();
                lm.requestLocationUpdates(
                        LocationManager.NETWORK_PROVIDER,
                        60000,
```

```
                    1000,
                    locationListener);

            //---abort the broadcast; SMS messages won¡¯t be broadcasted---
            this.abortBroadcast();
        }
    }
}

private class MyLocationListener implements LocationListener
{
    @
    public void onLocationChanged(Location loc) {
        if (loc != null) {
            //---send a SMS containing the current location---
            SmsManager sms = SmsManager.getDefault();
            sms.sendTextMessage(senderTel, null,
                    "http://maps.google.com/maps?q=" +
                    loc.getLatitude() + "," +
                        loc.getLongitude(), null, null);

            //---stop listening for location changes---
            lm.removeUpdates(locationListener);
        }
    }

    public void onProviderDisabled(String provider) {
    }

    public void onProviderEnabled(String provider) {
    }

    public void onStatusChanged(String provider, int status,
        Bundle extras) {
    }
}
}
```

(5) 为了测试应用程序，首先把它部署到一个真实的 Android 设备上。然后，使用另外一部手机(任意一部可以发送 SMS 消息的手机即可)向它发送一条 SMS 消息：Where are you?

(6) 在发送 SMS 消息后，等待该 Android 设备发回一条 SMS 消息。图 9-19 显示了 Android 设备的回复(这里使用了一部 iPhone)，其中包含了该设备的位置。

(7) 大多数智能手机能够识别 SMS 消息中的 URL 数据。因此，如果单击 SMS 消息中的 URL，可以在 Google Maps 中看到这个位置，如图 9-20 所示。

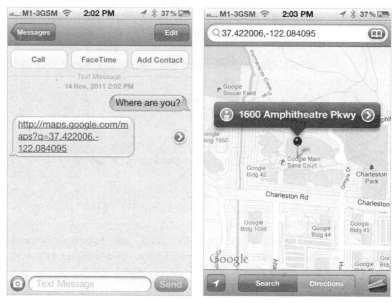

图 9-19　　　　　　　　图 9-20

示例说明

这个项目在一个完整的应用程序中结合运用了本章学习的概念和在第 8 章中学习的关于 SMS 消息传递的知识。在安装了这个应用程序后，它会侦听传入的 SMS 消息是否包含文本"Where are you?"。然后它会拦截这些 SMS 消息，这样用户就不会在 Android 设备上的 Messaging 应用程序中看到这些消息。

当收到 SMS 消息后，首先提取发送者的电话号码，这样以后就可以向这个号码发送包含设备位置的回复：

```java
//---retrieve the SMS message received---
senderTel = "";

//---retrieve the SMS message received---
Object[] pdus = (Object[]) bundle.get("pdus");
msgs = new SmsMessage[pdus.length];
for (int i=0; i<msgs.length; i++){
    msgs[i] = SmsMessage.createFromPdu((byte[])pdus[i]);
    if (i==0) {
        //---get the sender address/phone number---
        senderTel = msgs[i].getOriginatingAddress();
    }
    //---get the message body---
    str += msgs[i].getMessageBody().toString();
}
```

然后观察 SMS 消息的内容。如果它以句子 Where are you?开头，就使用 LocationManager 类请求位置更新：

```java
if (str.startsWith("Where are you?")) {
```

```
            //---use the LocationManager class to obtain locations data---
            lm = (LocationManager)
                    context.getSystemService(Context.LOCATION_SERVICE);

            //---request location updates---
            locationListener = new MyLocationListener();
            lm.requestLocationUpdates(
                    LocationManager.NETWORK_PROVIDER,
                    60000,
                    1000,
                    locationListener);

            this.abortBroadcast();
        }
```

注意，这个示例中使用了网络位置服务提供商来获取位置信息，因为它不需要在一个空旷的位置(这是 GPS 位置服务服务提供商的要求)。但是，使用网络位置服务提供商需要设备连接到 Internet，所以如果设备没有网络连接，应用程序将无法工作。

获得位置后，使用一个指向 Google Maps 的 URL 发回一条包含设备位置的 SMS 消息：

```
        public void onLocationChanged(Location loc) {
            if (loc != null) {
                //---send a SMS containing the current location---
                SmsManager sms = SmsManager.getDefault();
                sms.sendTextMessage(senderTel, null,
                        "http://maps.google.com/maps?q="+loc.getLatitude()+","+
                            loc.getLongitude(), null, null);

                //---stop listening for location changes---
                lm.removeUpdates(locationListener);
            }
        }
```

发送 SMS 消息后，立即删除位置更新应用程序，这样就不会继续侦听位置变化。

注意，代码中分别为 requestLocationUpdates()方法的 minTime 和 minDistance 参数使用了 60 000 毫秒和 1000 米。回忆一下，本章前面为这两个参数使用的值是 0，这是为了能够不断地收到位置变化的信息。但是，在这个项目中不应该这么做，否则应用程序会不断地一次性向 SMS 消息发送者发回多条 SMS 消息。这是因为当停止侦听位置更新时，Location Manager 将会调用几次 onLocationChanged()方法，报告设备最小的位置变化，从而导致发送多条 SMS 消息。在实际的测试中，SMS 消息的数量在 10 条到 60 条之间。因此，应该将 minTime 和 minDistance 参数设为更加合理的值，这样 Location Manager 就不能重复地触发 onLocationChanged()方法。

9.5 本章小结

本章对在 Android 应用程序中用于显示 Google Maps 的 MapView 对象进行了大致的了解。我们学习了操作地图的不同方式，还学习了如何使用不同的网络服务提供商来获取地理位置数据：采用 GPS、Cell-ID 和 Wi-Fi 三角测量法。最后，学习了如何构建一个使用 SMS 消息传递跟踪用户位置的位置跟踪应用程序。

> **练 习**
>
> 1. 如果您在 Android 应用程序中嵌入了 Google Maps API，但在应用程序加载时并没有显示出地图，那么最可能的原因有哪些？
> 2. 地理编码和反向地理编码有何区别？
> 3. 说出可用于获得位置数据的两个位置服务提供商。
> 4. 监控一个位置的方法是什么？
>
> 练习答案参见附录 C。

本章主要内容

主　　题	关　键　概　念

显示 MapView	`<com.google.android.maps.MapView`
	　　`android:id="@+id/mapView"`
	　　`android:layout_width="fill_parent"`
	　　`android:layout_height="fill_parent"`
	　　`android:enabled="true"`
	　　`android:clickable="true"`
	　　`android:apiKey="YOUR_MAPS_API_KEY" />`

引用 Map 库	`<uses-library android:name="com.google.android.maps" />`
显示缩放控件	`mapView.setBuiltInZoomControls(true);`
以编程方式对地图进行缩放	`mc.zoomIn();` `mc.zoomOut();`
改变视图	`mapView.setSatellite(true);` `mapView.setTraffic(true);`

动画显示到一个特定位置	`mc = mapView.getController();`
	`String coordinates[] = {"1.352566007", "103.78921587"};`
	`double lat = Double.parseDouble(coordinates[0]);`
	`double lng = Double.parseDouble(coordinates[1]);`
	`p = new GeoPoint(`
	　　`(int) (lat * 1E6),`
	　　`(int) (lng * 1E6));`
	`mc.animateTo(p);`

添加标记	实现一个 Overlay 类并重写 draw()方法

获取在地图上触摸的位置	`GeoPoint p = mapView.getProjection().fromPixels(`
	　　`(int) event.getX(),`
	　　`(int) event.getY());`

(续表)

主　题	关　键　概　念
地理编码和反向地理编码	使用 Geocoder 类
| 获取位置数据 | ```java
private LocationManager lm;
//...
        lm = (LocationManager)
            getSystemService(Context.LOCATION_SERVICE);
        locationListener = new MyLocationListener();
        lm.requestLocationUpdates(
                LocationManager.GPS_PROVIDER,
                0,
                0,
                locationListener);
//...
    private class MyLocationListener implements LocationListener
    {
        public void onLocationChanged(Location loc){
            if (loc != null) {
            }
        }

        public void onProviderDisabled(String
            provider) {
        }

        public void onProviderEnabled(String
            provider){
        }

        public void onStatusChanged(String
            provider, int status,Bundle extras) {
        }
    }
``` |
| 监控一个位置 | `lm.addProximityAlert(37.422006, -122.084095,5,-1, pendingIntent);` |
```

第 10 章

联 网

本章将介绍以下内容：
- 如何利用 HTTP 连接 Web。
- 如何使用 XML Web 服务。
- 如何使用 JSON Web 服务。
- 如何与套接字服务器建立连接。

在第 8 章，已经学习了应用程序如何使用 SMS 消息传递和邮件来与外界环境通信。另一种与外界环境进行通信的方式是通过 Android 设备可用的无线网络。因此，在本章中将学习如何使用 HTTP 协议与 Web 服务器通信，以下载文本和二进制数据，也将学习如何分析 XML 文件来提取 XML 文档的相关部分——这是一种在访问 Web 服务时非常有用的技术。除了 XML Web 服务以外，本章将涵盖 JSON(JavaScript Object Notation，JavaScript 对象符号)，这是 XML 的一种轻量级替代形式。本章中将使用 Android SDK 提供的类来操作 JSON 内容。

最后，本章还将演示如何编写一个使用 TCP 套接字连接服务器的 Android 应用程序。利用套接字编程，可以编写复杂但有趣的网络应用程序。

10.1 通过 HTTP 使用 Web 服务

与外界进行通信的一种常用方法是通过 HTTP。大多数人对 HTTP 并不会感到陌生，它是推动 Web 走向成功的一种协议。通过使用 HTTP 协议，可以执行各种广泛的任务，比如从 Web 服务器下载网页、下载二进制数据等。

下面的"试一试"创建了一个 Android 项目，从而可以利用 HTTP 协议连接到 Web 来下载各种内容。

试一试 为 HTTP 连接创建基础项目

代码文件 Networking.zip 可从 Wrox.com 下载

(1) 利用 Eclipse 创建一个新的 Android 项目,将其命名为 Networking。
(2) 在 AndroidManifest.xml 文件中添加下列粗体显示的语句。

```xml
<?xml version="1.0" encoding="utf-8"?>
<manifest xmlns:android="http://schemas.android.com/apk/res/android"
    package="net.learn2develop.Networking"
    android:versionCode="1"
    android:versionName="1.0" >

    <uses-sdk android:minSdkVersion="14" />
    <uses-permission android:name="android.permission.INTERNET"/>

    <application
        android:icon="@drawable/ic_launcher"
        android:label="@string/app_name" >
        <activity
            android:label="@string/app_name"
            android:name=".NetworkingActivity" >
            <intent-filter >
                <action android:name="android.intent.action.MAIN" />

                <category android:name="android.intent.category.LAUNCHER" />
            </intent-filter>
        </activity>
    </application>

</manifest>
```

(3) 把下列包导入到 NetworkingActivity.java 文件中。

```java
package net.learn2develop.Networking;

import android.app.Activity;
import android.os.Bundle;
import java.io.IOException;
import java.io.InputStream;
import java.net.HttpURLConnection;
import java.net.URL;
import java.net.URLConnection;
import android.util.Log;
public class NetworkingActivity extends Activity {
    /** Called when the activity is first created. */
    @Override
    public void onCreate(Bundle savedInstanceState) {
        super.onCreate(savedInstanceState);
        setContentView(R.layout.main);
    }
}
```

(4) 在 NetworkingActivity.java 文件中定义 OpenHttpConnection()方法。

```java
public class NetworkingActivity extends Activity {
    private InputStream OpenHttpConnection(String urlString) throws
    IOException
    {
        InputStream in = null;
        int response = -1;

        URL url = new URL(urlString);
        URLConnection conn = url.openConnection();

        if (!(conn instanceof HttpURLConnection))
            throw new IOException("Not an HTTP connection");
        try{
            HttpURLConnection httpConn = (HttpURLConnection) conn;
            httpConn.setAllowUserInteraction(false);
            httpConn.setInstanceFollowRedirects(true);
            httpConn.setRequestMethod("GET");
            httpConn.connect();
            response = httpConn.getResponseCode();
            if (response == HttpURLConnection.HTTP_OK) {
                in = httpConn.getInputStream();
            }
        }
        catch (Exception ex)
        {
            Log.d("Networking", ex.getLocalizedMessage());
            throw new IOException("Error connecting");
        }
        return in;
    }

    /** Called when the activity is first created. */
    @Override
    public void onCreate(Bundle savedInstanceState) {
        super.onCreate(savedInstanceState);
        setContentView(R.layout.main);
    }
}
```

示例说明

因为是使用 HTTP 协议连接到 Web，应用程序需要 INTERNET 许可，因此，首先将许可协议添加到 AndroidManifest.xml 文件中。

然后定义 OpenHttpConnection 方法，它以一个 URL 字符串作为参数，并返回一个 InputStream 对象。为下载数据，通过使用一个 InputStream 对象从流对象读取数据。在这个方法中，可以利用 HttpURLConnection 对象建立与远程 URL 的 HTTP 连接。可以设置连接的各种特性，比如请求方法等等。

```java
HttpURLConnection httpConn = (HttpURLConnection) conn;
```

```
httpConn.setAllowUserInteraction(false);
httpConn.setInstanceFollowRedirects(true);
httpConn.setRequestMethod("GET");
```

在尝试建立与服务器的连接之后，会返回 HTTP 响应代码。如果连接成功建立(响应代码为 HTTP_OK)，就可以继续从连接中得到一个 InputStream 对象：

```
httpConn.connect();
response = httpConn.getResponseCode();
if (response == HttpURLConnection.HTTP_OK) {
    in = httpConn.getInputStream();
}
```

接下来可以使用 InputStreams 对象开始从服务器下载数据。

10.1.1 下载二进制数据

需要执行的一件常见任务是从 Web 下载二进制数据。比如，可能想要从服务器下载一张图片，以便可以在应用程序中显示。下面的"试一试"展示了下载二进制数据的方法。

试一试　下载二进制数据

(1) 利用之前创建的相同项目，在 main.xml 文件中，将默认的 TextView 替换为下列粗体显示的语句。

```xml
<?xml version="1.0" encoding="utf-8"?>
<LinearLayout xmlns:android="http://schemas.android.com/apk/res/android"
    android:layout_width="fill_parent"
    android:layout_height="fill_parent"
    android:orientation="vertical" >

    <ImageView
        android:id="@+id/img"
        android:layout_width="wrap_content"
        android:layout_height="wrap_content"
        android:layout_gravity="center" />

</LinearLayout>
```

(2) 在 NetworkingActivity.java 文件中添加下列粗体显示的语句。

```java
import android.widget.ImageView;
import android.graphics.Bitmap;
import android.graphics.BitmapFactory;
import android.os.AsyncTask;

public class NetworkingActivity extends Activity {
    ImageView img;

    private InputStream OpenHttpConnection(String urlString) throws IOException
```

```java
{
    InputStream in = null;
    int response = -1;

    URL url = new URL(urlString);
    URLConnection conn = url.openConnection();

    if (!(conn instanceof HttpURLConnection))
        throw new IOException("Not an HTTP connection");
    try{
        HttpURLConnection httpConn = (HttpURLConnection) conn;
        httpConn.setAllowUserInteraction(false);
        httpConn.setInstanceFollowRedirects(true);
        httpConn.setRequestMethod("GET");
        httpConn.connect();
        response = httpConn.getResponseCode();
        if (response == HttpURLConnection.HTTP_OK) {
            in = httpConn.getInputStream();
        }
    }
    catch (Exception ex)
    {
        Log.d("Networking", ex.getLocalizedMessage());
        throw new IOException("Error connecting");
    }
    return in;
}

private Bitmap DownloadImage(String URL)
{
    Bitmap bitmap = null;
    InputStream in = null;
    try {
        in = OpenHttpConnection(URL);
        bitmap = BitmapFactory.decodeStream(in);
        in.close();
    } catch (IOException e1) {
        Log.d("NetworkingActivity", e1.getLocalizedMessage());
    }
    return bitmap;
}

private class DownloadImageTask extends AsyncTask<String,Void,Bitmap>{
    protected Bitmap doInBackground(String... urls) {
        return DownloadImage(urls[0]);
    }

    protected void onPostExecute(Bitmap result) {
        ImageView img = (ImageView) findViewById(R.id.img);
        Img.setImageBitmap(result);
```

```java
        }
    }

    /** Called when the activity is first created. */
    @Override
    public void onCreate(Bundle savedInstanceState) {
        super.onCreate(savedInstanceState);
        setContentView(R.layout.main);
        new DownloadImageTask().execute(
            "http://www.mayoff.com/5-01cablecarDCP01934.jpg");
    }
}
```

(3) 按 F11 键在 Android 模拟器上调试应用程序，图 10-1 显示从 Web 下载并在 ImageView 中展示的一幅图片。

示例说明

DownLoadImage()方法接受要下载的图片的 URL 作为参数，并使用之前定义的 OpenHttpConnection()方法来打开与服务器的连接。通过使用 BitmapFactory 类的 decodeStream()方法和连接返回的 InputStream 对象下载数据并把数据解码为 Bitmap 对象。DownloadImage()方法返回一个 Bitmap 对象。

为了下载一张图片并在活动中显示它，可以调用 DownloadImage()方法，但是，从 Android3.0 版本开始，不能直接在 UI 线程中直接执行同步操作。如果尝试在 onCreate()方式中直接调用 DowmloadImage()方法(如以下代码段所示)，那么当应用程序运行在 Android3.0 及更高版本的设备上时会崩溃。

```java
    /** Called when the activity is first created. */
    @Override
    public void onCreate(Bundle savedInstanceState) {
        super.onCreate(savedInstanceState);
        setContentView(R.layout.main);
        //---download an image---
        //---code will not run in Android 3.0 and beyond---
        Bitmap bitmap =
            DownloadImage("http://www.mayoff.com/5-01cablecarDCP01934.
               jpg");
        img = (ImageView) findViewById(R.id.img);
        img.setImageBitmap(bitmap);
    }
```

因为 DownloadImage()方法是同步的——这意味着，在图片下载完之前不会返回控制权——所以直接调用它会导致活动 UI 冻结，这在 Android 3.0 及更高版本上是不允许的，必须使用 AsyncTask 类封装所有同步代码。使用 AsyncTask 允许在单独的线程中执行后台任务，然后在 UI 线程中返回结果。这样的话，不需要处理复杂的线程问题就可以执行后台操作。

图 10-1

如要异步访问 DownloadImage()方法,需要将代码段封装在 AsyncTask 类的子类中,如下所示:

```
private class DownloadImageTask extends AsyncTask<String,Void,Bitmap>{
    protected Bitmap doInBackground(String... urls) {
        return DownloadImage(urls[0]);
    }
    protected void onPostExecute(Bitmap result) {
        ImageView img = (ImageView) findViewById(R.id.img);
        img.setImageBitmap(result);
    }
}
```

这里基本上就是定义一个扩展了 AsyncTask 类的类(DownloadImageTask)。在本例中,DownloadImageTask 类中有两个方法:doInBackground()和 onPostExcute()。

把所有需要异步运行的代码都置于 doInBackGround()方法中。当任务完成后,通过 onPostExcute()方法传回结果。在本例中,利用 ImageView 展示下载的图片。

在 UI 线程中运行同步操作

具体来说,如果将 AndroidManifest.xml 文件中 android:minSdkVersion 属性的值设置为小于或等于 9,那么在 Android 3.0 或更高版本的设备上运行应用程序时,同步代码仍然会在 UI 线程中运行(尽管不推荐)。但是,如果将 android:minSdkVersion 属性的值设置为大于或等于 10,同步代码不会在 UI 线程中运行。

 注意:第 11 章会详细讨论 AsyncTask 类。

为调用 DownloadImageTask 类,创建它的一个实例并调用它的 execute()方法,传入要下载的图片的 URL。

```java
@Override
public void onCreate(Bundle savedInstanceState) {
    super.onCreate(savedInstanceState);
    setContentView(R.layout.main);
    new DownloadImageTask().execute(
        "http://www.mayoff.com/5-01cablecarDCP01934.jpg");
}
```

如果想要异步下载一系列图片,可以按如下方法修改 DownloadImageTask 类。

```java
...
import android.widget.Toast;
...
    private class DownloadImageTask extends AsyncTask
    <String, Bitmap, Long> {
        //---takes in a list of image URLs in String type---
        protected Long doInBackground(String... urls) {
            long imagesCount = 0;
            for (int i = 0; i < urls.length; i++) {
                //---download the image---
                Bitmap imageDownloaded = DownloadImage(urls[i]);
                if (imageDownloaded != null) {
                    //---increment the image count---
                    imagesCount++;
                    try {
                        //---insert a delay of 3 seconds---
                        Thread.sleep(3000);
                    } catch (InterruptedException e) {
                        e.printStackTrace();
                    }
                    //---return the image downloaded---
                    publishProgress(imageDownloaded);
                }
            }
            //---return the total images downloaded count---
            return imagesCount;
        }

        //---display the image downloaded---
        protected void onProgressUpdate(Bitmap... bitmap) {
            img.setImageBitmap(bitmap[0]);
        }

        //---when all the images have been downloaded---
        protected void onPostExecute(Long imagesDownloaded) {
            Toast.makeText(getBaseContext(),
                "Total " + imagesDownloaded + " images downloaded",
```

```
            Toast.LENGTH_LONG).show();
    }
}
```

注意在本例中，DownloadImageTask 类有另一个方法：onProgressUpdate()。因为要在 AsyncTask 类内部执行的任务可能耗时很长，所以调用 publishProgess()方法来更新操作的进度。这会触发 onProgressUpdate()方法，在本例中它会显示要下载的图片。onProgress-Update()方法在 UI 线程中执行，因此，使用从服务器下载的位图更新 ImageView 是线程安全的。

为了在后台异步下载一系列图片，需要创建一个 BackgroundTask 类的实例并访问它的 execute()方法，如下所示：

```
@Override
public void onCreate(Bundle savedInstanceState) {
    super.onCreate(savedInstanceState);
    setContentView(R.layout.main);

    /* new DownloadImageTask().execute(
        "http://www.mayoff.com/5-01cablecarDCP01934.jpg");
    */

    img = (ImageView) findViewById(R.id.img);
    new DownloadImageTask().execute(
            "http://www.mayoff.com/5-01cablecarDCP01934.jpg",
            "http://www.hartiesinfo.net/greybox/Cable_Car_
                Hartbeespoort.jpg",
            "http://mcmanuslab.ucsf.edu/sites/default/files/
                imagepicker/m/mmcmanus/
                CaliforniaSanFranciscoPaintedLadiesHz.jpg",
            "http://www.fantom-xp.com/wallpapers/63/San_Francisco
                _-_Sunset.jpg",
            "http://travel.roro44.com/europe/france/
                Paris_France.jpg",
            "http://wwp.greenwichmeantime.com/time-zone/usa/nevada
                /las-vegas/hotel/the-strip/paris-las-vegas/paris-
                las-vegas-hotel.jpg",
            "http://designheaven.files.wordpress.com/2010/04/
                eiffel_tower_paris_france.jpg");
}
```

当运行上述代码段时，图片在后台下载并且以每 3 秒一次的频率显示。当下载完最后一张图片后，Toast 类显示已下载的图片总数。

从模拟器指向 localhost

当使用 Android 模拟器时，可能经常需要利用 localhost 访问本地 Web 服务器上存储的数据。例如，你自己的 Web 服务很可能在开发过程中驻留在本地计算机上，而你想要在编写 Android 应用程序所用的相同开发机器上测试它们。在这种情况下，应该利用特殊的 IP

地址 10.0.2.2(而不是 127.0.0.1)来指向主机的回环接口。从 Android 模拟器的角度来看，localhost(127.0.0.1)指的是它本身的回环接口。

10.1.2 下载文本内容

除了下载二进制数据以外，还可以下载纯文本内容。例如，想要访问一个返回随机引文字符串的 Web 服务。下面的"试一试"展示如何在应用程序中从 Web 服务下载一个字符串。

试一试　下载纯文本内容

(1) 利用之前创建相同项目，在 NetworkingActivity.java 文件中添加下列粗体显示的语句。

```java
import java.io.InputStreamReader;

    private String DownloadText(String URL)
    {
        int BUFFER_SIZE = 2000;
        InputStream in = null;
        try {
            in = OpenHttpConnection(URL);
        } catch (IOException e) {
            Log.d("Networking", e.getLocalizedMessage());
            return "";
        }

        InputStreamReader isr = new InputStreamReader(in);
        int charRead;
        String str = "";
        char[] inputBuffer = new char[BUFFER_SIZE];
        try {
            while ((charRead = isr.read(inputBuffer))>0) {
                //---convert the chars to a String---
                String readString =
                    String.copyValueOf(inputBuffer, 0, charRead);
                str += readString;
                inputBuffer = new char[BUFFER_SIZE];
            }
            in.close();
        } catch (IOException e) {
            Log.d("Networking", e.getLocalizedMessage());
            return "";
        }
        return str;
    }

    private class DownloadTextTask extends AsyncTask<String, Void, String>
    {
        protected String doInBackground(String... urls) {
```

```
            return DownloadText(urls[0]);
        }

        @Override
        protected void onPostExecute(String result) {
            Toast.makeText(getBaseContext(), result, Toast.LENGTH_LONG).
                show();
        }
    }

    /** Called when the activity is first created. */
    @Override
    public void onCreate(Bundle savedInstanceState) {
        super.onCreate(savedInstanceState);
        setContentView(R.layout.main);

        //---download text---
        new DownloadTextTask().execute(
            "http://iheartquotes.com/api/v1/random?max_characters=256&
              max_lines=10");
    }
}
```

(2) 按 F11 键在 Android 模拟器中调试应用程序。图 10-2 展示了利用 Toast 类下载和展示的随机引文字符串。

图 10-2

示例说明

DownloadText()方法接受要下载的文本文件的 URL 作为参数,返回所下载文本文件的字符串。它基本上就是打开一个与服务器的 HTTP 连接,然后使用一个 InputStreamReader 对象读取流中的每个字符,并将其保存在一个 String 对象中。正如前一节所示,必须创建一个 AsyncTask 类的子类才能异步调用 DownloadText()方法。

10.1.3 通过 GET 方法访问 Web 服务

到现在为止,已经介绍如何从 Web 下载图像和文本。前一节介绍如何从服务器下载一些纯文本。常见的一种情况是需要下载 XML 文件并解析其内容(使用 Web 服务是一个很好的例子)。因此,本节将介绍如何使用 HTTP GET 方法连接 Web 服务。一旦 Web 服务返回 XML 格式的结果,就提取相关的部分并使用 Toast 类显示它的内容。

本例中,使用了 http://services.aonaware.com/DictService/DictService.asmx?op=Define 提供的 Web 方法。这个 Web 方法来自一个词典 Web 服务,它会返回给定词的定义。

该 Web 方法接受如下格式的请求。

```
GET /DictService/DictService.asmx/Define?word=string HTTP/1.1
Host: services.aonaware.com
HTTP/1.1 200 OK
Content-Type: text/xml; charset=utf-8
Content-Length: length
```

它以下列格式返回响应。

```xml
<?xml version="1.0" encoding="utf-8"?>
<WordDefinition xmlns="http://services.aonaware.com/webservices/">
  <Word>string</Word>
  <Definitions>
    <Definition>
      <Word>string</Word>
      <Dictionary>
        <Id>string</Id>
        <Name>string</Name>
      </Dictionary>
      <WordDefinition>string</WordDefinition>
    </Definition>
    <Definition>
      <Word>string</Word>
      <Dictionary>
        <Id>string</Id>
        <Name>string</Name>
      </Dictionary>
      <WordDefinition>string</WordDefinition>
    </Definition>
  </Definitions>
</WordDefinition>
```

因此，要获取一个词的定义，必须先建立一个连接到该 Web 方法的 HTTP 连接，然后解析返回的 XML 结果。下面的"试一试"展示了具体做法。

试一试　　使用 Web 服务

(1) 利用之前创建的相同项目，在 NetworkingActivity.java 文件中添加下列粗体显示的语句。

```
import javax.xml.parsers.DocumentBuilder;
import javax.xml.parsers.DocumentBuilderFactory;
import javax.xml.parsers.ParserConfigurationException;

import org.w3c.dom.Document;
import org.w3c.dom.Element;
import org.w3c.dom.Node;
import org.w3c.dom.NodeList;

    private String WordDefinition(String word) {
        InputStream in = null;
        String strDefinition = "";
        try {
            in = OpenHttpConnection(
  "http://services.aonaware.com/DictService/DictService.asmx/Define?word=
    " + word);
            Document doc = null;
            DocumentBuilderFactory dbf =
                DocumentBuilderFactory.newInstance();
            DocumentBuilder db;
            try {
                db = dbf.newDocumentBuilder();
                doc = db.parse(in);
            } catch (ParserConfigurationException e) {
                // TODO Auto-generated catch block
                e.printStackTrace();
            } catch (Exception e) {
                // TODO Auto-generated catch block
                e.printStackTrace();
            }
            doc.getDocumentElement().normalize();

            //---retrieve all the <Definition> elements---
            NodeList definitionElements =
                doc.getElementsByTagName("Definition");

            //---iterate through each <Definition> elements---
            for (int i = 0; i < definitionElements.getLength(); i++) {
                Node itemNode = definitionElements.item(i);
                if (itemNode.getNodeType() == Node.ELEMENT_NODE)
                {
                    //---convert the Definition node into an Element---
                    Element definitionElement = (Element) itemNode;
```

```java
                    //---get all the <WordDefinition> elements under
                    // the <Definition> element---
                    NodeList wordDefinitionElements =
                        (definitionElement).getElementsByTagName(
                        "WordDefinition");

                    strDefinition = "";
                    //---iterate through each <WordDefinition> elements---
                    for (int j = 0;j<wordDefinitionElements.getLength();j++){
                        //---convert a <WordDefinition> node into an
                            Element---
                        Element wordDefinitionElement =
                            (Element) wordDefinitionElements.item(j);

                        //---get all the child nodes under the
                        // <WordDefinition> element---
                        NodeList textNodes =
                            ((Node)wordDefinitionElement).getChildNodes();

                        strDefinition +=
                            ((Node)textNodes.item(0)).getNodeValue()+".\n";
                    }

                }
            }
        } catch (IOException e1) {
            Log.d("NetworkingActivity", e1.getLocalizedMessage());
        }
        //---return the definitions of the word---
        return strDefinition;
    }

    private class AccessWebServiceTask extends AsyncTask<String, Void,
        String> {
        protected String doInBackground(String... urls) {
            return WordDefinition(urls[0]);
        }

        protected void onPostExecute(String result) {
            Toast.makeText(getBaseContext(),result,Toast.LENGTH_LONG).show();
        }
    }

    /** Called when the activity is first created. */
    @Override
    public void onCreate(Bundle savedInstanceState) {
        super.onCreate(savedInstanceState);
        setContentView(R.layout.main);
```

```
//---access a Web Service using GET---
new AccessWebServiceTask().execute("apple");
}
```

(2) 按 F11 键在 Android 模拟器中调试应用程序。图 10-3 展示了 Web 服务调用被解析的结果，然后利用 Toast 类进行显示。

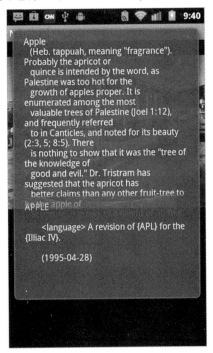

图 10-3

示例说明

WordDefinition()方法首先建立一个与 Web 服务的 HTTP 连接，需要传入感兴趣的词：

```
in = OpenHttpConnection(
"http://services.aonaware.com/DictService/DictService.asmx/Define
?word=" + word);
```

然后利用 DocumentBuliderFactory 和 DocumentBulider 对象从一个 XML(从 Web 服务返回的 XML 结果)文件中获取一个 Document (DOM)对象。

```
Document doc = null;
DocumentBuilderFactory dbf =
    DocumentBuilderFactory.newInstance();
DocumentBuilder db;
try {
    db = dbf.newDocumentBuilder();
    doc = db.parse(in);
} catch (ParserConfigurationException e) {
    // TODO Auto-generated catch block
```

```
        e.printStackTrace();
    } catch (Exception e) {
        // TODO Auto-generated catch block
        e.printStackTrace();
    }
    doc.getDocumentElement().normalize();
```

一旦获取 Document 对象，可以在<Definition>标签中发现所有元素。

```
//---retrieve all the <Definition> elements---
NodeList definitionElements =
    doc.getElementsByTagName("Definition");
```

图 10-4 展示了从 Web 服务返回的 XML 文档的结构。

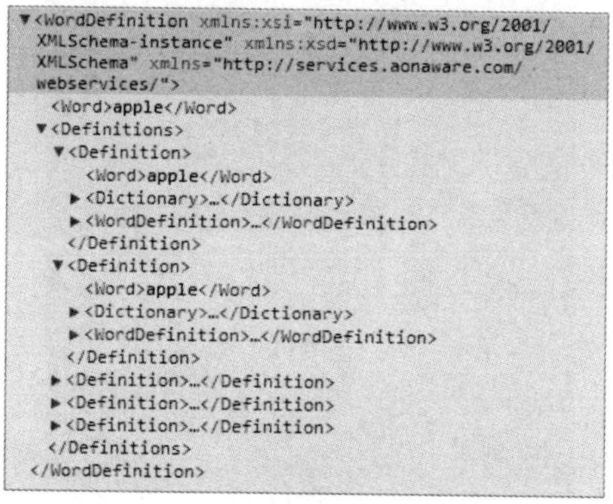

```
▼<WordDefinition xmlns:xsi="http://www.w3.org/2001/
XMLSchema-instance" xmlns:xsd="http://www.w3.org/2001/
XMLSchema" xmlns="http://services.aonaware.com/
webservices/">
    <Word>apple</Word>
    ▼<Definitions>
        ▼<Definition>
            <Word>apple</Word>
            ▶<Dictionary>…</Dictionary>
            ▶<WordDefinition>…</WordDefinition>
        </Definition>
        ▼<Definition>
            <Word>apple</Word>
            ▶<Dictionary>…</Dictionary>
            ▶<WordDefinition>…</WordDefinition>
        </Definition>
        ▶<Definition>…</Definition>
        ▶<Definition>…</Definition>
        ▶<Definition>…</Definition>
    </Definitions>
</WordDefinition>
```

图 10-4

因为词的定义包含在<WordDefinition>元素中，接下来可以提取所有定义。

```
//---iterate through each <Definition> elements---
for (int i = 0; i < definitionElements.getLength(); i++) {
    Node itemNode = definitionElements.item(i);
    if (itemNode.getNodeType() == Node.ELEMENT_NODE)
    {
        //---convert the Definition node into an Element---
        Element definitionElement = (Element) itemNode;

        //---get all the <WordDefinition> elements under
        // the <Definition> element---
        NodeList wordDefinitionElements =
            (definitionElement).getElementsByTagName(
            "WordDefinition");

        strDefinition = "";
        //---iterate through each <WordDefinition> elements---
        for (int j = 0; j < wordDefinitionElements.getLength(); j++){
            //---convert a <WordDefinition> node into an Element---
            Element wordDefinitionElement =
                (Element) wordDefinitionElements.item(j);
```

```
                //---get all the child nodes under the
                // <WordDefinition> element---
                NodeList textNodes =
                    ((Node) wordDefinitionElement).getChildNodes();

                strDefinition +=
                    ((Node) textNodes.item(0)).getNodeValue() + ". \n";
            }
        }
    }
```

上述的代码段遍历所有<Definition>元素，查找名为<WordDefinition>的子元素。<WordDefinition>元素的文本内容包含词的定义，词的定义会连接起来，并由 WordDefinition()方法返回。

```
    //---return the definitions of the word---
    return strDefinition;
```

与往常一样，需要创建 AsyncTask 类的一个子类来异步调用 WordDefinition()方法。

```
    private class AccessWebServiceTask extends AsyncTask<String, Void,
        String> {
        protected String doInBackground(String... urls) {
            return WordDefinition(urls[0]);
        }

        protected void onPostExecute(String result) {
            Toast.makeText(getBaseContext(), result, Toast.LENGTH_LONG).
            show();
        }
    }
```

最后，使用 execute()方法异步访问 Web 服务。

```
    //---access a Web Service using GET---
    new AccessWebServiceTask().execute("apple");
```

10.2 使用 JSON 服务

在前一节中，通过利用 HTTP 连接到 Web 服务器和获取 XML 格式的结果，学习了如何使用 XML Web 服务。还学习了如何使用 DOM 解析 XML 文档的结果。但是，就计算量而言，操作 XML 文档对于移动设备来说是一项代价高昂的操作，原因如下：
- XML 文档是冗长的，它利用标签来嵌入信息，其规模可能会很快变大。一个很大的 XML 文档意味着设备需要更大的带宽来下载它，开销自然更高。
- XML 文档更难处理。正如之前所示，必须利用 DOM 遍历树，以便定位想要的信息。另外，在遍历树之前，DOM 本身需要在内存中建立整个文档的树结构。两者都是 CPU 和内存密集型的操作。

表示信息更加有效的一种方法是使用 JSON(JavaScript Object Notation)格式。JSON 是一种轻量级的数据交换格式，便于人们读写，同样它也易于机器解析和生成。下列代码段展示了一段 JSON 消息。

```
[
    {
        "appeId":"1",
        "survId":"1",
        "location":"",
        "surveyDate":"2008-03 14",
        "surveyTime":"12:19:47",
        "inputUserId":"1",
        "inputTime":"2008-03-14 12:21:51",
        "modifyTime":"0000-00-00 00:00:00"
    },
    {
        "appeId":"2",
        "survId":"32",
        "location":"",
        "surveyDate":"2008-03-14",
        "surveyTime":"22:43:09",
        "inputUserId":"32",
        "inputTime":"2008-03-14 22:43:37",
        "modifyTime":"0000-00-00 00:00:00"
    },
    {
        "appeId":"3",
        "survId":"32",
        "location":"",
        "surveyDate":"2008-03-15",
        "surveyTime":"07:59:33",
        "inputUserId":"32",
        "inputTime":"2008-03-15 08:00:44",
        "modifyTime":"0000-00-00 00:00:00"
    },
    {
        "appeId":"4",
        "survId":"1",
        "location":"",
        "surveyDate":"2008-03-15",
        "surveyTime":"10:45:42",
        "inputUserId":"1",
        "inputTime":"2008-03-15 10:46:04",
        "modifyTime":"0000-00-00 00:00:00"
    },
    {
        "appeId":"5",
        "survId":"32",
        "location":"",
```

```
        "surveyDate":"2008-03-16",
        "surveyTime":"08:04:49",
        "inputUserId":"32",
        "inputTime":"2008-03-16 08:05:26",
        "modifyTime":"0000-00-00 00:00:00"
    },
    {
        "appeId":"6",
        "survId":"32",
        "location":"",
        "surveyDate":"2008-03-20",
        "surveyTime":"20:19:01",
        "inputUserId":"32",
        "inputTime":"2008-03-20 20:19:32",
        "modifyTime":"0000-00-00 00:00:00"
    }
]
```

上述的代码段表示一组参与调查的数据。注意信息被表示成键/值对的集合，每个键/值对组成对象的有序列表。不同于 XML，JSON 中没有冗长的标签名，只有方括号和花括号。

下面的"试一试"演示了如何利用 Android SDK 提供的 JSONArray 类和 JSONObject 类轻松地处理 JSON 消息。

试一试　　**使用 JSON 服务**

(1) 利用 Eclipse 创建一个新的 Android 项目，将其命名为 JSON。

(2) 在 AndroidManifest.xml 文件中添加下列粗体显示的语句。

```xml
<?xml version="1.0" encoding="utf-8"?>
<manifest xmlns:android="http://schemas.android.com/apk/res/android"
    package="net.learn2develop.JSON"
    android:versionCode="1"
    android:versionName="1.0" >

    <uses-sdk android:minSdkVersion="14" />
    <uses-permission android:name="android.permission.INTERNET"/>

    <application
        android:icon="@drawable/ic_launcher"
        android:label="@string/app_name" >
        <activity
            android:label="@string/app_name"
            android:name=".JSONActivity" >
            <intent-filter >
                <action android:name="android.intent.action.MAIN" />
                <category android:name="android.intent.category.
                    LAUNCHER" />
            </intent-filter>
        </activity>
    </application>
```

```
</manifest>
```

(3) 在 JSONActivity.java 文件中添加下列粗体显示的语句。

```java
package net.learn2develop.JSON;

import java.io.BufferedReader;
import java.io.IOException;
import java.io.InputStream;
import java.io.InputStreamReader;

import org.apache.http.HttpEntity;
import org.apache.http.HttpResponse;
import org.apache.http.StatusLine;
import org.apache.http.client.ClientProtocolException;
import org.apache.http.client.HttpClient;
import org.apache.http.client.methods.HttpGet;
import org.apache.http.impl.client.DefaultHttpClient;
import org.json.JSONArray;
import org.json.JSONObject;

import android.app.Activity;
import android.os.AsyncTask;
import android.os.Bundle;
import android.util.Log;
import android.widget.Toast;

public class JSONActivity extends Activity {

    public String readJSONFeed(String URL) {
        StringBuilder stringBuilder = new StringBuilder();
        HttpClient client = new DefaultHttpClient();
        HttpGet httpGet = new HttpGet(URL);
        try {
            HttpResponse response = client.execute(httpGet);
            StatusLine statusLine = response.getStatusLine();
            int statusCode = statusLine.getStatusCode();
            if (statusCode == 200) {
                HttpEntity entity = response.getEntity();
                InputStream content = entity.getContent();
                BufferedReader reader = new BufferedReader(
                        new InputStreamReader(content));
                String line;
                while ((line = reader.readLine()) != null) {
                    stringBuilder.append(line);
                }
            } else {
                Log.e("JSON", "Failed to download file");
            }
        } catch (ClientProtocolException e) {
```

```
            e.printStackTrace();
        } catch (IOException e) {
            e.printStackTrace();
        }
        return stringBuilder.toString();
    }

    private class ReadJSONFeedTask extends AsyncTask<String, Void, String>
    {
        protected String doInBackground(String... urls) {
            return readJSONFeed(urls[0]);
        }

        protected void onPostExecute(String result) {
            try {
            JSONArray jsonArray = new JSONArray(result);
            Log.i("JSON", "Number of surveys in feed: " +
                    jsonArray.length());

            //---print out the content of the json feed---
            for (int i = 0; i < jsonArray.length(); i++) {
                JSONObject jsonObject = jsonArray.getJSONObject(i);
                Toast.makeText(getBaseContext(),jsonObject.getString
                   ("appeId") +
                        " - " + jsonObject.getString("inputTime"),
                        Toast.LENGTH_SHORT).show();
            }
        } catch (Exception e) {
            e.printStackTrace();
        }
      }
    }

    /** Called when the activity is first
     created. */
    @Override
    public void onCreate(Bundle
     savedInstanceState) {
        super.onCreate(savedInstanceState);
        setContentView(R.layout.main);
        new ReadJSONFeedTask().execute(
            "http://extjs.org.cn/extjs/ex
            amples/grid/survey.html");
    }
}
```

图 10-5

(4) 按 F11 键在 Android 模拟器中调试应用程序。你会看到 Toast 类多次出现,显示该信息(如图 10-5 所示)。

示例说明

在这个项目中,首先定义了 readJSONFeed()方法:

```java
public String readJSONFeed(String URL) {
    StringBuilder stringBuilder = new StringBuilder();
    HttpClient client = new DefaultHttpClient();
    HttpGet httpGet = new HttpGet(URL);
    try {
        HttpResponse response = client.execute(httpGet);
        StatusLine statusLine = response.getStatusLine();
        int statusCode = statusLine.getStatusCode();
        if (statusCode == 200) {
            HttpEntity entity = response.getEntity();
            InputStream content = entity.getContent();
            BufferedReader reader = new BufferedReader(
                new InputStreamReader(content));
            String line;
            while ((line = reader.readLine()) != null) {
                stringBuilder.append(line);
            }
        } else {
            Log.e("JSON", "Failed to download file");
        }
    } catch (ClientProtocolException e) {
        e.printStackTrace();
    } catch (IOException e) {
        e.printStackTrace();
    }
    return stringBuilder.toString();
}
```

该方法简单地连接到指定的 URL,然后从 Web 服务器读取响应。它返回字符串作为结果。为了异步访问 readJSONFeed()方法,需要创建 AsyncTask()类的子类:

```java
Private class ReadJSONFeedTask extends yncTask<String,Void,String>{
    protected String doInBackground(String... urls) {
        return readJSONFeed(urls[0]);
    }

    protected void onPostExecute(String result) {
        try {
            JSONArray jsonArray = new JSONArray(result);
            Log.i("JSON", "Number of surveys in feed: " +
                jsonArray.length());

            //---print out the content of the json feed---
            for (int i = 0; i < jsonArray.length(); i++) {
                JSONObject jsonObject = jsonArray.getJSONObject(i);
                Toast.makeText(getBaseContext(), jsonObject.getString
                    ("appeId") +
```

```
                        " - " + jsonObject.getString("inputTime"),
                        Toast.LENGTH_SHORT).show();
            }
        } catch (Exception e) {
            e.printStackTrace();
        }
    }
}
```

在 doInBackground()方法中调用 readJSONFeed()方法，已经提取的 JSON 字符串通过 onPostExcute()方法传递。本例中使用的 JSON 字符串(之前介绍过的)来自 http://extjs.org.cn/extjs/examples/grip/survey.html。

为在 JSON 字符串里获取对象列表，使用了 JSONArray 类，并向其传入 JSON 源作为该类的构造函数。

```
JSONArray jsonArray = new JSONArray(result);
Log.i("JSON", "Number of surveys in feed: " +
        jsonArray.length());
```

length()方法在 jsonArray 对象里返回对象的数目。通过迭代存储在 jsonArray 对象里的对象列表，使用 getJOSNObject()方法来获取每个对象：

```
//---print out the content of the json feed---
for (int i = 0; i < jsonArray.length(); i++) {
    JSONObject jsonObject = jsonArray.getJSONObject(i);

    Toast.makeText(this, jsonObject.getString("appeId") +
            " - " + jsonObject.getString("inputTime"),
            Toast.LENGTH_SHORT).show();
}
```

getJSONObject()方法返回一个类型为 JSONObject 的对象。为了获取存储在对象内部的键/值对的值，使用了 getString()方法(对于其他数据类型，也可以使用 getInt()、getLong()以及 getBoolean()方法)。

最后，使用 execute()方法异步访问 JSON 源。

```
new ReadJSONFeedTask().execute(
        "http://extjs.org.cn/extjs/examples/grid/survey.html");
```

本例展示如何使用一个 JSON 服务并快速解析其结果。更有意思的一个例子是使用一个真实生活的场景：Twitter。下列修改能使应用程序从 Twitter 中提取最新的 tweet 并在 Toast 类中显示它们(如图 10-6 所示)：

图 10-6

```java
    private class ReadJSONFeedTask extends AsyncTask<String, Void, String> {
        protected String doInBackground(String... urls) {
            return readJSONFeed(urls[0]);
        }

        protected void onPostExecute(String result) {
            try {
                JSONArray jsonArray = new JSONArray(result);
                Log.i("JSON", "Number of surveys in feed: " +
                        jsonArray.length());

                //---print out the content of the json feed---
                for (int i = 0; i < jsonArray.length(); i++) {
                    JSONObject jsonObject = jsonArray.getJSONObject(i);
                    /*
                      Toast.makeText(getBaseContext(),jsonObject.getString
                        ("appeId") +
                                " - " + jsonObject.getString("inputTime"),
                                Toast.LENGTH_SHORT).show();
                     */

                    Toast.makeText(getBaseContext(), jsonObject.getString
                      ("text") +
                            " - " + jsonObject.getString("created_at"),
                            Toast.LENGTH_SHORT).show();
                }
            } catch (Exception e) {
                e.printStackTrace();
            }
        }
    }

    /** Called when the activity is first created. */
```

```
@Override
public void onCreate(Bundle savedInstanceState) {
    super.onCreate(savedInstanceState);
    setContentView(R.layout.main);
    /*
    new ReadJSONFeedTask().execute(
        "http://extjs.org.cn/extjs/examples/grid/survey.html");
    */
    new ReadJSONFeedTask().execute(
        "https://twitter.com/statuses/user_timeline/weimenglee.json");
}
```

10.3 套接字编程

至此,已经见过如何通过 HTTP 使用 XML 服务和 JSON Web 服务。大多数 Web 服务利用 HTTP 进行通信,它们本身存在一个巨大的缺点:这些 Web 服务是无状态的。当利用 HTTP 与 Web 服务建立连接时,每个连接都被当作一个新连接——Web 服务器不维护与客户端的持久连接。

设想一个应用程序与电影院订票 Web 服务建立连接的场景。当某个客户端在服务器上订票时,其他客户端直到再次连接到 Web 服务以获取最新的座位信息时才意识到这一点。客户端不断轮询 Web 服务增加了不必要的带宽,并降低了应用程序的工作效率。一个更好的解决方案是,让服务器维持和每个客户端之间的单独连接,一旦其他客户端预定某个座位后,就将此消息发送给所有客户端。

如果想要应用程序维护一个与服务器的持续连接,并且在修改发生时接到服务器的通知,需要使用一种叫做"套接字编程"的编程技术。通过套接字编程,可以在服务器和客户端之间建立连接。下面的"试一试"展示了如何建立一个连接到套接字服务器的 Android 聊天客户端应用程序。多个应用程序可以同时与服务器建立连接并进行聊天。

试一试　　**与套接字服务器建立连接**

(1) 利用 Eclipse 创建一个新的 Android 项目,将其命名为 Sockets。
(2) 在 AndroidManifest.xml 文件中添加下列粗体显示的语句:

```xml
<?xml version="1.0" encoding="utf-8"?>
<manifest xmlns:android="http://schemas.android.com/apk/res/android"
    package="net.learn2develop.Sockets"
    android:versionCode="1"
    android:versionName="1.0" >

    <uses-sdk android:minSdkVersion="14" />
    <uses-permission android:name="android.permission.INTERNET"/>

    <application
        android:icon="@drawable/ic_launcher"
        android:label="@string/app_name" >
```

```xml
        <activity
            android:label="@string/app_name"
            android:name=".SocketsActivity" >
            <intent-filter >
                <action android:name="android.intent.action.MAIN" />

                <category android:name="android.intent.category.
                    LAUNCHER" />
            </intent-filter>
        </activity>
    </application>

</manifest>
```

(3) 在 main.xml 文件中添加下列粗体显示的语句，用来代替 TextView：

```xml
<?xml version="1.0" encoding="utf-8"?>
<LinearLayout xmlns:android="http://schemas.android.com/apk/res/android"
    android:layout_width="fill_parent"
    android:layout_height="fill_parent"
    android:orientation="vertical" >

<EditText
    android:id="@+id/txtMessage"
    android:layout_width="fill_parent"
    android:layout_height="wrap_content" />

<Button
    android:layout_width="fill_parent"
    android:layout_height="wrap_content"
    android:text="Send Message"
    android:onClick="onClickSend"/>

<TextView
    android:id="@+id/txtMessagesReceived"
    android:layout_width="fill_parent"
    android:layout_height="200dp"
    android:scrollbars = "vertical" />

</LinearLayout>
```

(4) 将一个新的 Java 类文件添加到包里，并将其命名为 CommsThread。使用如下代码填充 CommsThread.java 文件：

```java
package net.learn2develop.Sockets;

import java.io.IOException;
import java.io.InputStream;
import java.io.OutputStream;
import java.net.Socket;
import android.util.Log;
```

```java
public class CommsThread extends Thread {
    private final Socket socket;
    private final InputStream inputStream;
    private final OutputStream outputStream;

    public CommsThread(Socket sock) {
        socket = sock;
        InputStream tmpIn = null;
        OutputStream tmpOut = null;
        try {
            //---creates the inputstream and outputstream objects
            // for reading and writing through the sockets---
            tmpIn = socket.getInputStream();
            tmpOut = socket.getOutputStream();
        } catch (IOException e) {
            Log.d("SocketChat", e.getLocalizedMessage());
        }
        inputStream = tmpIn;
        outputStream = tmpOut;
    }

    public void run() {
        //---buffer store for the stream---
        byte[] buffer = new byte[1024];

        //---bytes returned from read()---
        int bytes;

        //---keep listening to the InputStream until an
        // exception occurs---
        while (true) {
            try {
                //---read from the inputStream---
                bytes = inputStream.read(buffer);

                //---update the main activity UI---
                SocketsActivity.UIupdater.obtainMessage(
                    0,bytes, -1, buffer).sendToTarget();
            } catch (IOException e) {
                break;
            }
        }
    }

    //---call this from the main activity to
    // send data to the remote device---
    public void write(byte[] bytes) {
        try {
            outputStream.write(bytes);
```

```java
        } catch (IOException e) { }
    }

    //---call this from the main activity to
    // shutdown the connection---
    public void cancel() {
        try {
            socket.close();
        } catch (IOException e) { }
    }
}
```

(5) 在 SocketsActivity.java 文件中添加下列粗体显示的语句：

```java
package net.learn2develop.Sockets;

import java.io.IOException;
import java.net.InetAddress;
import java.net.Socket;
import java.net.UnknownHostException;

import android.app.Activity;
import android.os.AsyncTask;
import android.os.Bundle;
import android.os.Handler;
import android.os.Message;
import android.view.View;
import android.widget.EditText;
import android.widget.TextView;

import android.util.Log;

public class SocketsActivity extends Activity {
    static final String NICKNAME = "Wei-Meng";
    InetAddress serverAddress;
    Socket socket;

    //---all the Views---
    static TextView txtMessagesReceived;
    EditText txtMessage;

    //---thread for communicating on the socket---
    CommsThread commsThread;

    //---used for updating the UI on the main activity---
    static Handler UIupdater = new Handler() {
        @Override
        public void handleMessage(Message msg) {
            int numOfBytesReceived = msg.arg1;
            byte[] buffer = (byte[]) msg.obj;
```

```
            //---convert the entire byte array to string---
            String strReceived = new String(buffer);

            //---extract only the actual string received---
            strReceived = strReceived.substring(
                    0, numOfBytesReceived);

            //---display the text received on the TextView---
            txtMessagesReceived.setText(
                    txtMessagesReceived.getText().toString() +
                    strReceived);
        }
    };

    private class CreateCommThreadTask extends AsyncTask
    <Void, Integer, Void> {
        @Override
        protected Void doInBackground(Void... params) {
            try {
                //---create a socket---
                serverAddress =
                    InetAddress.getByName("192.168.1.142");
                    //--remember to change the IP address above to match your
                        own--
                socket = new Socket(serverAddress, 500);
                commsThread = new CommsThread(socket);
                commsThread.start();
                //---sign in for the user; sends the nick name---
                sendToServer(NICKNAME);
            } catch (UnknownHostException e) {
                Log.d("Sockets", e.getLocalizedMessage());
            } catch (IOException e) {
                Log.d("Sockets", e.getLocalizedMessage());
            }
            return null;
        }
    }

    private class WriteToServerTask extends AsyncTask
    <byte[], Void, Void> {
        protected Void doInBackground(byte[]...data) {
            commsThread.write(data[0]);
            return null;
        }
    }

    private class CloseSocketTask extends AsyncTask
    <Void, Void, Void> {
        @Override
```

```java
    protected Void doInBackground(Void... params) {
        try {
            socket.close();
        } catch (IOException e) {
            Log.d("Sockets", e.getLocalizedMessage());
        }
        return null;
    }
}

/** Called when the activity is first created. */
@Override
public void onCreate(Bundle savedInstanceState) {
    super.onCreate(savedInstanceState);
    setContentView(R.layout.main);

    //---get the views---
    txtMessage = (EditText) findViewById(R.id.txtMessage);
    txtMessagesReceived = (TextView)
            findViewById(R.id.txtMessagesReceived);
}

public void onClickSend(View view) {
    //---send the message to the server---
    sendToServer(txtMessage.getText().toString());
}

private void sendToServer(String message) {
    byte[] theByteArray =
            message.getBytes();
    new WriteToServerTask().execute(theByteArray);
}

@Override
public void onResume() {
    super.onResume();
    new CreateCommThreadTask().execute();
}

@Override
public void onPause() {
    super.onPause();
    new CloseSocketTask().execute();
}
}
```

(6) 为了进行测试，需要使用作者编写的一个套接字服务器应用程序(可从 Wrox.com 上下载的本书源代码中提供了这个应用程序)。该应用程序是一个多用户控制台应用程序(Windows)，它侦听本地计算机的 500 端口，并将接收的消息传播给连接到该服务器的其他所有客户端。要运行该服务器，先要在 Windowns 环境下打开一个命令窗口，并键入以下命令：C:\>Server.exe Your_IP_Address。例如，如果你的 IP 地址为 192.168.1.142，就需要输入以下语句：

```
C:\>Server.exe 192.168.1.142
```

(7) 在一台真实设备上配置应用程序之前，先确保将设备连接到前一步介绍的运行服务器的计算机所在的网络。在常见的设置中，计算机连接到无线路由器上(可以是有线或无线)，而设备无线连接到同一个无线路由器上。一旦完成这些步骤，就可以按 F11 键在 Android 设备上部署应用程序。

(8) 输入一条消息并点击 Send Message 按钮(如图 10-7 所示)。

(9) 将能看到服务器接收的信息，如图 10-8 所示。

图 10-7

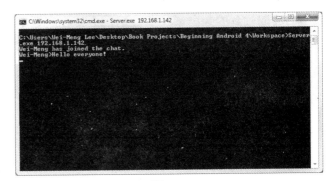

图 10-8

示例说明

为了处理套接字通信错综复杂的问题，创建了一个单独的类并将其命名为 CommsThread(用于通信线程)。该类扩展了 Thread 类，使所有套接字通信能在一个不同于主 UI 线程的线程上执行。

```
public class CommsThread extends Thread {
}
```

在该类中声明 3 种对象：

```
private final Socket socket;
private final InputStream inputStream;
private final OutputStream outputStream;
```

第一种是 Socket 对象，它提供了一个客户端 TCP 套接字。InputStream 对象用于从套接字连接读取数据。OutputStream 对象用于向套接字连接写数据。

CommsThread 类的构造函数接受一个 Socket 实例作为参数，然后尝试从套接字连接获取一个 InputStream 对象和一个 OutputStream 对象：

```
public CommsThread(Socket sock) {
    socket = sock;
    InputStream tmpIn = null;
    OutputStream tmpOut = null;
    try {
        //---creates the inputstream and outputstream objects
        // for reading and writing through the sockets---
        tmpIn = socket.getInputStream();
        tmpOut = socket.getOutputStream();
    } catch (IOException e) {
        Log.d("SocketChat", e.getLocalizedMessage());
    }
    inputStream = tmpIn;
    outputStream = tmpOut;
}
```

run()方法(当调用该线程的 start()方法时会访问 run()方法)通过使用 InputStream 对象不断读取收到的数据来侦听数据。当接收到数据时，它会更新主要活动的 UI，将一个包含接收到数据的消息传递给主要活动的 UI：

```
public void run() {
    //---buffer store for the stream---
    byte[] buffer = new byte[1024];

    //---bytes returned from read()---
    int bytes;

    //---keep listening to the InputStream until an
    // exception occurs---
    while (true) {
        try {
            //---read from the inputStream---
            bytes = inputStream.read(buffer);

            //---update the main activity UI---
            SocketsActivity.UIupdater.obtainMessage(
                0,bytes, -1, buffer).sendToTarget();
        } catch (IOException e) {
            break;
        }
    }
}
```

write()方法将数据写入到套接字连接中：

```
//---call this from the main activity to
// send data to the remote device---
public void write(byte[] bytes) {
```

```
    try {
        outputStream.write(bytes);
    } catch (IOException e) { }
}
```

最后，cancel()方法关闭了套接字连接：

```
//---call this from the main activity to
// shutdown the connection---
public void cancel() {
    try {
        socket.close();
    } catch (IOException e) { }
}
```

在 SocketsActivity.java 文件中，创建的 3 个子类扩展了 AsyncTask 类：

```
private class CreateCommThreadTask extends AsyncTask
<Void, Integer, Void> {
    @Override
    protected Void doInBackground(Void... params) {
        try {
            //---create a socket---
            serverAddress =
                InetAddress.getByName("192.168.1.142");
            socket = new Socket(serverAddress, 500);
            commsThread = new CommsThread(socket);
            commsThread.start();
            //---sign in for the user; sends the nick name---
            sendToServer(NICKNAME);
        } catch (UnknownHostException e) {
            Log.d("Sockets", e.getLocalizedMessage());
        } catch (IOException e) {
            Log.d("Sockets", e.getLocalizedMessage());
        }
        return null;
    }
}

private class WriteToServerTask extends AsyncTask
<byte[], Void, Void> {
    protected Void doInBackground(byte[]...data) {
        commsThread.write(data[0]);
        return null;
    }
}

private class CloseSocketTask extends AsyncTask
<Void, Void, Void> {
    @Override
    protected Void doInBackground(Void... params) {
```

```
        try {
            socket.close();
        } catch (IOException e) {
            Log.d("Sockets", e.getLocalizedMessage());
        }
        return null;
    }
}
```

CreateCommThreadTask 类异步创建了与服务器相连的套接字连接。对于套接字服务器来说，连接建立后，客户端发送的第一个字符串作为该客户端的名称。因此，在套接字连接建立后，立即向服务器发送一条包含想使用的客户端名称的消息：

```
//---sign in for the user; sends the nick name---
sendToServer(NICKNAME);
```

当 CloseSocketTask 类关闭一条套接字连接时，WriteToServerTask 类允许向服务器异步发送数据。

sendToServer()方法接受一个 String 参数，并将其转换为一个字节数组。然后它调用 WriteToServerTask 类的 execute()方法将数据异步发送到服务器：

```
private void sendToServer(String message) {
    byte[] theByteArray =
            message.getBytes();
    new WriteToServerTask().execute(theByteArray);
}
```

最后，当活动暂停时，关闭套接字连接；当活动恢复时，重新建立连接：

```
@Override
public void onPause() {
    super.onPause();
    new CloseSocketTask().execute();
}

@Override
public void onResume() {
    super.onResume();
    new CreateCommThreadTask().execute();
}
```

10.4 本章小结

本章中介绍了应用程序如何使用 HTTP 协议与外界建立连接。通过使用 HTTP 协议，可以从 Web 服务器下载各种类型的数据。其中一个优秀的应用就是与 Web 服务进行通信，这需要解析 XML 文件。除了 XML Web 服务以外，还学习了如何使用 JSON

第 10 章 联　　网

服务，相比 XML Web 服务，它更简便。最后介绍了 HTTP 的替代方式：使用套接字进行通信。套接字使应用程序与服务器持续保持连接，以使应用程序在数据可用时接收数据。从本章的学习中应该重点知道：所有同步操作都必须利用 AsyncTask 类封装，否则应用程序不会在运行 Honeycomb 或更高版本的设备上工作。

练　习

1. 为了建立 HTTP 连接，需要在 AndroidManifest.xml 文件中声明哪些许可？
2. 有哪些类可用于处理 JSON 消息？
3. 列举用于执行后台异步任务的类。

附录 C 提供了练习的答案。

本章主要内容

主　　题	关　键　概　念
建立 HTTP 连接	使用 HttpURLConnection 类
访问 XML Web 服务	使用 Document、DocumentBuilderFactory 和 DocumentBuilder 类解析 Web 服务返回的 XML 结果
处理 JSON 消息	使用 JSONArray 和 JSONObject 类
套接字编程	使用 Socket 类建立一个 TCP 连接。使用 InputStream 对象和 OutputStream 对象来分别接收和发送数据
AsyncTask 类中的 3 个方法	3 个方法分别是 doInBackground()、onProgressUpdate() 和 onPostExecute()

第 11 章

开发 Android 服务

本章将介绍以下内容:
- 如何创建一个在后台运行的服务
- 如何在一个单独的线程中执行长时间运行的任务
- 如何在一个服务中执行重复的任务
- 如何在活动和服务间进行通信

服务是 Android 中一个在后台运行的应用程序,它无须与用户进行交互。例如,在使用一个应用程序时,您可能想要在同一时间播放背景音乐。在这种情况下,播放背景音乐的代码就不需要与用户进行交互,因此它可以作为一个服务来运行。对于不需要向用户展示用户界面的情况下,使用服务也是一个理想的选择。用来持续记录设备所在地理坐标的应用程序就是一个很好的示例。在这种情况下,可以写一个服务在后台执行任务。在本章中,将介绍如何创建自己的服务以及利用它们来异步地执行后台任务。

11.1 创建自己的服务

理解服务如何工作的最好方法就是亲自创建一个服务。下面的"试一试"向您展示了创建一个简单服务的步骤。后续小节的内容将为这个服务增加更多的功能。至于现在,先学习如何启动和停止服务。

试一试　创建一个简单服务

Services.zip 代码文件可以在 Wrox.com 上下载

(1) 打开 Eclipse,创建一个新的 Android 项目,命名为 Services。
(2) 在该项目中添加一个新的名为 MyService 的 Java 类文件。在 MyService.java 类文件中填充如下代码:

```java
package net.learn2develop.Services;

import android.app.Service;
import android.content.Intent;
import android.os.IBinder;
import android.widget.Toast;

public class MyService extends Service {

    @Override
    public IBinder onBind(Intent arg0) {
        return null;
    }

    @Override
    public int onStartCommand(Intent intent, int flags, int startId) {
        // We want this service to continue running until it is explicitly
        // stopped, so return sticky.
        Toast.makeText(this, "Service Started", Toast.LENGTH_LONG).show();
        return START_STICKY;
    }

    @Override
    public void onDestroy() {
        super.onDestroy();
        Toast.makeText(this, "Service Destroyed", Toast.LENGTH_LONG).
        show();
    }
}
```

(3) 在 AndroidManifest.xml 文件中添加下列粗体显示的语句：

```xml
<?xml version="1.0" encoding="utf-8"?>
<manifest xmlns:android="http://schemas.android.com/apk/res/android"
    package="net.learn2develop.Services"
    android:versionCode="1"
    android:versionName="1.0" >

    <uses-sdk android:minSdkVersion="14" />

    <application
        android:icon="@drawable/ic_launcher"
        android:label="@string/app_name" >
        <activity
            android:label="@string/app_name"
            android:name=".ServicesActivity" >
            <intent-filter >
                <action android:name="android.intent.action.MAIN" />

                <category android:name="android.intent.category.
                    LAUNCHER"/>
```

```xml
        </intent-filter>
    </activity>
    <service android:name=".MyService" />
</application>

</manifest>
```

(4) 在 main.xml 文件中添加下列粗体显示的语句，替换掉 TextView：

```xml
<?xml version="1.0" encoding="utf-8"?>
<LinearLayout xmlns:android="http://schemas.android.com/apk/res/android"
    android:layout_width="fill_parent"
    android:layout_height="fill_parent"
    android:orientation="vertical" >

<Button android:id="@+id/btnStartService"
    android:layout_width="fill_parent"
    android:layout_height="wrap_content"
    android:text="Start Service"
    android:onClick="startService"/>

<Button android:id="@+id/btnStopService"
    android:layout_width="fill_parent"
    android:layout_height="wrap_content"
    android:text="Stop Service"
    android:onClick="stopService" />

</LinearLayout>
```

(5) 在 ServicesActivity.java 文件中添加下列粗体显示的语句：

```java
package net.learn2develop.Services;

import android.app.Activity;
import android.content.Intent;
import android.os.Bundle;
import android.view.View;

public class ServicesActivity extends Activity {
    /** Called when the activity is first created. */
    @Override
    public void onCreate(Bundle savedInstanceState) {
        super.onCreate(savedInstanceState);
        setContentView(R.layout.main);
    }

    public void startService(View view) {
        startService(new Intent(getBaseContext(), MyService.class));
    }

    public void stopService(View view) {
```

```
            stopService(new Intent(getBaseContext(), MyService.class));
    }

}
```

(6) 按 F11 键在 Android 模拟器上调试应用程序。

(7) 单击 Start Service 按钮将启动服务(如图 11-1 所示)。要停止服务,可单击 Stop Service 按钮。

示例说明

这个示例展示了您可以创建的最简单的服务。当然,服务本身是没有做什么有用的事情,但它足以说明整个创建的过程。

首先,它定义了一个扩展 Service 基类的类。所有服务都扩展 Service 类:

```
public class MyService extends Service {
}
```

图 11-1

在 MyService 类中,实现了 3 个方法:

```
@Override
public IBinder onBind(Intent arg0) { ... }

@Override
public int onStartCommand(Intent intent, int flags, int startId) { ... }

@Override
public void onDestroy() { ... }
```

onBind()方法可以用来将一个活动绑定到一个服务。这反过来又使活动可以直接访问服务内部的成员和方法。目前,只是为这个方法返回 null。在本章的后面,您将学习更多有关绑定的内容。

当使用 startService()方法(稍后讨论)显式启动服务时将调用 onStartCommand( )方法。这个方法意味着服务的启动,并为服务做一些需要做的事情。此方法返回常量 START_STICKY,因此只要服务不被显式地停止,它都将继续运行。

当使用 stopService()方法停止服务时将调用 onDestroy()方法。这一方法将清除服务所使用的资源。

所有已创建的服务必须在 AndroidManifest.xml 文件中声明,如下所示:

```
<service android:name=".MyService" />
```

如果想让其他应用程序也可使用您的服务,可以添加一个带有动作名称的意图筛选器,如下所示:

第 11 章 开发 Android 服务

```xml
<service android:name=".MyService">
    <intent-filter>
        <action android:name="net.learn2develop.MyService" />
    </intent-filter>
</service>
```

要启动一个服务，使用 startService()方法，如下所示：

```
startService(new Intent(getBaseContext(), MyService.class));
```

如果调用一个外部服务，要按如下所示调用 startService()方法：

```
startService(new Intent("net.learn2develop.MyService"));
```

要停止一个服务，使用 stopService()方法，如下所示：

```
stopService(new Intent(getBaseContext(), MyService.class));
```

11.1.1 在服务中执行长时间运行的任务

由于在上一节中所创建的服务没有做任何有用的事情，因此在本节中将修改这一服务，使其可以执行一个任务。在下面的"试一试"中，将模拟一个从 Internet 上下载文件的服务。

试一试 使服务变得有用

（1）使用在第一个例子中所创建的同一个 Services 项目，在 ServicesActivity.java 文件中添加下列粗体显示的语句：

```java
package net.learn2develop.Services;

import java.net.MalformedURLException;
import java.net.URL;

import android.app.Service;
import android.content.Intent;
import android.os.IBinder;
import android.widget.Toast;

public class MyService extends Service {

    @Override
    public IBinder onBind(Intent arg0) {
        return null;
    }

    @Override
    public int onStartCommand(Intent intent, int flags, int startId) {
        // We want this service to continue running until it is explicitly
        // stopped, so return sticky.
        //Toast.makeText(this, "Service Started", Toast.LENGTH_LONG).
```

```
        show();

    try {
        int result = DownloadFile(new URL("http://www.amazon.com/
         somefile.pdf"));
        Toast.makeText(getBaseContext(),
            "Downloaded " + result + " bytes",
            Toast.LENGTH_LONG).show();
    } catch (MalformedURLException e) {
        // TODO Auto-generated catch block
        e.printStackTrace();
    }
    return START_STICKY;
}

private int DownloadFile(URL url) {
    try {
        //---simulate taking some time to download a file---
        Thread.sleep(5000);
    } catch (InterruptedException e) {
        e.printStackTrace();
    }
    //---return an arbitrary number representing
    // the size of the file downloaded---
    return 100;
}

@Override
public void onDestroy() {
    super.onDestroy();
    Toast.makeText(this,"Service
    Destroyed",Toast.LENGTH_LONG).
    show();
}
```

(2) 按 F11 键在 Android 模拟器上调试应用程序。

(3) 单击 Start Service 按钮启动服务来下载文件。可以看到活动在 Toast 类显示 Downloaded 100 bytes 消息之前被"冻结"了几秒钟(如图 11-2 所示)。

示例说明

在这个示例中，服务调用 DownloadFile()方法来模拟从一个给定的 URL 下载文件。此方法返回下载的总字节数(已经硬编码为 100)。为了模拟下载文件时服务所经历的延迟，使用 Thread.Sleep()方法使服务暂停 5 秒(5000 毫秒)。

图 11-2

当启动服务时，可以注意到活动被暂停了大约 5 秒钟，这正是从 Internet 上下载文件

第 11 章 开发 Android 服务

的时间。在这段时间内,整个活动没有响应,这证明了很重要的一点:服务和活动在相同的线程上运行。在这种情况下,因为服务暂停了 5 秒钟,所以活动也一样。

因此,对于一个长时间运行的服务,将所有长时间运行的代码放入一个单独的线程中是很重要的,这样服务就不会阻塞调用它的应用程序了。下面的"试一试"将告诉您如何做到这一点。

试一试 在服务中异步执行任务

Services.zip 代码文件可以在 Wrox.com 上下载

(1) 使用在第一个例子中创建的 Services 项目,在 MyService.java 文件中添加下列粗体显示的语句:

```
package net.learn2develop.Services;

import java.net.MalformedURLException;
import java.net.URL;

import android.app.Service;
import android.content.Intent;
import android.os.AsyncTask;
import android.os.IBinder;
import android.util.Log;
import android.widget.Toast;

public class MyService extends Service {

    @Override
    public IBinder onBind(Intent arg0) {
        return null;
    }

    @Override
    public int onStartCommand(Intent intent, int flags, int startId) {
        // We want this service to continue running until it is explicitly
        // stopped, so return sticky.
        //Toast.makeText(this, "Service Started", Toast.LENGTH_LONG).
        show();

        try {
            new DoBackgroundTask().execute(
                    new URL("http://www.amazon.com/somefiles.pdf"),
                    new URL("http://www.wrox.com/somefiles.pdf"),
                    new URL("http://www.google.com/somefiles.pdf"),
                    new URL("http://www.learn2develop.net/somefiles.
                      pdf"));
```

```java
            } catch (MalformedURLException e) {
                // TODO Auto-generated catch block
                e.printStackTrace();
            }
            return START_STICKY;
    }

    private int DownloadFile(URL url) {
        try {
            //---simulate taking some time to download a file---
            Thread.sleep(5000);
        } catch (InterruptedException e) {
             e.printStackTrace();
        }
        //---return an arbitrary number representing
        // the size of the file downloaded---
        return 100;
    }

    private class DoBackgroundTask extends AsyncTask<URL,Integer,Long> {
        protected Long doInBackground(URL... urls) {
            int count = urls.length;
            long totalBytesDownloaded = 0;
            for (int i = 0; i < count; i++) {
                totalBytesDownloaded += DownloadFile(urls[i]);
                //---calculate percentage downloaded and
                // report its progress---
                publishProgress((int) (((i+1) / (float) count) * 100));
            }
            return totalBytesDownloaded;
        }

        protected void onProgressUpdate(Integer... progress) {
            Log.d("Downloading files",
                    String.valueOf(progress[0]) + "% downloaded");
            Toast.makeText(getBaseContext(),
                String.valueOf(progress[0]) + "% downloaded",
                Toast.LENGTH_LONG).show();
        }

        protected void onPostExecute(Long result) {
            Toast.makeText(getBaseContext(),
                    "Downloaded " + result + " bytes",
                    Toast.LENGTH_LONG).show();
            stopSelf();
        }
    }

    @Override
    public void onDestroy() {
```

```
        super.onDestroy();
        Toast.makeText(this,"ServiceDestroyed",Toast.
        LENGTH_LONG).show();
    }
}
```

(2) 按 F11 键在 Android 模拟器上调试应用程序。

(3) 单击 Start Service 按钮。Toast 类将显示一个消息表明下载完成的百分比。可以看到 4 个值：25%、50%、75%和 100%。

(4) 还可以看到在 LogCat 窗口中的如下输出内容：

```
12-06 01:58:24.967: D/Downloading files(6020): 25% downloaded
12-06 01:58:30.019: D/Downloading files(6020): 50% downloaded
12-06 01:58:35.078: D/Downloading files(6020): 75% downloaded
12-06 01:58:40.096: D/Downloading files(6020): 100% downloaded
```

示例说明

这个示例说明了可以在您的服务中异步执行任务的一种方法。通过创建一个扩展 AsyncTask 类的内部类来做到这一点。AsyncTask 类使您能够在后台执行，而无须手动操纵线程和处理程序。

DoBackgroundTask 类通过指定 3 个泛型类型来扩展 AsyncTask 类：

```
private class DoBackgroundTask extends AsyncTask<URL, Integer, Long>{
```

这里指定的 3 种类型为 URL、Integer 和 Long。这 3 种类型指定了以下 3 种方法所用到的数据类型，这些方法将在一个 AsyncTask 类中来实现：

- doInBackground()——这个方法接受一个先前指定的第一个泛型类型的数组。在这里，类型是 URL。这个方法用于放置需要长时间运行的代码并在后台线程中执行。为了报告任务的进度，调用 publishProgress()方法，它将调用第二个方法 onProgressUpdate()，这个方法在一个 AsyncTask 类中实现。此方法的返回类型使用先前指定的第三个泛型类型，本例中是 Long。

- onProgressUpdate()——这一方法在 UI 线程中启动并在调用 publishProgress()方法时调用。它接受一个先前指定的第二个泛型类型的数组。在这里，类型是 Integer。使用这一方法向用户报告后台任务的进度。

- onPostExecute()——这一方法在 UI 线程中启动并在 doInBackground()方法执行完毕后调用。这一方法接受一个先前指定的第三个泛型类型的参数，本例中是 Long。

图 11-3 总结了在 AsyncTask 类的子类中指定的类型和它们与 3 个方法的关系。

要在后台下载多个文件，则创建 DoBackgroundTask 类的一个实例，然后通过传入一个 URL 数组来调用其 execute()方法：

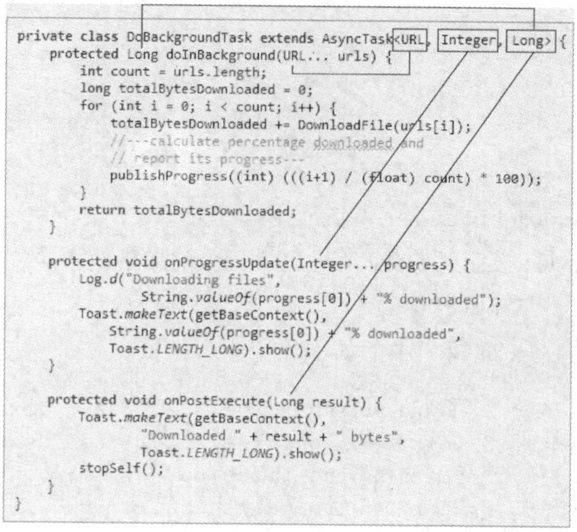

```
private class DoBackgroundTask extends AsyncTask<URL, Integer, Long> {
    protected Long doInBackground(URL... urls) {
        int count = urls.length;
        long totalBytesDownloaded = 0;
        for (int i = 0; i < count; i++) {
            totalBytesDownloaded += DownloadFile(urls[i]);
            //---calculate percentage downloaded and
            // report its progress---
            publishProgress((int) (((i+1) / (float) count) * 100));
        }
        return totalBytesDownloaded;
    }

    protected void onProgressUpdate(Integer... progress) {
        Log.d("Downloading files",
                String.valueOf(progress[0]) + "% downloaded");
        Toast.makeText(getBaseContext(),
                String.valueOf(progress[0]) + "% downloaded",
                Toast.LENGTH_LONG).show();
    }

    protected void onPostExecute(Long result) {
        Toast.makeText(getBaseContext(),
                "Downloaded " + result + " bytes",
                Toast.LENGTH_LONG).show();
        stopSelf();
    }
}
```

图 11-3

```
try {
    new DoBackgroundTask().execute(
            new URL("http://www.amazon.com/somefiles.pdf"),
            new URL("http://www.wrox.com/somefiles.pdf"),
            new URL("http://www.google.com/somefiles.pdf"),
            new URL("http://www.learn2develop.net/somefiles.pdf"));
} catch (MalformedURLException e) {
    // TODO Auto-generated catch block
    e.printStackTrace();
}
```

前述的代码使服务在后台下载文件，并用文件下载的百分比形式报告进度。更重要的是，当在后台通过一个单独的线程下载文件时，活动仍然保持响应。

注意，当后台线程执行完毕时，需要手动调用 stopSelf()方法来停止服务：

```
protected void onPostExecute(Long result) {
    Toast.makeText(getBaseContext(),
            "Downloaded " + result + " bytes",
            Toast.LENGTH_LONG).show();
    stopSelf();
}
```

stopSelf()方法相当于调用 stopService()方法来停止服务。

11.1.2 在服务中执行重复的任务

除了在服务中执行长时间运行的任务，一些重复的任务也会在服务中执行。例如，您可能编写了一个一直在后台运行的闹钟服务。在这种情况下，服务可能需要定期执行一些代码来检查是否到了一个预设的时间以便发出提醒。要运行一个按固定的时间间隔执行的代码块，可以在服务中使用 Timer 类。下面的"试一试"将告诉您怎么做。

第 11 章 开发 Android 服务

试一试 使用 Timer 类来运行重复的任务

Services.zip 代码文件可以在 Wrox.com 上下载

(1) 再次使用 Services 项目，在 MyService.java 文件中添加下列粗体显示的语句：

```java
package net.learn2develop.Services;

import java.net.MalformedURLException;
import java.net.URL;
import java.util.Timer;
import java.util.TimerTask;

import android.app.Service;
import android.content.Intent;
import android.os.AsyncTask;
import android.os.IBinder;
import android.util.Log;
import android.widget.Toast;

public class MyService extends Service {
    int counter = 0;
    static final int UPDATE_INTERVAL = 1000;
    private Timer timer = new Timer();

    @Override
    public IBinder onBind(Intent arg0) {
        return null;
    }

    @Override
    public int onStartCommand(Intent intent, int flags, int startId) {
        // We want this service to continue running until it is explicitly
        // stopped, so return sticky.
        //Toast.makeText(this, "Service Started", Toast.LENGTH_LONG).show();

        doSomethingRepeatedly();

        try {
            new DoBackgroundTask().execute(
                new URL("http://www.amazon.com/somefiles.pdf"),
                new URL("http://www.wrox.com/somefiles.pdf"),
                new URL("http://www.google.com/somefiles.pdf"),
                new URL("http://www.learn2develop.net/somefiles.
                   pdf"));
        } catch (MalformedURLException e) {
            // TODO Auto-generated catch block
            e.printStackTrace();
```

```java
        }
        return START_STICKY;
    }

    private void doSomethingRepeatedly() {
        timer.scheduleAtFixedRate(new TimerTask() {
            public void run() {
                Log.d("MyService", String.valueOf(++counter));
            }
        }, 0, UPDATE_INTERVAL);
    }

    private int DownloadFile(URL url) {
        try {
            //---simulate taking some time to download a file---
            Thread.sleep(5000);
        } catch (InterruptedException e) {
            e.printStackTrace();
        }
        //---return an arbitrary number representing
        // the size of the file downloaded---
        return 100;
    }

    private class DoBackgroundTask extends AsyncTask<URL, Integer, Long>{
        protected Long doInBackground(URL... urls) {
            int count = urls.length;
            long totalBytesDownloaded = 0;
            for (int i = 0; i < count; i++) {
                totalBytesDownloaded += DownloadFile(urls[i]);
                //---calculate percentage downloaded and
                // report its progress---
                publishProgress((int) (((i+1) / (float) count) * 100));
            }
            return totalBytesDownloaded;
        }

        protected void onProgressUpdate(Integer... progress) {
            Log.d("Downloading files",
                    String.valueOf(progress[0]) + "% downloaded");
            Toast.makeText(getBaseContext(),
                    String.valueOf(progress[0]) + "% downloaded",
                    Toast.LENGTH_LONG).show();
        }

        protected void onPostExecute(Long result) {
            Toast.makeText(getBaseContext(),
                    "Downloaded " + result + " bytes",
                    Toast.LENGTH_LONG).show();
            stopSelf();
```

```
        }
    }

    @Override
    public void onDestroy() {
        super.onDestroy();

        if (timer != null){
            timer.cancel();
        }

        Toast.makeText(this, "Service Destroyed", Toast.LENGTH_LONG).
            show();
    }
}
```

(2) 按 F11 键在 Android 模拟器上调试应用程序。
(3) 单击 Start Service 按钮。
(4) 观察在 LogCat 窗口中显示的输出内容，如下所示：

```
12-06 02:37:54.118: D/MyService(7752): 1
12-06 02:37:55.109: D/MyService(7752): 2
12-06 02:37:56.120: D/MyService(7752): 3
12-06 02:37:57.111: D/MyService(7752): 4
12-06 02:37:58.125: D/MyService(7752): 5
12-06 02:37:59.137: D/MyService(7752): 6
```

示例说明

在这一示例中，创建了一个 Timer 对象并在您已经定义的 doSomethingRepeatedly()方法中调用这一对象的 scheduleAtFixedRate()方法：

```java
private void doSomethingRepeatedly() {
    timer.scheduleAtFixedRate( new TimerTask() {
        public void run() {
            Log.d("MyService", String.valueOf(++counter));
        }
    }, 0, UPDATE_INTERVAL);
}
```

将一个 TimerTask 类的实例传递给 scheduleAtFixedRate()方法，以便可以重复执行位于 run()方法中的代码块。scheduleAtFixedRate()方法的第二个参数指定了在第一次执行之前的时间间隔，以毫秒为单位。第三个参数以毫秒为单位，指定了后继执行之间的时间间隔。

在前面的示例中，基本上每 1 秒钟(1000 毫秒)打印一次计数器的值。这一服务重复打印 counter 的值，直到服务终止。

```java
    @Override
    public void onDestroy() {
        super.onDestroy();
```

```
        if (timer != null){
            timer.cancel();
        }

        Toast.makeText(this, "Service Destroyed",Toast.LENGTH_LONG).show();
    }
```

对于 scheduleAtFixedRate()方法，不管每个任务需要多长时间，代码都是按照固定的时间间隔执行。例如，如果 run()方法内的代码需要两秒钟来完成，那么第二个任务将在第一个任务结束后立即开始。同样，如果延迟时间设置为 3 秒，而完成任务只需要 2 秒，那么第二个任务在开始前将等待 1 秒钟。

另外，注意在 onStartCommand()方法中直接调用了 doSomethingRepeatedly()方法，而不需要将其封装到 AsyncTask 类的一个子类中。这是因为 TimerTask 类本身实现了 Runnable 接口，该接口允许在一个单独的线程上运行。

11.1.3 使用 IntentService 在单独的线程上执行异步任务

本章的前面部分讲述了如何使用 startService()方法启动服务以及使用 stopService()方法停止服务。我们还学习了如何在一个单独的线程上执行长时间运行的任务——与主调活动不是同一个线程。重要的是需要注意，一旦服务执行完任务，应尽快停止，防止它继续占用宝贵的资源。这就是为什么当任务完成时需要使用 stopSelf()方法来停止它的原因。遗憾的是，很多开发人员在任务已经完成后常常忘记将服务终止。为了能较容易地创建一个服务，使其异步运行一个任务并在结束时自行终止，可以使用 IntentService 类。

IntentService 类是可以按需处理异步请求的 Service 的基类。它像一个普通服务那样启动，在一个工作者线程中执行其任务并在任务完成时自行终止。下面的"试一试"演示了如何使用 IntentService 类。

试一试 使用 IntentService 类自动停止一个服务

Services 代码文件可以在 Wrox.com 上下载

(1) 使用第一个示例中创建的 Services 项目，添加一个新的类文件 MyIntentService.java。
(2) 在 MyIntentService.java 文件中输入以下内容：

```
package net.learn2develop.Services;

import java.net.MalformedURLException;
import java.net.URL;

import android.app.IntentService;
import android.content.Intent;
import android.util.Log;
```

```java
public class MyIntentService extends IntentService {

    public MyIntentService() {
        super("MyIntentServiceName");
    }

    @Override
    protected void onHandleIntent(Intent intent) {
        try {
            int result =
                DownloadFile(new URL("http://www.amazon.com/somefile.
                    pdf"));
            Log.d("IntentService", "Downloaded " + result + " bytes");
        } catch (MalformedURLException e) {
            e.printStackTrace();
        }
    }

    private int DownloadFile(URL url) {
        try {
            //---simulate taking some time to download a file---
            Thread.sleep(5000);
        } catch (InterruptedException e) {
            e.printStackTrace();
        }
        return 100;
    }
}
```

(3) 在 AndroidManifest.xml 文件中添加下列粗体显示的语句：

```xml
<?xml version="1.0" encoding="utf-8"?>
<manifest xmlns:android="http://schemas.android.com/apk/res/android"
    package="net.learn2develop.Services"
    android:versionCode="1"
    android:versionName="1.0" >

    <uses-sdk android:minSdkVersion="14" />

    <application
        android:icon="@drawable/ic_launcher"
        android:label="@string/app_name" >
        <activity
            android:label="@string/app_name"
            android:name=".ServicesActivity" >
            <intent-filter >
                <action android:name="android.intent.action.MAIN" />

                <category android:name="android.intent.category.LAUNCHER"/>
            </intent-filter>
        </activity>
```

```xml
<service android:name=".MyService">
    <intent-filter>
        <action android:name="net.learn2develop.MyService" />
    </intent-filter>
</service>
<service android:name=".MyIntentService" />

    </application>

</manifest>
```

(4) 在 ServicesActivity.java 文件中添加下列粗体显示的语句：

```java
public void startService(View view) {
    //startService(new Intent(getBaseContext(), MyService.class));
    //OR
    //startService(new Intent("net.learn2develop.MyService"));
    startService(new Intent(getBaseContext(), MyIntentService.class));
}
```

(5) 按 F11 键在 Android 模拟器上调试应用程序。

(6) 单击 Start Service 按钮。大约 5 秒钟之后，应该可以在 LogCat 窗口中观察到以下语句：

```
12-06 13:35:32.181: D/IntentService(861): Downloaded 100 bytes
```

示例说明

首先，定义了 MyIntentService 类，它扩展 IntentService 类而非 Service 类：

```java
public class MyIntentService extends IntentService {
}
```

需要实现这个类的构造函数并利用意图服务的名称(设置为一个字符串)调用其父类：

```java
public MyIntentService() {
    super("MyIntentServiceName");
}
```

然后，实现 onHandleIntent()方法，它在一个工作者线程上执行：

```java
@Override
protected void onHandleIntent(Intent intent) {
    try {
        int result =
            DownloadFile(new URL("http://www.amazon.com/somefile.pdf"));
        Log.d("IntentService", "Downloaded " + result + " bytes");
    } catch (MalformedURLException e) {
        e.printStackTrace();
    }
}
```

需要在一个单独的线程中执行的代码可以放在 onHandleIntent()方法中，如从服务器下载一个文件。当代码执行完毕，线程被终止，并且服务自动停止。

11.2 在服务和活动之间通信

服务通常只是在自己的线程中执行，独立于调用它的活动。如果您只是想让服务定期执行某些任务并且不需要将服务的状态通知给活动的话，这并不造成任何问题。例如，您可能有一个将设备的地理位置定期记录到数据库中的服务。在这种情况下，您的服务没有必要与任何活动进行交互，因为其主要目的是将坐标保存到数据库中。然而，假设您想监控一个特定的位置。当服务记录一个靠近您正在监控的位置的地址时，它可能需要将这一信息传递给活动。在这种情况下，就需要为服务和活动的交互设计一种方法。

下面的"试一试"展示了服务是如何使用一个 BroadcastReceiver 来与活动通信的。

试一试 从一个服务来启动一个活动

Services 代码文件可以在 Wrox.com 上下载

(1) 使用前面创建的 Services 项目，在 MyIntentService.java 文件中添加下列粗体显示的语句：

```java
package net.learn2develop.Services;

import java.net.MalformedURLException;
import java.net.URL;

import android.app.IntentService;
import android.content.Intent;
import android.util.Log;

public class MyIntentService extends IntentService {

    public MyIntentService() {
        super("MyIntentServiceName");
    }

    @Override
    protected void onHandleIntent(Intent intent) {
        try {
            int result =
                DownloadFile(new URL("http://www.amazon.com/somefile.
                    pdf"));
            Log.d("IntentService", "Downloaded " + result + " bytes");

            //---send a broadcast to inform the activity
```

```java
            // that the file has been downloaded---
            Intent broadcastIntent = new Intent();
            broadcastIntent.setAction("FILE_DOWNLOADED_ACTION");
            getBaseContext().sendBroadcast(broadcastIntent);
        } catch (MalformedURLException e) {
            e.printStackTrace();
        }
    }

    private int DownloadFile(URL url) {
        try {
            //---simulate taking some time to download a file---
            Thread.sleep(5000);
        } catch (InterruptedException e) {
            // TODO Auto-generated catch block
            e.printStackTrace();
        }
        return 100;
    }
}
```

(2) 在 ServicesActivity.java 文件中添加下列粗体显示的语句：

```java
package net.learn2develop.Services;

import android.app.Activity;
import android.content.BroadcastReceiver;
import android.content.Context;
import android.content.Intent;
import android.content.IntentFilter;
import android.os.Bundle;
import android.view.View;
import android.widget.Toast;

public class ServicesActivity extends Activity {
    IntentFilter intentFilter;

    /** Called when the activity is first created. */
    @Override
    public void onCreate(Bundle savedInstanceState) {
        super.onCreate(savedInstanceState);
        setContentView(R.layout.main);
    }

    @Override
    public void onResume() {
        super.onResume();

        //---intent to filter for file downloaded intent---
```

```java
        intentFilter = new IntentFilter();
        intentFilter.addAction("FILE_DOWNLOADED_ACTION");

        //---register the receiver---
        registerReceiver(intentReceiver, intentFilter);
    }

    @Override
    public void onPause() {
        super.onPause();

        //---unregister the receiver---
        unregisterReceiver(intentReceiver);
    }

    public void startService(View view) {
        //startService(new Intent(getBaseContext(), MyService.class));
        //OR
        //startService(new Intent("net.learn2develop.MyService"));
        startService(new Intent(getBaseContext(),MyIntentService.class));
    }

    public void stopService(View view) {
        stopService(new Intent(getBaseContext(), MyService.class));
    }

    private BroadcastReceiver intentReceiver = new BroadcastReceiver() {
        @Override
        public void onReceive(Context context, Intent intent) {
            Toast.makeText(getBaseContext(), "File downloaded!",
                Toast.LENGTH_LONG).show();
        }
    };
}
```

(3) 按 F11 键在 Android 模拟器上调试应用程序。

(4) 单击 Start Service 按钮。大约 5 秒钟后，Toast 类将显示一个消息，表明文件已经被下载(如图 11-4 所示)。

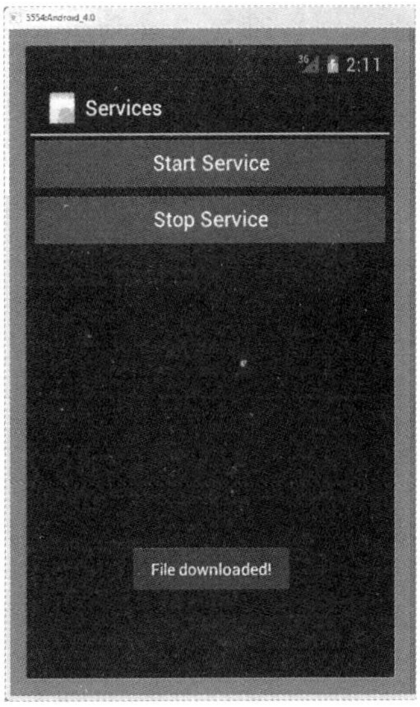

示例说明

为了在一个服务执行结束后可以通知一个活动，需要使用 sendBroadcast()方法广播一个意图：

```java
    @Override
    protected void onHandleIntent(Intent intent)
    {
        try {
            int result =
                DownloadFile(new URL("http://www.amazon.com/somefile.
```

图 11-4

```
                pdf"));
        Log.d("IntentService", "Downloaded " + result + " bytes");

        //---send a broadcast to inform the activity
        // that the file has been downloaded---
        Intent broadcastIntent = new Intent();
        broadcastIntent.setAction("FILE_DOWNLOADED_ACTION");
        getBaseContext().sendBroadcast(broadcastIntent);

    } catch (MalformedURLException e) {
        e.printStackTrace();
    }
}
```

广播的这一意图的动作被设置为 FILE_DOWNLOADED_ACTION，这意味着侦听这一意图的任何活动都将被调用。因此，在 ServicesActivity.java 文件中，使用 IntentFilter 类的 registerReceiver()方法来侦听这个意图：

```
@Override
public void onResume() {
    super.onResume();

    //---intent to filter for file downloaded intent---
    intentFilter = new IntentFilter();
    intentFilter.addAction("FILE_DOWNLOADED_ACTION");

    //---register the receiver---
    registerReceiver(intentReceiver, intentFilter);
}
```

当收到这一意图后，它将启动一个已定义的 BroadcastReceiver 类的实例：

```
private BroadcastReceiver intentReceiver = new BroadcastReceiver() {
    @Override
    public void onReceive(Context context, Intent intent) {
        Toast.makeText(getBaseContext(), "File downloaded!",
                Toast.LENGTH_LONG).show();
    }
};
```

注意：第 8 章详细讨论了 BroadcastReceiver 类。

在本例中，显示了消息"File downloaded!"。当然，如果需要从服务传递一些数据给活动，可以使用 Intent 对象。下一节将讨论这个问题。

11.3 将活动绑定到服务

到目前为止，您已经了解了如何创建服务，如何调用服务以及任务完成后如何终止服务。所有已经看到的服务都很简单——要么启动一个计数器并以固定的时间间隔递增，要么从 Internet 上下载一组固定的文件。然而，现实世界中的服务通常更为复杂，需要传递数据来使它们为您正确地工作。

使用先前展示的下载一组文件的服务，假设现在想让主调活动来决定下载什么样的文件，而不是在服务中靠硬编码实现，那么就需要按以下方式来做。

首先，在主调活动中，创建一个 Intent 对象，指定服务名称：

```java
public void startService(View view) {
    Intent intent = new Intent(getBaseContext(), MyService.class);
}
```

然后，创建一个 URL 对象的数组，并通过 Intent 对象的 putExtra()方法将其赋给 Intent 对象。最后，使用 Intent 对象启动服务：

```java
public void startService(View view) {
    Intent intent = new Intent(getBaseContext(), MyService.class);
    try {
        URL[] urls = new URL[] {
                new URL("http://www.amazon.com/somefiles.pdf"),
                new URL("http://www.wrox.com/somefiles.pdf"),
                new URL("http://www.google.com/somefiles.pdf"),
                new URL("http://www.learn2develop.net/somefiles.
                    pdf")};
        intent.putExtra("URLs", urls);
    } catch (MalformedURLException e) {
        e.printStackTrace();
    }
    startService(intent);
}
```

注意，URL 数组作为一个 Object 数组被赋给了 Intent 对象。

在服务端，需要通过 onStartCommand()方法中的 Intent 对象提取传入的数据：

```java
@Override
public int onStartCommand(Intent intent, int flags, int startId) {
    // We want this service to continue running until it is explicitly
    // stopped, so return sticky.
    Toast.makeText(this, "Service Started", Toast.LENGTH_LONG).show();
    Object[] objUrls = (Object[]) intent.getExtras().get("URLs");
    URL[] urls = new URL[objUrls.length];
    for (int i=0; i<objUrls.length-1; i++) {
        urls[i] = (URL) objUrls[i];
    }
```

```
        new DoBackgroundTask().execute(urls);
        return START_STICKY;
}
```

上述代码首先使用 getExtras()方法提取数据来返回一个 Bundle 对象。然后使用 get()方法以 Object 数组形式提取出 URL 数组。由于在 Java 中，不能直接将数组从一个类型转换到另一个类型，因此必须创建一个循环，对数组中的每个成员单独进行转换。最后，通过将 URL 数组传递给 execute()方法来执行后台任务。

这是活动可以将值传递给服务的一种方式。正如您看到的，如果要将比较复杂的数据传递给服务，就必须做一些额外的工作，以确保数据被正确传递。传递数据的一个更好的办法是直接将活动绑定到服务上，这样活动可以直接调用服务的任何公共成员和方法。下面的"试一试"展示了如何将活动绑定到服务上。

试一试　直接通过绑定访问属性成员

Services.zip 代码文件可以在 Wrox.com 上下

(1) 使用先前创建的 Services 项目，在 MyService.java 文件中添加下列粗体显示的语句(注意这是在修改现有的 onStartCommand()方法):

```
import android.os.Binder;

import android.os.IBinder;

public class MyService extends Service {
    int counter = 0;
    URL[] urls;
    static final int UPDATE_INTERVAL = 1000;
    private Timer timer = new Timer();
    private final IBinder binder = new MyBinder();

    public class MyBinder extends Binder {
        MyService getService() {
            return MyService.this;
        }
    }

    @Override
    public IBinder onBind(Intent arg0) {
        return binder;
    }

    @Override
    public int onStartCommand(Intent intent, int flags, int startId) {
        // We want this service to continue running until it is explicitly
        // stopped, so return sticky.
        Toast.makeText(this, "Service Started", Toast.LENGTH_LONG).show();
```

```
        new DoBackgroundTask().execute(urls);
        return START_STICKY;
    }


    private void doSomethingRepeatedly() { ;- }

    private int DownloadFile(URL url) { ... }

    private class DoBackgroundTask extends AsyncTask<URL,Integer,Long> { ... }

    @Override
    public void onDestroy() { ... }
}
```

(2) 在 ServicesActivity.java 文件中添加下列粗体显示的语句(注意对现有 startService() 方法的修改):

```
import android.content.ComponentName;
import android.os.IBinder;
import android.content.ServiceConnection;
import java.net.MalformedURLException;
import java.net.URL;

public class ServicesActivity extends Activity {
    IntentFilter intentFilter;

    MyService serviceBinder;
    Intent i;

    private ServiceConnection connection = new ServiceConnection() {
        public void onServiceConnected(
            ComponentName className, IBinder service) {
            //--called when the connection is made--
            serviceBinder = ((MyService.MyBinder)service).getService();
            try {
                URL[] urls = new URL[] {
                    new URL("http://www.amazon.com/somefiles.pdf"),
                    new URL("http://www.wrox.com/somefiles.pdf"),
                    new URL("http://www.google.com/somefiles.pdf"),
                    new URL("http://www.learn2develop.net/somefiles.
                       pdf")};
                //---assign the URLs to the service through the
                // serviceBinder object---
                serviceBinder.urls = urls;
            } catch (MalformedURLException e) {
                e.printStackTrace();
            }
            startService(i);
        }
```

```java
        public void onServiceDisconnected(ComponentName className) {
            //---called when the service disconnects---
            serviceBinder = null;
        }
    } ;

    public void startService(View view) {
        i = new Intent(ServicesActivity.this, MyService.class);
        bindService(i, connection, Context.BIND_AUTO_CREATE);
    }

    @Override
    public void onCreate(Bundle savedInstanceState) { ... }

    @Override
    public void onResume() { ... }

    @Override
    public void onPause() { ... }

    public void stopService(View view) { ... }

    private BroadcastReceiver intentReceiver = new BroadcastReceiver() {
        ...
    };

}
```

(3) 按 F11 键调试应用程序。单击 Start Service 按钮将正常启动服务。

示例说明

为了将活动绑定到一个服务，必须首先在服务中声明一个扩展 Binder 类的内部类：

```java
public class MyBinder extends Binder {
    MyService getService() {
        return MyService.this;
    }
}
```

在这个类中实现 getService()方法，这一方法返回服务的一个实例。然后创建一个 MyBinder 类的实例：

```java
private final IBinder binder = new MyBinder();
```

还要修改 onBind()方法来返回 MyBinder 实例：

```java
@Override
public IBinder onBind(Intent arg0) {
    return binder;
}
```

在 onStartCommand()方法中，使用先前在服务中作为公共成员声明的 urls 数组调用 execute()方法：

```java
public class MyService extends Service {
    int counter = 0;
    URL[] urls;
...
...
    @Override
    public int onStartCommand(Intent intent, int flags, int startId) {
        // We want this service to continue running until it is explicitly
        // stopped, so return sticky.
        Toast.makeText(this, "Service Started", Toast.LENGTH_LONG).show();
        new DoBackgroundTask().execute(urls);
        return START_STICKY;
    }
}
```

下一步要做的就是从活动中直接对这个 URL 数组进行设置。

在 ServicesActivity.java 文件中，首先声明服务的一个实例和一个 Intent 对象：

```java
MyService serviceBinder;
Intent i;
```

serviceBinder 对象将用作指向服务的引用，可以直接访问它。

然后创建一个 ServiceConnection 类的实例以便监控服务的状态：

```java
private ServiceConnection connection = new ServiceConnection() {
    public void onServiceConnected(
        ComponentName className, IBinder service) {
        //---called when the connection is made---
        ServiceBinder = ((MyService.MyBinder)service).getService();
        try {
            URL[] urls = new URL[] {
                new URL("http://www.amazon.com/somefiles.pdf"),
                new URL("http://www.wrox.com/somefiles.pdf"),
                new URL("http://www.google.com/somefiles.pdf"),
                new URL("http://www.learn2develop.net/somefiles.
                pdf")};
                //---assign the URLs to the service through the
                // serviceBinder object---
                serviceBinder.urls = urls;
        } catch (MalformedURLException e) {
            e.printStackTrace();
        }
        startService(i);
    }
    public void onServiceDisconnected(ComponentName className) {
        //---called when the service disconnects---
        serviceBinder = null;
```

```
        }
    };
```

需要实现两个方法：onServiceConnected()和 onServiceDisconnected()。onServiceConnected() 方法是当活动连接到服务时调用的。当服务与活动断开时调用 onServiceDisconnected()方法。

在 onServiceConnected()方法中，当活动连接到服务时，通过使用 service 参数的 getService()方法，然后将其赋给 serviceBinder 对象来获取服务的一个实例。serviceBinder 对象是一个指向服务的引用，可以通过这一对象访问服务的所有成员和方法。这里，创建了一个 URL 数组并将其直接赋给服务中的公共成员：

```
URL[] urls = new URL[] {
    new URL("http://www.amazon.com/somefiles.pdf"),
    new URL("http://www.wrox.com/somefiles.pdf"),
    new URL("http://www.google.com/somefiles.pdf"),
    new URL("http://www.learn2develop.net/somefiles.pdf") };
    //---assign the URLs to the service through the
    // serviceBinder object---
    serviceBinder.urls = urls;
```

然后使用一个 Intent 对象启动服务：

```
startService(i);
```

在可以启动服务之前，需要将活动绑定到服务。这在 Start Service 按钮的 startservice() 方法中完成：

```
public void startService(View view) {
    i = new Intent(ServicesActivity.this, MyService.class);
    bindService(i, connection, Context.BIND_AUTO_CREATE);
}
```

bindService()方法使活动与服务建立了连接。它接受 3 个参数：一个 Inent 对象、一个 ServiceConnection 对象以及一个用来指示服务绑定方式的标志。

11.4 理解线程

到目前为止，您已经看到了如何创建服务，以及为什么确保长时间运行的任务得到恰当的处理是非常重要的，特别是在更新 UI 线程时。在本章前面(和第 10 章)看到了如何使用 AsyncTask 类在后台执行长时间运行的代码。本节将简要总结一下使用各个可用的方法正确处理长时间运行的任务的几种方式。

这里的讨论假定有一个名为 Threading 的 Android 项目。main.xml 文件中包含一个 Button 和一个 TextView：

```
<?xml version="1.0" encoding="utf-8"?>
<LinearLayout xmlns:android="http://schemas.android.com/apk/res/android"
```

```xml
    android:layout_width="fill_parent"
    android:layout_height="fill_parent"
    android:orientation="vertical" >

    <TextView
        android:layout_width="fill_parent"
        android:layout_height="wrap_content"
        android:text="@string/hello" />

    <Button
        android:id="@+id/btnStartCounter"
        android:layout_width="match_parent"
        android:layout_height="wrap_content"
        android:text="Start"
        android:onClick="startCounter" />

<TextView
    android:id="@+id/textView1"
    android:layout_width="match_parent"
    android:layout_height="wrap_content"
    android:text="TextView" />

</LinearLayout>
```

假设想要在活动中显示一个从 1 到 1000 的计数器。在 ThreadingActivity 类中有如下代码：

```java
package net.learn2develop.Threading;

import android.app.Activity;
import android.os.Bundle;
import android.util.Log;
import android.view.View;
import android.widget.TextView;

public class ThreadingActivity extends Activity {
    TextView txtView1;

    /** Called when the activity is first created. */
    @Override
    public void onCreate(Bundle savedInstanceState) {
        super.onCreate(savedInstanceState);
        setContentView(R.layout.main);

        txtView1 = (TextView) findViewById(R.id.textView1);
    }

    public void startCounter(View view) {
        for (int i=0; i<=1000; i++) {
            txtView1.setText(String.valueOf(i));
            try {
                Thread.sleep(1000);
```

```
            } catch (InterruptedException e) {
                Log.d("Threading", e.getLocalizedMessage());
            }
        }
    }
}
```

运行应用程序并单击 Start 按钮时,应用程序将被短暂地冻结,一段时间后可以看到如图 11-5 所示的消息。

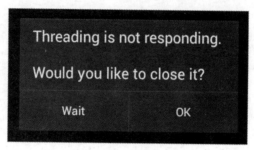

图 11-5

用户界面之所以会冻结,是因为在应用程序显示了计数器的一个值后会暂停 1 秒钟,而与此同时想让应用程序不断地显示计数器的另一个值。UI 会等待数字完成显示,所以来不及处理。结果导致应用程序不能响应,让用户感到失望。

为了解决这个问题,一种方法是使用一个 Thread 类和一个 Runnable 类来封装包含了循环的代码部分,如下所示:

```
public void startCounter(View view) {
    new Thread(new Runnable() {
        public void run() {
            for (int i=0; i<=1000; i++) {
                txtView1.setText(String.valueOf(i));
                try {
                    Thread.sleep(1000);
                } catch (InterruptedException e) {
                    Log.d("Threading", e.getLocalizedMessage());
                }
            }
        }
    }).start();
}
```

在前面的代码中,首先创建一个实现了 Runnable 接口的类。利用该类将长时间运行的代码放到 run()方法中。然后,Runnable 块开始使用 Thread 类。

 注意:一个 Runnable 是可被一个线程执行的代码块。

但是，前面的应用程序不能工作，试图运行它会使它崩溃。放到 Runnable 块中的代码在一个单独的线程上运行，在前面的示例中，试图从另外一个线程来更新 UI。由于 Android 用户界面不是线程安全的，所以这么做并不安全。为了解决这个问题，需要使用一个 View 的 post()方法来创建另一个 Runnable()块来添加到消息队列中。简言之，新创建的 Runnable 块将在 UI 线程中执行，所以现在执行应用程序是安全的：

```java
public void startCounter(View view) {
    new Thread(new Runnable() {
        @Override
        public void run() {
            for (int i=0; i<=1000; i++) {
                final int valueOfi = i;

                //---update UI---
                txtView1.post(new Runnable() {
                    public void run() {
                        //---UI thread for updating---
                        txtView1.setText(String.valueOf(valueOfi));
                    }
                });

                //---insert a delay
                try {
                    Thread.sleep(1000);
                } catch (InterruptedException e) {
                    Log.d("Threading", e.getLocalizedMessage());
                }
            }
        }
    }).start();
}
```

现在应用程序可以正确运行，但是其代码十分复杂，不易维护。

另外一种在其他线程中更新 UI 的方法是使用 Handler 类。Handler 类允许发送和处理消息，就像使用一个 View 的 post()方法一样。下面的代码片段显示了一个名为 UIupdater 的 Handler 类，它使用接收到的消息更新 UI：

 注意：需要导入 android.os.Handler 包，并向 txtView1 添加 static 修饰符才能使这段代码工作。

```java
//---used for updating the UI on the main activity---
static Handler UIupdater = new Handler() {
    @Override
    public void handleMessage(Message msg) {
        byte[] buffer = (byte[]) msg.obj;
```

```
            //---convert the entire byte array to string---
            String strReceived = new String(buffer);

            //---display the text received on the TextView---
            txtView1.setText(strReceived);
            Log.d("Threading", "running");
        }
    };

    public void startCounter(View view) {
        new Thread(new Runnable() {
            @Override
            public void run() {
                for (int i=0; i<=1000; i++) {
                    //---update the main activity UI---
                    ThreadingActivity.UIupdater.obtainMessage(
                        0,String.valueOf(i).getBytes() ).sendToTarget();
                    //---insert a delay
                    try {
                        Thread.sleep(1000);
                    } catch (InterruptedException e) {
                        Log.d("Threading", e.getLocalizedMessage());
                    }
                }
            }
        }).start();
    }

}
```

对 Handler 类的详细讨论不在本书的范围内。更多信息请查看位于以下地址的文档：http://developer.android.com/reference/android/os/Handler.html。

刚才讨论的这两种方法可以在一个单独的线程中更新 UI。在 Android 中，可以使用更简单的 AsyncTask 类来实现相同的操作。当使用 AsyncTask 时，按如下所示，重写前面的代码：

```
private class DoCountingTask extends AsyncTask<Void, Integer, Void> {
protected Void doInBackground(Void... params) {
        for (int i = 0; i < 1000; i++) {
            //---report its progress---
            publishProgress(i);
            try {
                Thread.sleep(1000);
            } catch (InterruptedException e) {
                Log.d("Threading", e.getLocalizedMessage());
            }
        }
        return null;
    }
```

```java
        protected void onProgressUpdate(Integer... progress) {
            txtView1.setText(progress[0].toString());
            Log.d("Threading", "updating...");
        }
    }

    public void startCounter(View view) {
        new DoCountingTask().execute();
    }
```

前面的代码将从另一个线程安全地更新 UI。那么如何停止任务呢？如果运行这个应用程序，然后单击 Start 按钮，计数器将从 0 开始进行显示。但是，如果按模拟器/设备上的 Back 按钮，即使活动已被销毁，任务也会继续运行。通过 LogCat 窗口可以验证这一点。如果想要停止任务，可以使用下面的代码片段：

```java
public class ThreadingActivity extends Activity {
    static TextView txtView1;

    DoCountingTask task;

    /** Called when the activity is first created. */
    @Override
    public void onCreate(Bundle savedInstanceState) {
        super.onCreate(savedInstanceState);
        setContentView(R.layout.main);


        txtView1 = (TextView) findViewById(R.id.textView1);
    }

    public void startCounter(View view) {
        task = (DoCountingTask) new DoCountingTask().execute();
    }

    public void stopCounter(View view) {
        task.cancel(true);
    }

    private class DoCountingTask extends AsyncTask<Void, Integer, Void> {
        protected Void doInBackground(Void... params) {
            for (int i = 0; i < 1000; i++) {
                //---report its progress---
                publishProgress(i);
                try {
                    Thread.sleep(1000);
                } catch (InterruptedException e) {
                    Log.d("Threading", e.getLocalizedMessage());
                }
                if (isCancelled()) break;
            }
```

```
            return null;
        }

        protected void onProgressUpdate(Integer... progress) {
            txtView1.setText(progress[0].toString());
            Log.d("Threading", "updating...");
        }
    }

    @Override
    protected void onPause() {
        super.onPause();
        stopCounter(txtView1);
    }
}
```

为停止 AsyncTask 子类,首先需要获得它的一个实例。调用任务的 cancel()方法可以停止任务。在任务内,调用 isCancelled()方法来检查任务是否应该终止。

11.5 本章小结

在本章中,我们学习了如何在 Android 项目中创建一个服务来执行长时间运行的任务。了解了可以用来确保后台任务以异步方式执行而不阻塞主调活动的许多方法。我们还学习了活动是如何给服务传递数据的,以及如何绑定到一个活动,使之可以更直接地访问服务。

练 习

1. 为什么说将一个服务中的长时间运行的代码放在一个单独的线程中是重要的?
2. IntentService 类的作用是什么?
3. 说出需要在一个 AsyncTask 类中实现的 3 个方法。
4. 服务如何通知活动发生了一个事件?
5. 对于线程,推荐用哪种方法来确保代码在运行时不会阻塞应用程序的 UI?

练习答案参见附录 C。

本章主要内容

主 题	关 键 概 念
创建一个服务	创建一个类并扩展 Service 类
在一个服务中实现方法	实现以下方法:onBind()、onStartCommand()和 onDestroy()
启动一个服务	使用 startService()方法
停止一个服务	使用 stopService()方法
执行长时间运行的任务	使用 AsyncTask 类并实现 3 个方法:doInBackground()、onProgressUpdate()和 onPostExecute()
执行重复的任务	使用 Timer 类并调用它的 scheduleAtFixedRate()方法

(续表)

主　题	关　键　概　念
在一个单独的线程中执行任务并自动停止一个服务	使用 IntentService 类
在活动和服务之间通信	使用 Intent 对象给服务传递数据。对于一个服务，广播一个 Intent 来通知一个活动
将活动绑定到服务	在服务中使用 Binder 类并在主调活动中实现 ServiceConnection 类
在一个 Runnable 块中更新 UI	使用视图的 post()方法来更新 UI。或者，可以使用 Handler 类。推荐的方法是使用 AsyncTask 类

第 12 章

发布 Android 应用程序

本章将介绍以下内容：
- 如何为部署应用程序做准备
- 如何将应用程序导出为一个 APK 文件并用新的证书对其签名
- 如何分发 Android 应用程序
- 如何在 Android Market 上发布应用程序

到目前为止，您已经了解到了使用 Android 可以做很多有趣的事情。然而，为了使您的应用程序可以在用户的设备上运行，需要一个方法来部署和分发它。在本章中，将学习如何为部署 Android 应用程序做准备并将它们转移到客户设备上。此外，还将学习如何将您的应用程序发布到 Android Market 上，在那里您可以通过出售应用程序来赚钱！

12.1 为发布做准备

Google 已经使得 Android 应用程序的发布变得相当容易，因此可以很迅速地将其分发给终端用户。发布 Android 应用程序的步骤通常包含以下几步：

(1) 将应用程序导出为一个 APK(Android Package)文件。
(2) 生成自己的自签名证书并用它对应用程序进行数字签名。
(3) 部署签名后的应用程序。
(4) 利用 Android Market 对应用程序进行托管和出售。

在接下来的小节中，将学习如何为签署应用程序做准备，以及学习部署应用程序的不同方法。

本章中将使用在第 9 章创建的 LBS 项目来说明如何部署 Android 应用程序。

12.1.1 版本化

从 Android SDK 1.0 版本开始，每一个 Android 应用程序的 AndroidManifest.xml 文件

都包括了 android:versionCode 和 android:versionName 属性:

```xml
<?xml version="1.0" encoding="utf-8"?>
<manifest xmlns:android="http://schemas.android.com/apk/res/android"
    package="net.learn2develop.LBS"
    android:versionCode="1"
    android:versionName="1.0" >

    <uses-sdk android:minSdkVersion="14" />
    <uses-permission android:name="android.permission.INTERNET"/>
    <uses-permission android:name="android.permission.ACCESS_FINE_
        LOCATION"/>
    <uses-permission android:name="android.permission.ACCESS_COARSE_
        LOCATION"/>

    <application
        android:icon="@drawable/ic_launcher"
        android:label="@string/app_name" >
        <uses-library android:name="com.google.android.maps" />
        <activity
            android:label="@string/app_name"
            android:name=".LBSActivity" >
            <intent-filter >
                <action android:name="android.intent.action.MAIN" />

                <category android:name="android.intent.category.
                    LAUNCHER" />
            </intent-filter>
        </activity>
    </application>

</manifest>
```

android:versionCode 属性表示了应用程序的版本号。对于对应用程序所做的每一次修改,都应该将这个值加 1,以便可以以编程方式来区分先前版本和最新版本。Android 系统永远不会使用这个值。但对于开发人员来说,这是一个获得应用程序版本号的很有用的方法。不过,Android Market 使用 android:versionCode 属性来确定应用程序是否有新的版本可用。

可以使用 PackageManager 类的 getPackageInfo()方法以编程方式检索 android:versionCode 属性的值,如下所示:

```java
import android.content.pm.PackageInfo;
import android.content.pm.PackageManager;
import android.content.pm.PackageManager.NameNotFoundException;

    private void checkVersion() {
        PackageManager pm = getPackageManager();
        try {
```

```
        //---get the package info---
        PackageInfo pi =
            pm.getPackageInfo("net.learn2develop.LBS", 0);
        //---display the versioncode---
        Toast.makeText(getBaseContext(),
            "VersionCode: " +Integer.toString(pi.versionCode),
            Toast.LENGTH_SHORT).show();
    } catch (NameNotFoundException e) {
        // TODO Auto-generated catch block
        e.printStackTrace();
    }
}
```

android:versionName 属性包含了对用户可见的版本信息。它应该以<major>.<minor>.<point>格式包含值。如果应用程序进行了重要的升级，那么应该将<major>加1。对于微小的增量更新，可以将<minor>或<point>加1。例如，一个新的应用程序的版本名称是1.0.0。对于较小的增量更新，可以将版本名称修改为1.1.0 或 1.0.1。如果下一次有较大的更新，可以将其变为2.0.0。

如果计划在 Android Market(www.android.com/market/)上发布应用程序，AndroidManifest.xml 文件必须具有以下属性：

- android:versionCode (位于<manifest>元素中)
- android:versionName (位于<manifest>元素中)
- android:icon (位于<application>元素中)
- android:label (位于<application>元素中)

android:label 属性指定了应用程序的名称。这个名称将显示在 Android 设备的 Settings | Apps 部分中。对于 LBS 项目，我们给应用程序起名为 Where Am I：

```xml
<?xml version="1.0" encoding="utf-8"?>
<manifest xmlns:android="http://schemas.android.com/apk/res/android"
    package="net.learn2develop.LBS"
    android:versionCode="1"
    android:versionName="1.0" >

    <uses-sdk android:minSdkVersion="14" />
    <uses-permission android:name="android.permission.INTERNET"/>
    <uses-permission android:name="android.permission.ACCESS_FINE_
        LOCATION"/>
    <uses-permission android:name="android.permission.ACCESS_COARSE_
        LOCATION"/>

    <application
        android:icon="@drawable/ic_launcher"
        android:label="Where Am I" >
        <uses-library android:name="com.google.android.maps" />
        <activity
            android:label="@string/app_name"
            android:name=".LBSActivity" >
```

```xml
            <intent-filter >
                <action android:name="android.intent.action.MAIN" />
                <category android:name="android.intent.category.
                    LAUNCHER" />
            </intent-filter>
        </activity>
    </application>

</manifest>
```

另外，如果运行应用程序需要一个最低版本的 Android 操作系统，那么可以在 AndroidManifest.xml 文件中使用<uses-sdk>元素来指定：

```xml
<?xml version="1.0" encoding="utf-8"?>
<manifest xmlns:android="http://schemas.android.com/apk/res/android"
    package="net.learn2develop.LBS"
    android:versionCode="1"
    android:versionName="1.0" >

    <uses-sdk android:minSdkVersion="13" />
    <uses-permission android:name="android.permission.INTERNET"/>
    <uses-permission android:name="android.permission.ACCESS_FINE_
        LOCATION"/>
    <uses-permission android:name="android.permission.ACCESS_COARSE_
        LOCATION"/>

    <application
        android:icon="@drawable/ic_launcher"
        android:label="Where Am I" >
        <uses-library android:name="com.google.android.maps" />
        <activity
            android:label="@string/app_name"
            android:name=".LBSActivity" >
            <intent-filter >
                <action android:name="android.intent.action.MAIN" />

                <category android:name="android.intent.category.LAUNCHER"/>
            </intent-filter>
        </activity>
    </application>

</manifest>
```

在前面的示例中，应用程序需要的最低 SDK 版本为 13，即 Android 3.2.1。一般来说，将这个版本号设置为应用程序可支持的最低版本是明智的。这可以保证有更广泛的用户可以运行您的应用程序。

12.1.2 对 Android 应用程序进行数字签名

所有 Android 应用程序在被允许部署到设备(或模拟器)上之前必须经过数字签名。与

第 12 章　发布 Android 应用程序

一些手机平台不同，您不需要从认证机构(CA)购买数字证书来进行签名。相反，您可以生成自己的自签名证书并用它为 Android 应用程序签名。

如果使用 Eclipse 开发 Android 应用程序，然后按 F11 键将其部署到一个模拟器上，Eclipse 将自动为您对其进行签名。可以在 Eclipse 中选择 Window | Preferences，展开 Android 项，选择 Build(如图 12-1 所示)来验证这一点。Eclipse 使用一个默认的调试密钥库(被命名为 debug.keystore)来对应用程序签名。密钥库常被称为数字证书。

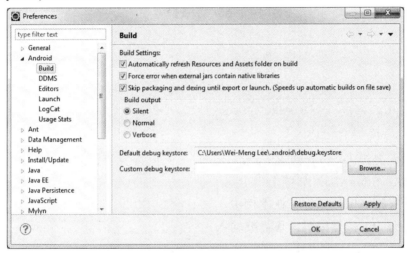

图 12-1

如果您正在发布一个 Android 应用程序，就必须使用您自己的证书来对其进行签名。使用调试证书签名的应用程序是不能被发布的。虽然您可以使用 Java SDK 提供的 keytool.exe 实用程序来手动生成自己的证书，但 Eclipse 通过提供一个向导使这一过程变得更容易了，它可以引导您一步步地生成证书。它将使用生成的证书对应用程序进行签名(也可以使用 Java SDK 的 jarsigner.exe 工具进行手动签名)。

下面的"试一试"展示了如何使用 Eclipse 导出一个 Android 应用程序并使用新生成的证书进行签名。

试一试　导出 Android 应用程序并对其签名

在这个"试一试"中，将使用第 9 章创建的 LBS 项目。

(1) 在 Eclipse 中选择 LBS 项目并选择 File | Export...。

(2) 在 Export 对话框中，展开 Android 项，并选择 Export Android Application(如图 12-2 所示)，单击 Next 按钮。

(3) 现在 LBS 项目应该显示出来了(如图 12-3 所示)，单击 Next 按钮。

(4) 选择 Create new keystore 选项来创建一个新的证书(密钥库)用于应用程序签名(如图 12-4 所示)。输入保存新密钥库的路径并输入一个密码来保护此密钥库。在本例中，输入 keystorepassword 作为密码，单击 Next 按钮。

图 12-2

图 12-3

(5) 为私钥提供一个别名(命名为 DistributionKeyStoreAlias,如图 12-5 所示)并输入一个密码来保护私钥。在本例中,输入 keypassword 作为密码。还需要输入密钥的有效期。根据 Google 的规定,您的应用程序必须用一个加密的私钥进行签名,其有效期要到 2033 年 10 月 22 日以后才能结束。因此,应该输入一个大于 2033 减去当前年份的数。最后,在 First and Last Name 字段中输入自己的姓名。单击 Next 按钮。

(6) 输入存储 APK 目标文件的路径(如图 12-6 所示)。单击 Finish 按钮,将生成 APK 文件。

图 12-4

图 12-5

图 12-6

(7) 在第 9 章中，LBS 应用程序需要使用 Google Maps API 密钥。这个密钥是通过使用您的 debug.keystore 的 MD5 指纹申请到的。这意味着 Google Maps API 密钥与用来对应用程序进行签名的 debug.keystore 是捆绑在一起的。因为您现在生成了新的密钥库来对应用程序进行签名和部署，所以需要使用新密钥库的 MD5 指纹再次申请 Google Maps API 密钥。要做到这一点，在命令提示符窗口中输入以下命令(keytool.exe 实用程序的位置可能略微不同，如图 12-7 所示)：

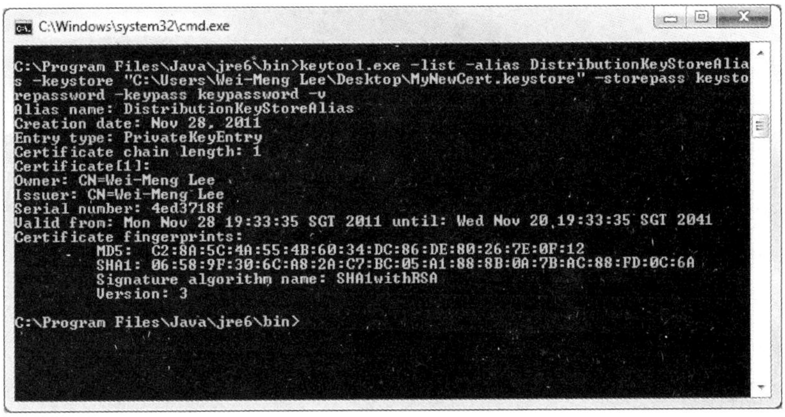

图 12-7

```
C:\Program Files\Java\jre6\bin>keytool.exe -list -v -alias
DistributionKeyStoreAlias
-keystore "C:\Users\Wei-Meng Lee\Desktop\MyNewCert.keystore"
-storepass keystorepassword -keypass keypassword -v
```

(8) 使用从前面的步骤获得的 MD5 指纹，转到 http://code.google.com/android/add-ons/google-apis/maps-api-signup.html 并注册一个新的 Maps API 密钥。

(9) 在 main.xml 文件中输入新的 Maps API 密钥：

```xml
<?xml version="1.0" encoding="utf-8"?>
<LinearLayout xmlns:android="http://schemas.android.com/apk/res/android"
    android:layout_width="fill_parent"
    android:layout_height="fill_parent"
    android:orientation="vertical" >

<com.google.android.maps.MapView
    android:id="@+id/mapView"
    android:layout_width="fill_parent"
    android:layout_height="fill_parent"
    android:enabled="true"
    android:clickable="true"
    android:apiKey="your_key_here" />

</LinearLayout>
```

(10) 由于在 main.xml 文件中输入了新的 Maps API 密钥，现在需要您再一次导出应用程序并重新对其签名。重复第 2~4 步。当要求您选择一个密钥库时，选择 Use existing keystore 选项(如图 12-8 所示)并输入一个先前用于保护您的密钥库的密码(在这里，密码是 keystorepassword)。单击 Next 按钮。

(11) 选择 Use existing key 选项(如图 12-9 所示)并输入先前设置的用于保护私钥的密码(输入 keypassword)。单击 Next 按钮。

图 12-8　　　　　　　　　　　　　　　图 12-9

(12) 单击 Finish 按钮(如图 12-10 所示)再次生成 APK 文件。

到此为止，APK 已经生成了，它包含了捆绑到新密钥库的新的 Maps API 密钥。

图 12-10

示例说明

Eclipse 提供了 Export Android Application 选项，可以帮助您将 Android 应用程序导出为一个 APK 文件并生成一个用于对 APK 文件进行签名的新的密钥库。对于使用 Maps API 的应用程序，要注意 Maps API 密钥必须和为您的 APK 文件签名的新密钥库相关联。

12.2 部署 APK 文件

一旦对 APK 文件进行了签名，就需要有个方法将它们转移到用户设备上。下面的小节将描述部署 APK 文件的不同方法，主要包括以下 3 个方法：

- 使用 adb.exe 工具手动部署
- 在 Web 服务器上托管应用程序
- 通过 Android Market 发布

除了以上方法，还可以通过电子邮件、SD 卡等方式将您的应用程序安装到用户设备上。只要能将 APK 文件转移到用户设备上，您就能够安装这些应用程序。

12.2.1 使用 adb.exe 工具

一旦 Android 应用程序经过签名，就可以使用 abd.exe(Android Debug Bridge)工具(位于 Android SDK 的 platform-tools 文件夹中)将其部署到模拟器和设备上。

使用 Windows 中的命令提示符，转到<Android_SDK>\platform-tools 文件夹下。要在模拟器/设备上安装应用程序(假设模拟器当前已经启动并运行或设备已连接)，可输入以下命令：

```
adb install "C:\Users\Wei-Meng Lee\Desktop\LBS.apk"
```

adb.exe 工具介绍

adb.exe 工具是一个多功能的工具，可以用来控制与您的计算机相连的 Android 设备(和模拟器)。

默认情况下，当使用 adb 命令时，假定当前只连接了一台设备/模拟器。如果连接了多台设备，那么 adb 命令会返回一个错误消息：

```
error: more than one device and emulator
```

可以使用 adb 的 devices 选项来查看当前连接到计算机上的设备，如下所示：

```
D:\Android 4.0\android-sdk-windows\platform-tools>adb devices
List of devices attached
HT07YPY09335    device
emulator-5554   device
emulator-5556   device
```

正如上面的示例所显示的，这个命令返回一个当前已连接设备的列表。为了给一个特定的设备发出命令，需要使用-s 选项来指明此设备，如下所示：

```
adb -s emulator-5556 install LBS.apk
```

如果试图在一台已有 APK 文件的设备上安装该文件，将显示以下错误消息：

```
Failure [INSTALL_FAILED_ALREADY_EXISTS]
```

如果先前的 LBS 应用程序仍然在设备或模拟器上，可以通过 Settings | Apps | LBS | Uninstall 来删除它。

有时候 ADB 会失败(当同时打开了过多的 ADV 时，您会注意到无法再把应用程序从 Eclipse 部署到真实的设备或者模拟器上)。此时，需要关闭服务器并重启它：

```
adb kill-server
adb start-server
```

如果查看 Android 设备/模拟器上的 Launcher，可以看到 LBS 图标(在图 12-11 的顶部)。如果在 Android 设备/模拟器上选择 Settings | Apps，将可以看到 Where Am I 应用程序(在图 12-11 的底部)。

adb.exe 工具除了可以用来安装应用程序外，还可以用来移除已安装的应用程序。要做到这一点，可以使用 uninstall 选项将一个应用程序从其安装文件夹下移除：

```
adb uninstall net.learn2develop.LBS
```

部署应用程序的另外一种方法是使用 Eclipse 中的 DDMS 工具(如图 12-12 所示)。选择一个模拟器(或设备)后，使用 DDMS 中的 File Explorer 进入到/data/app 文件夹并通过 Push a file onto the device 按钮将 APK 文件复制到设备上。

图 12-11

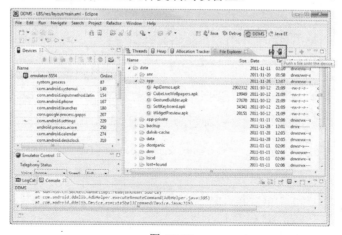

图 12-12

12.2.2 使用 Web 服务器

如果希望自己驻留应用程序,可以使用一台 Web 服务器来完成。如果您有自己的 Web 托管服务并且打算为您的用户免费提供应用程序(或者将访问限制为特定用户群),这是个不错的选择。

> **注意**:在下载了 APK 文件后,即使您将对应用程序的访问权限限制为某个特定的用户群,也还是没有什么能阻止这些用户将您的应用程序重新分发给其他用户。

为了说明这一点,作者将使用 Windows 7 计算机上的 IIS(Internet Information Server),将已签名的 LBS.apk 文件复制到 c:\inetpub\wwwroot\。另外,创建一个新的名为 index.html 的 HTML 文件,其内容如下:

```
<html>
<title>Where Am I application</title>
<body>
Download the Where Am I application <a href="LBS.apk">here</a>
</body>
</html>
```

> **注意**:如果不清楚如何在 Windows 7 计算机上设置 IIS,可查阅以下链接:http://technet.microsoft.com/en-us/library/cc725762.aspx。

在您的 Web 服务器上,需要为 APK 文件注册一个新的 MIME 类型。.apk 扩展名对应的 MIME 类型是 application/vnd.android.package-archive。

> **注意**:如果不知道如何在 IIS 中设置 MIME 类型,可查阅以下链接:http://technet.microsoft.com/en-us/library/cc725608(WS.10).aspx。

> **注意**:要在 Web 上安装 APK 文件,需要在您的模拟器或设备上安装一个 SD 卡。这是由于下载的 APK 文件将保存在 SD 卡的 download 文件夹下。为了在模拟器中对此进行测试,确保 SD 卡至少有 128MB。有报道称,一些开发人员在小于 128MB 的 SD 卡安装应用程序时遇到了问题。

默认情况下,对于在线安装 Android 应用程序,Android 模拟器或设备只允许安装来自 Android Market(www.android.com/market/)的应用程序。因此,对于在 Web 服务器上的安装,需要配置您的 Android 模拟器/设备以接受来自非 Android Market 源的应用程序。

在 Settings 菜单中,选中 Security 项,并滚动到屏幕的底部。选中 Unknown sources 项(如图 12-13 所示),会出现一个警告消息,单击 OK 按钮。选中这一项将允许模拟器/设备安装来自非 Android Market 源的应用程序(例如来自一台 Web 服务器)。

为了从运行在您计算机上的 IIS Web 服务器上安装 LBS.apk 应用程序,可在 Android 模拟器/设备上启动 Browser 应用程序,并导航到指向 APK 文件的 URL。为了指向运行模拟器的计算机,应该使用计算机的 IP 地址。图 12-14 展示了 Web 浏览器上加载的 index.html 文件。单击 here 链接将会把 APK 文件下载到您的设备上。单击屏幕顶部的状态栏可以显示下载的状态。

图 12-13

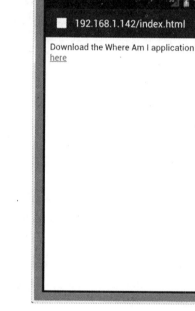

图 12-14

为了安装下载的应用程序,只要轻轻地单击应用程序,将显示出该应用程序所需要的权限。单击 Install 按钮继续安装。当应用程序安装完毕,可以单击 Open 按钮来启动它。

除了使用 Web 服务器,还可以用电子邮件将应用程序作为附件发给用户。当用户收到邮件时,他们可以下载附件并在自己的设备上直接安装。

12.2.3 在 Android Market 上发布

到目前为止,您已经学习了如何打包 Android 应用程序以及分发应用程序所能使用的多种方法——通过 Web 服务器、adb.exe 文件、电子邮件、SD 卡等。

然而,这些方法不能为用户提供一个可以很容易地发现您的应用程序的途径。更好的方法是由 Android Market 托管您的应用程序,这是一个可以使用户为其 Android 设备发现和下载(购买)应用程序变得非常容易的由 Google 托管的服务。用户只需要在他们的 Android

设备上启动 Market 应用程序,就能够发现可以在他们设备上进行安装的各种应用程序。

在本节中,将学习到如何在 Android Market 上发布 Android 应用程序。我将一步步地引导您完成发布,包括在提交给 Android Market 前应用程序所需要做的各项准备。

1. 创建一个开发者简介

在 Android Market 上发布应用程序的第一步是在 http://market.android.com/publish/Home 上创建一个开发者简介。这需要一个 Google 账户(例如您的 Gmail 账户)。一旦登录进 Android Market,首先创建开发者简介(如图 12-15 所示)。在输入所要求的信息后单击 Continue。

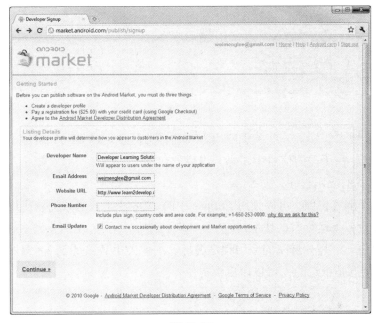

图 12-15

为了在 Android Market 上发布应用程序,需要支付一次性的注册费用,目前是 25 美元。单击 Google Checkout 按钮会被重定向到一个可以支付注册费用的页面。付费之后,单击 Continue 链接。

接下来,需要同意 Android Market Developer Distribution Agreement。选中 I agree 复选框并单击 I agree. Continue 链接。

2. 提交应用程序

一旦设置好简介后,就要准备向 Android Market 提交应用程序了。如果您的应用程序打算收费,单击位于屏幕底部的 Setup Merchant Account 链接。在这里输入诸如银行账号、税号等额外信息。

对于免费的应用程序,单击 Upload Application 链接,如图 12-16 所示。

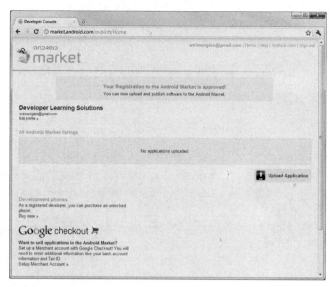

图 12-16

这时，将会要求您输入一些关于应用程序的详细信息。图 12-17 展示了需要您提供的第一组详细信息。在所需的信息中，必须提供的包括：

- APK 格式的应用程序。
- 至少两个屏幕快照。可以使用 Eclipse 中的 DDMS 透视图来捕获应用程序在模拟器或真实设备上运行时的快照。
- 需要提供一个高分辨率的应用程序图标。图像大小必须是 512×512 像素。

其他详细信息是可选的，可以以后再提供。

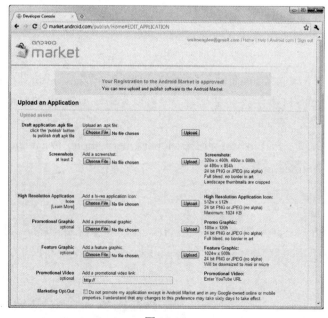

图 12-17

图 12-18 展示了作者已向 Android Market 站点上传了 LBS.apk 文件。特别要注意，基于您上传的 APK 文件，将会有警告消息，告诉用户所需的特定的权限，并且您的应用程序的功能将被用来作为搜索结果的筛选条件。例如，由于我的应用程序要求接入 GPS，对于那些没有 GPS 接收器的设备的用户，我的应用程序就不会在他们的搜索结果列表上显示。

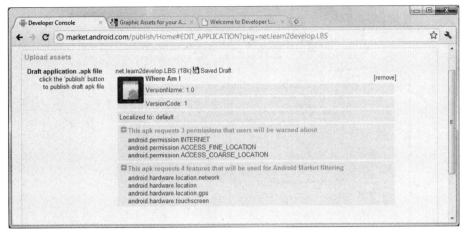

图 12-18

需要提供的下一组信息如图 12-19 所示，包含了应用程序的标题、其描述以及最新修改的细节(对于应用程序更新很有用)。还可以选择应用程序的类型和在 Android Market 中所属的类别。

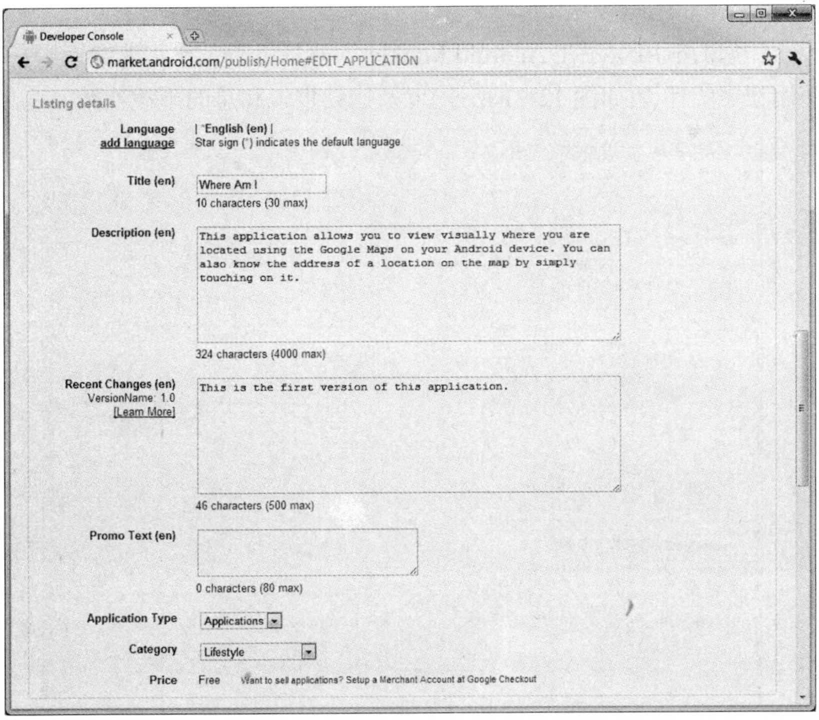

图 12-19

在最后一个对话框中，可以指明您的应用程序是否采用版权保护以及指定内容评级。还可以提供您的网站的 URL 以及您的联系信息(如图 12-20 所示)。当同意了两个准则和协议后，单击 Publish 按钮将您的应用程序发布到 Android Market 上。

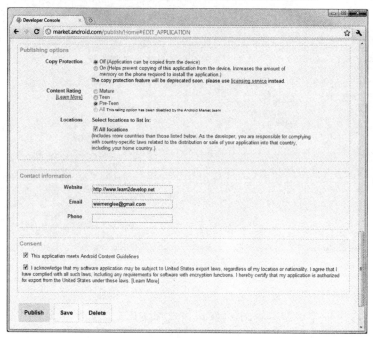

图 12-20

到此为止，您的应用程序在 Android Market 上就可用了。您将能够查看任何已提交的关于您的应用程序的评论(如图 12-21 所示)以及错误报告和总的下载次数。

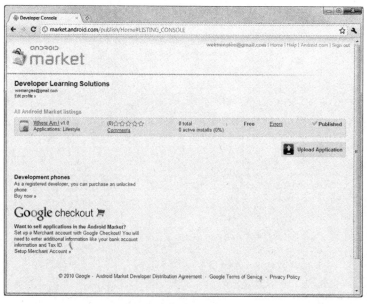

图 12-21

祝您好运！现在，您只要等好消息就行了。希望很快您就能一路笑着去银行！

12.3 本章小结

在本章中，我们学习了如何将 Android 应用程序导出为一个 APK 文件并使用自己创建的一个密钥库对其进行数字签名。然后，学习了分发应用程序的多种方法以及每一种方法的优点。最后，我们亲自体验了在 Android Market 上发布应用程序所需的步骤，这将使您可以出售您的应用程序，并且带来更多的用户。希望这样可以使您卖出更多的产品并因此获得不错的收益！

> **练 习**
>
> 1. 如何指定应用程序所需的 Android 最低版本？
> 2. 如何生成一个自签名证书来为 Android 应用程序签名？
> 3. 如何配置 Android 设备，使其可以接收非 Anroid Market 源的应用程序？
>
> 练习答案参见附录 C。

本章主要内容

主 题	关 键 概 念
用于发布应用程序的检查表	要在 Android Market 上发布一个应用程序，该应用程序必须在 AndroidManifest.xml 文件中具有以下 4 个属性：android：versionCode、android：versionName、android：icon、android：label
应用程序必须签名	要分发的所有应用程序必须使用一个自签名证书进行签名。调试密钥库对于分发来说是无效的
导出一个应用程序并对其签名	使用 Eclipse 的 Export 功能将应用程序导出为一个 APK 文件并使用一个自签名证书对其签名
部署 APK 文件	可以使用各种方式部署：Web 服务器、电子邮件、adb.exe 和 DDMS 等
在 Android Market 上发布应用程序	一次性花费 25 美元向 Android Market 进行申请，这样就可以在 Android Market 上托管和出售应用程序

附录 A

使用 Eclipse 进行 Android 开发

尽管 Google 支持使用诸如 IntelliJ 这样的 IDE 或者像 Emacs 这样的基本编辑器来进行 Android 应用程序的开发，但它还是推荐将 Eclipse IDE 和 ADT 插件一起使用。这样做可以使开发 Android 应用程序变得更容易也更高效。本附录描述了 Eclipse 中可用的一些可以使您的开发工作变得更加容易的极好功能。

 警告：如果您还没有下载 Eclipse，那么可以从第 1 章开始学起。在那里，您将学到如何获取 Eclipse 以及对它进行配置，使其可以和 Android SDK 一起使用。本附录假定您已经为 Android 开发设置好了 Eclipse 环境。

A.1 Eclipse 概览

Eclipse 是一个高度可扩展的多语言软件开发环境，可以支持各种应用程序开发。使用 Eclipse，可以利用多种语言来编写和测试应用程序，例如 Java、C、C++、PHP、Ruby 等。由于其可扩展性，Eclipse 的新手常常感到对其 IDE 无所适从。因此，以下小节的内容旨在帮助您在 Android 应用程序的开发过程中可以更加熟练地使用 Eclipse。

A.1.1 工作区

Eclipse 采用工作区(workspace)的概念。所谓工作区就是您所选择的用来保存所有项目的一个文件夹。

当第一次启动 Eclipse 时，将提示您选择一个工作区(如图 A-1 所示)。

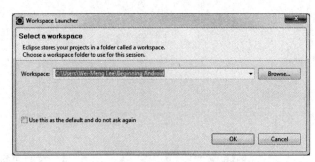

图 A-1

当 Eclipse 将位于工作区中的项目启动完之后，将在 IDE 中显示几个窗格(如图 A-2 所示)。

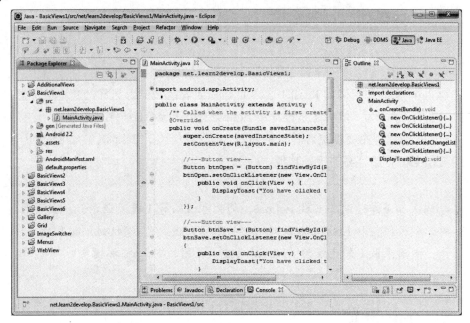

图 A-2

下面的小节将对在开发 Android 应用程序时必须了解的一些较重要的窗格进行重点讲解。

A.1.2 Package Explorer

如图 A-3 所示，Package Explorer 列出了当前位于工作区中的所有项目。要编辑项目中一个特定的项，可以双击这一项，文件将会显示在各自的编辑器中。

还可以右击在 Package Explorer 中列出的每一项，以显示与所选项有关的上下文敏感菜单。例如，如果希望在项目中添加一个新的.java 文件，可以在 Package Explorer 中右击包的名称，然后选择 New | Class (如图 A-4 所示)。

图 A-3

附录 A 使用 Eclipse 进行 Android 开发

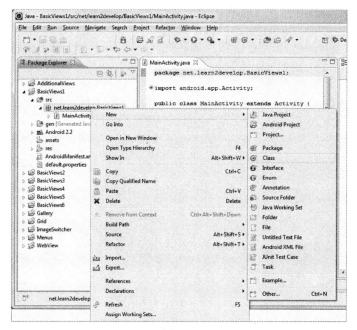

图 A-4

A.1.3 使用其他工作区的项目

也许您有时候会创建多个工作区来存储不同的项目。如果需要访问另一个工作区中的项目，通常有两种方法可以实现这一点。第一种方法是选择 File | Switch Workspace(如图 A-5 所示)切换到您所希望的工作区。指定新工作区后重新启动 Eclipse。

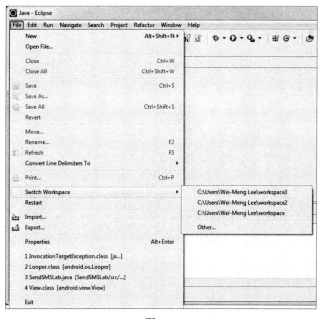

图 A-5

第二种方法是将项目从另一个工作区导入到当前工作区中。要做到这一点，选择 File |

Import...，然后选择 General | Existing Projects into Workspace(如图 A-6 所示)。单击 Next。

在 Select root directory 文本框中输入包含想要导入的项目的工作区的路径，并选中这些项目(如图 A-7 所示)。单击 Finish 按钮，导入所选择的项目。

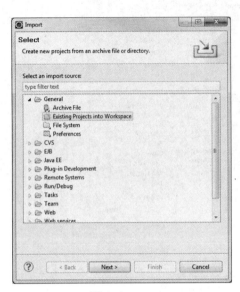

图 A-6

图 A-7

注意，当从另一个工作区将一个项目导入到当前工作区中时，导入项目的物理位置保持不变。也就是说，项目仍旧位于其原始目录下。要在当前工作区中保留项目的一个副本，可选中 Copy projects into workspace 选项。

A.1.4 在 Eclipse 中使用编辑器

根据在 Package Explorer 中双击的项目的类型，Eclipse 将为您打开对应的编辑器来编辑文件。举例来说，如果双击一个.java 文件，将打开用于编辑源文件的文本编辑器(如图 A-8 所示)。

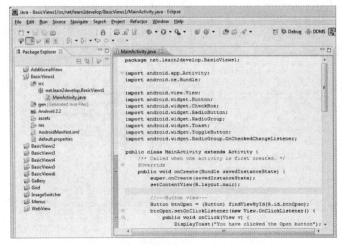

图 A-8

如果双击 res/drawable-mdpi 文件夹下的 ic_launcher.png 文件,将启动 Windows Photo Viewer 应用程序来显示图像(如图 A-9 所示)。

图 A-9

如果双击 res/layout 文件夹下的 main.xml 文件,Eclipse 将显示 UI 编辑器,在那里可以以图形化方式查看和构建 UI 布局(如图 A-10 所示)。

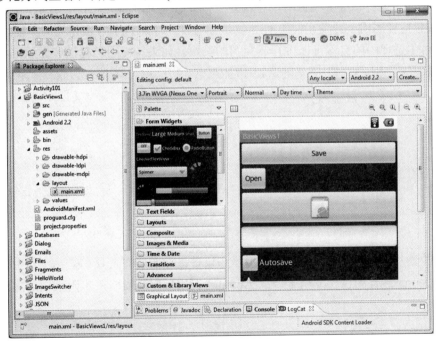

图 A-10

要使用 XML 手动编辑 UI,可以单击位于编辑器底部的 main.xml 选项卡,切换到 XML 视图(如图 A-11 所示)。

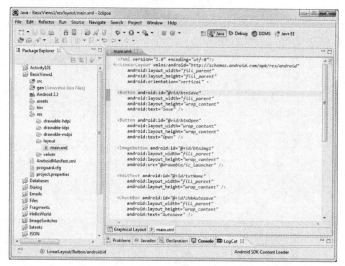

图 A-11

A.1.5 理解 Eclipse 透视图

在 Eclipse 中，透视图(perspective)是一个包含一组视图和编辑器的可视化容器。当在 Eclipse 中编辑 Android/Java 项目时，您就处于 Java 透视图中(如图 A-12 所示)。

Java EE 透视图用于开发企业级的 Java 应用程序，它包含了与之相关的其他模块。

通过单击透视图的名称可以进行透视图的切换。如果透视图名称没有显示，可以单击 Open Perspective 按钮添加一个新的透视图(如图 A-13 所示)。

图 A-12

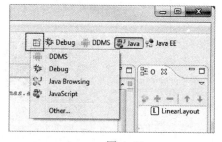

图 A-13

DDMS 透视图包含了与 Android 模拟器和设备进行通信的工具。在附录 B 中将进行详细介绍。Debug 透视图包含了用于调试 Android 应用程序的窗格，本附录稍后将对此进行详细介绍。

A.1.6 包的自动导入

Android 库中的各种类是以包的形式组织起来的。因此，当使用包的一个特定的类时，需要导入适当的包，如下所示：

```
import android.app.Activity;
import android.os.Bundle;
```

由于 Android 库中类的数目非常巨大，要记住每一个类所属的正确的包不是一件容易的事。幸运的是，Eclipse 可以帮您找到正确的包，使得您只要单击鼠标就可以导入它。

图 A-14 展示了本书作者所声明的一个 Button 类型的对象。由于作者没有为 Button 类导入正确的包，Eclipse 在语句下面提示一个错误。当将鼠标移动到 Button 类的上面时，Eclipse 会显示一个修改建议列表。在本例中，需要导入 android.widget.Button 包。单击 Import 'Button'（android.widget）链接将在文件开头添加导入语句。

图 A-14

或者，可以使用如下的组合键：Control+Shift+o。这一组合键将使 Eclipse 自动导入您的类所需要的所有包。

A.1.7 使用代码完成功能

Eclipse 另一个非常有用的功能就是支持代码完成。当您在代码编辑器中输入内容时，代码完成会显示一个上下文敏感的相关类、对象、方法以及属性名称的列表。例如，图 A-15 展示了起作用的代码完成功能。在我输入单词 fin 时，通过按下 Ctrl+Space 组合键能够激活代码完成功能，这时将出现一个以 fin 开头的名称列表。

图 A-15

要选择所需的名称，直接在其上双击或者使用光标高亮显示它并按 Enter 键就行了。在一个对象/类名之后输入 period(·)时，代码完成也可以起作用。图 A-16 展示了一个示例。

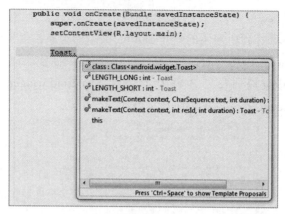

图 A-16

A.1.8 重构

重构是大多数现代 IDE 支持的一个非常有用的功能。Eclipse 支持大量的重构功能,可用来有效地进行应用程序开发。

在 Eclipse 中,当将光标放到一个特定对象/变量上时,编辑器将突出显示当前源中所选对象出现的所有地方(如图 A-17 所示)。

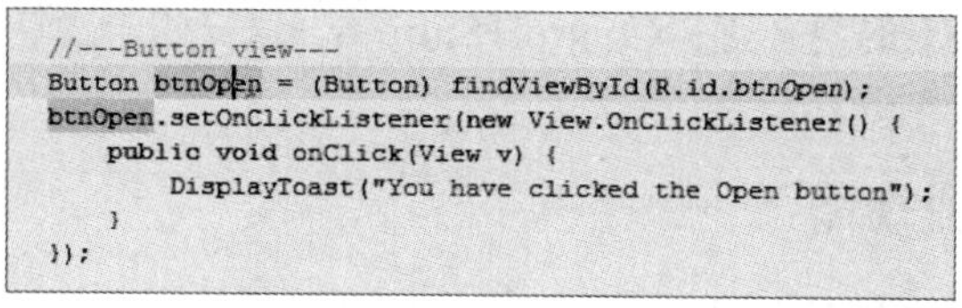

图 A-17

这一功能对于确定一个特定对象在代码中的位置是非常有用的。要改变一个对象的名称,可右击它并选择 Refactor | Rename...(如图 A-18 所示)。

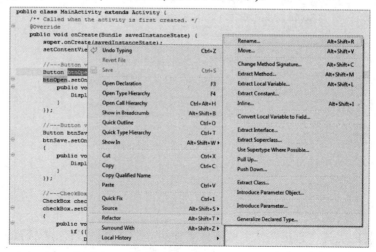

图 A-18

附录 A 使用 Eclipse 进行 Android 开发

在输入对象的新名称后，该对象出现的所有地方将动态变化(如图 A-19 所示)。注意为了使重构能够正确地工作，代码中不能有任何语法错误，而且必须能够被编译器正确地编译。

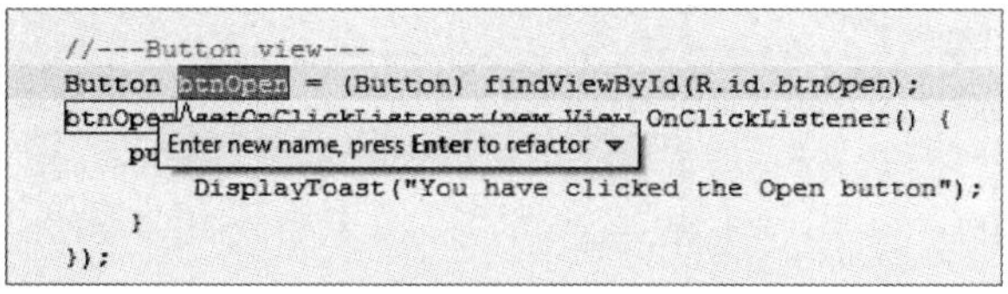

图 A-19

另外一个重构十分有用的领域是从 UI 文件中提取字符串常量。如第 1 章所述，在用户界面中使用的所有字符串常量最好都存储到 strings.xml 文件中，这样会方便以后进行本地化。然而在开发过程中，开发人员经常采取一种便捷的方法，即直接输入字符串常量。例如，您可能会使用一个字符串常量设置 Button 视图的 android:text 属性的值：

```
<Button android:id="@+id/btnSave"
    android:layout_width="fill_parent"
    android:layout_height="wrap_content"
    android:text="Save" />
```

在 Eclipse 中使用重构功能时，可以选择该字符串常量，然后选择 Refactor | Android | Extract Android String…，如图 A-20 所示。

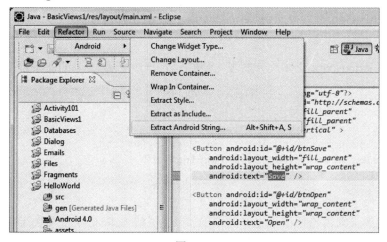

图 A-20

然后 Eclipse 将会提示您为该字符串常量指定一个名称，如图 A-21 所示。完成之后，单击 OK 按钮。

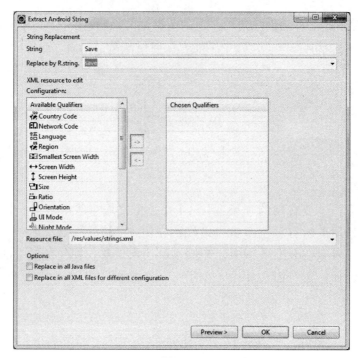

图 A-21

之后，android:text 属性的值将被替换为@string/save：

```
<Button android:id="@+id/btnSave"
    android:layout_width="fill_parent"
    android:layout_height="wrap_content"
    android:text="@string/save" />
```

如果观察 strings.xml 文件，会发现它现在包含一个名为 save 的新条目：

```
<?xml version="1.0" encoding="utf-8"?>
<resources>
    <string name="hello">Hello World, BasicViews1Activity!</string>
    <string name="app_name">BasicViews1</string>
    <string name="save">Save</string>
</resources>
```

对于重构的详细讨论超出了本书的范围。想要了解 Eclipse 中更多关于重构的信息，可以访问 www.ibm.com/developerworks/library/os-ecref/。

A.2 调试应用程序

Eclipse 支持在 Android 模拟器以及真实的 Android 设备上进行应用程序的调试。当在 Eclipse 中按下 F11 键时，Eclipse 将首先确定是否已经在运行一个 Android 模拟器的实例或是连接了一个真实设备。只要有一个模拟器(或设备)运行，Eclipse 就会将应用程序部署到运行中的模拟器或已连接的设备上。如果既没有模拟器运行又没有连接设备，Eclipse 将自动启动一个 Android 模拟器的实例并将应用程序部署在其上面。

如果有多个模拟器或设备连接，Eclipse 将提示您选择一个目标模拟器/设备，以便在其上部署应用程序(如图 A-22 所示)。选择一个您想使用的目标设备，然后单击 OK 按钮。没有运行应用程序所需的最低 OS 版本的设备将带有一个 X 标记。

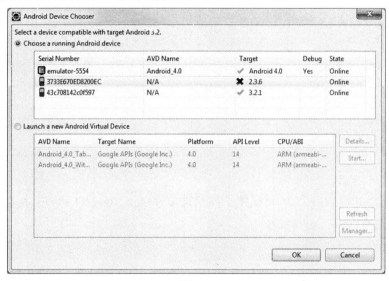

图 A-22

如果想启动一个新的模拟器实例来测试应用程序，可选择 Window | Android SDK and AVD Manager 来启动 AVD Manager。

A.2.1 设置断点

设置断点是临时暂停应用程序的执行，然后检查变量内容和对象内容的一个好方法。

要设置一个断点，可双击代码编辑器中的最左列。图 A-23 展示了设置在一个特定语句上的断点。

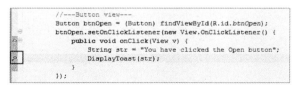

图 A-23

当应用程序运行到第一个断点时，Eclipse 将显示一个 Confirm Perspective Switch 对话框。从根本上说，它希望切换到 Debug 透视图。为了防止再次出现这一窗口，在底部选中 Remember my decision 复选框并单击 Yes。现在，Eclipse 突出显示了断点(如图 A-24 所示)。

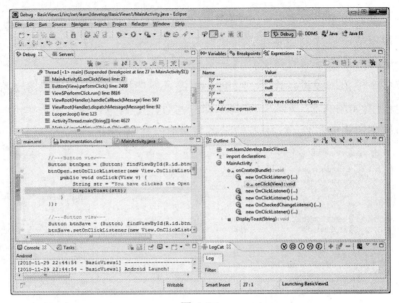

图 A-24

此时，可以利用如图 A-25 所示的不同的选项(Watch、Inspect 和 Display)，右击任意选择的对象/变量来查看它们的内容。

图 A-25

图 A-26 展示了使用 Inspect 选项显示 str 变量的内容。

图 A-26

此时，有以下几个选项可以继续执行：
- Step Into——按 F5 键步进到下一个方法调用/语句。
- Step Over——按 F6 键跳过下一个方法调用，不进入此方法中。
- Step Return——按 F7 键从已经进入的方法中返回。
- Resume Execution——按 F8 键继续执行。

A.2.2 异常

在 Android 中进行开发时,您将会碰到大量运行时异常,阻止您的程序继续运行。运行时异常的例子包括以下内容:
- 空引用异常(访问一个空对象)
- 没有指定应用程序所需的权限
- 算术运算异常

图 A-27 展示了一个应用程序产生异常时的当前状态。在这一示例中,我试图从我的应用程序中发送一条 SMS 消息。当消息刚要发送时,它崩溃了。

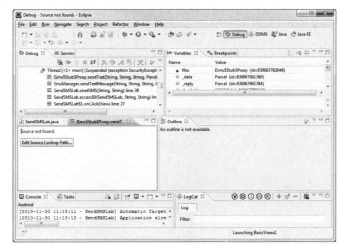

图 A-27

各式的窗口并不能真正识别异常发生的原因。要寻找更多信息,可以在 Eclipse 中按 F6 键跳过当前语句。Variables 窗口指出了异常的原因,如图 A-28 所示。在本例中,异常原因是缺少 SEND_SMS 权限。

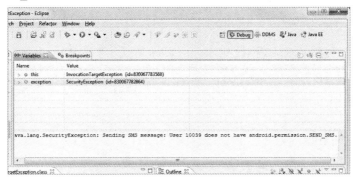

图 A-28

为了补救,只要在 AndroidManifest.xml 文件中添加以下权限声明就行了:

```
<uses-permission android:name="android.permission.SEND_SMS"/>
```

附录 B

使用 Android 模拟器

Android 模拟器自带了 Android SDK，它是一个很有用的工具，可以帮助您测试应用程序而不需要购买一台真实设备。虽然您应该在部署应用程序前在真实设备上对其进行彻底的测试，但模拟器还是可以模仿真实设备的大多数功能。模拟器是一个在项目开发阶段应该利用的非常方便的工具。本附录提供了有助于掌握 Android 模拟器的一些常见的提示和技巧。

B.1 Android 模拟器的使用

正如在第 1 章所讨论的，可以使用 Android 模拟器，通过创建 Android Virtual Device(AVD)来模拟不同的 Android 配置。

如果想模拟真实的设备，首先使用与真实设备相同的屏幕分辨率和 abstracted LCD 密度创建一个 AVD(具体做法参见本附录的"模拟具有不同屏幕尺寸的设备"一节)。然后，在 AVD Manager 窗口中直接启动创建好的 AVD(如图 B-1 所示)。直接选择 AVD 并单击 Start 按钮。如果想让模拟器显示的尺寸与真实设备的屏幕尺寸相同，则选中 Scale display to real size 选项，并将 Screen Size (in)选项设为真实设备的尺寸。输入当前使用的显示器的 dpi(如果不知道，可以单击?(问号)按钮，然后选择屏幕尺寸和分辨率)。Android 模拟器将显示一个接近真实设备的屏幕尺寸。这个选项十分有用，可以用来预览应用程序在具有不同屏幕尺寸的实际屏幕上显示的效果。

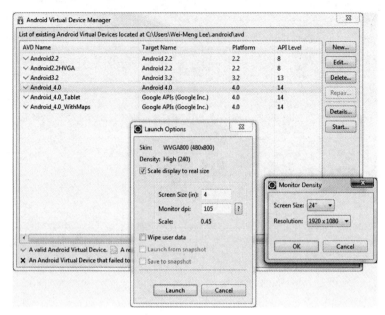

图 B-1

 注意：为了使 Android 模拟器具有最佳性能，可将屏幕尺寸设为可以允许的最小尺寸。这样可以让模拟器的运行速度加快。

或者，在 Eclipse 中运行一个 Android 项目时，Android 模拟器会自动启动来测试应用程序。可以在 Eclipse 中为每一个 Android 项目定制 Android 模拟器，只要选择 Run | Run Configurations。选择左边位于 Android Application 下的项目名称(如图 B-2 所示)，就会在右边看到 Target 选项卡。在其中可以选择用来进行应用程序测试的 AVD，以及选择模拟不同的场景，如网速和网络延迟。

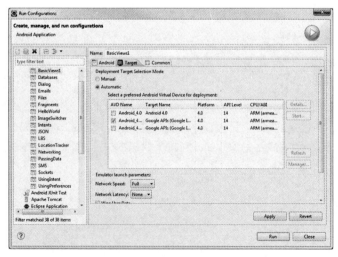

图 B-2

B.2 创建快照

在最新版的 AVD Manager 中，可以选择将模拟器的状态保存到一个快照文件中。这样下一次启动模拟器时就不会经历漫长的启动时间，所以启动速度变快。Android 3.0(及更高版本)的模拟器的启动时间可能长达 5 分钟，所以这么做尤为有用。

为使用快照功能，只要在创建新的 AVD 时，选中 Snapshot Enabled 复选框，如图 B-3 所示。

从 Start…按钮启动 AVD 时，选中 Launch from snapshot 和 Save to snapshot 复选框，如图 B-4 所示。第一次启动模拟器时，它将正常启动。关闭模拟器时，则会将状态保存到快照文件中。下一次启动模拟器时，它将立即显示，并从快照文件还原其状态。

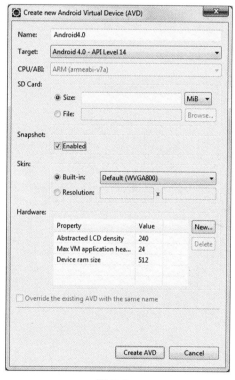

图 B-3   图 B-4

B.3 模拟 SD 卡

如果创建了一个新的 AVD，那么可以模拟存在着一张 SD 卡(如图 B-5 所示)。直接输入想模拟的 SD 卡的大小(在图中的大小为 200MB)。

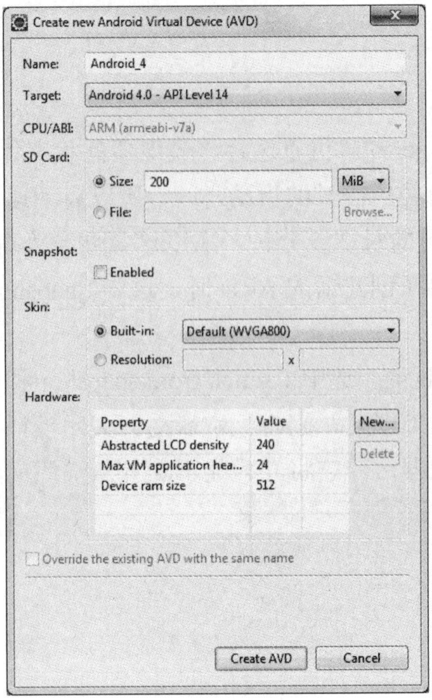

图 B-5

或者，可以通过首先创建一个磁盘映像并将其附加到 AVD 上来模拟 Android 模拟器中的一张 SD 卡。mksdcard.exe 实用程序(也位于 Android SDK 的 tools 文件夹下)可以用来创建一个 ISO 磁盘映像。下列命令创建一个大小为 2GB 的 ISO 映像(还可参见图 B-6)：

```
mksdcard 2048M sdcard.iso
```

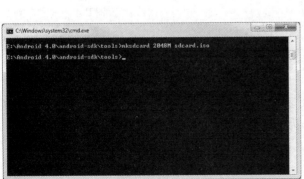

图 B-6

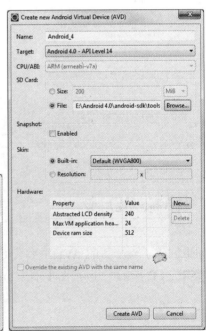

图 B-7

附录 B 使用 Android 模拟器

一旦创建了映像，就可以指定 ISO 文件的位置了，如图 B-7 所示。

B.4 模拟具有不同屏幕尺寸的设备

除了模拟 SD 卡外，还可以模拟具有不同屏幕大小的设备。图 B-8 展示了 AVD 在模拟 HVGA 外观，其分辨率是 320 × 480 像素。注意，抽象 LCD 的像素密度是 160，意思是该屏幕的像素密度为每英寸具有 160 个像素。

对于您选择的每一个目标，都有一个可用的外观列表。Android 支持以下屏幕分辨率：

- HVGA — 320 × 480
- QVGA — 240 × 320
- WQVGA400 — 240 × 400
- WQVGA432 — 240 × 432
- WVGA800 — 480 × 800
- WVGA854 — 480 × 854

除了使用内置的分辨率，还可以指定自定义分辨率。例如，可以通过使用图 B-9 所示的规范创建一个 AVD 来模拟 Samsung Galaxy Tab 10.1。

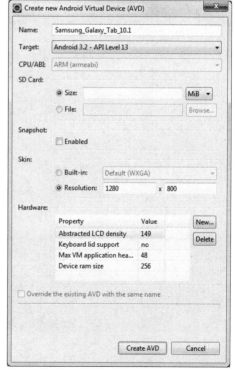

图 B-8   图 B-9

当 AVD 启动后，可以看到 Android 模拟器在模拟 Honeycomb 平板电脑，如图 B-10。

487

图 B-10

B.5 模拟物理功能

除了模拟具有不同屏幕大小的设备之外，还可以选择模拟不同的硬件功能。当创建一个新 AVD 时，单击 New…按钮将显示一个可用于选择打算模拟的硬件的类型的对话框(如图 B-11 所示)。

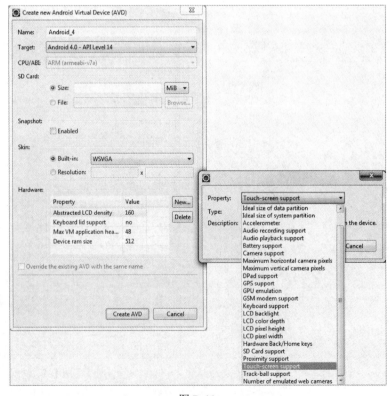

图 B-11

例如，如果想模拟一个没有触摸屏的 Android 设备，可以选择 Touch-screen support 属性并单击 OK 按钮。回到 AVD 对话框，将该属性值从 yes 变为 no(如图 B-12 所示)。

这将创建一个不带有触摸屏支持的 AVD(也就是说，用户不能使用鼠标在屏幕上单击)。

还可以使用 Android 模拟器来模拟位置数据。第 9 章详细讨论过这一点。

图 B-12

键盘快捷键

Android 模拟器支持多个键盘快捷键，可以使您模仿一个真实手机的行为。以下列表展示了可与模拟器一起使用的一组快捷键：

- Esc—返回
- Home—主屏幕
- F2—切换上下文敏感菜单
- F3—通话记录
- F4—结束通话按钮
- F7—电源按钮
- F5—搜索
- F6—切换跟踪球模式
- F8—切换数据网络(3G)
- Ctrl+F5—提高铃声音量
- Ctrl+F6—降低铃声音量
- Ctrl+F11/Ctrl+F12—切换方向

例如，通过按下 Ctrl+F11 组合键，可以将模拟器的方向改为纵向模式(如图 B-13 所示)。

图 B-13

使您的开发更有效率的一个有用的诀窍就是在开发过程中一直使 Android 模拟器处于运行状态——不要关闭和重启它。因为模拟器启动要花费时间,当您在调试应用程序时,最好让它一直运行着。

B.6 给模拟器发送 SMS 消息

使用 Eclipse 中可用的 Dalvik Debug Monitor Service(DDMS)工具或者 Telnet 客户端,可以模拟发送 SMS 消息到 Android 模拟器。

> **注意**:Telnet 客户端不是 Windows 7 的默认安装项。要安装它,可在 Windows 命令提示符下输入以下命令行:
> ```
> pkgmgr /iu: "TelnetClient "
> ```

现在看一看在 Telnet 中如何做到这一点。首先,确保 Android 模拟器正在运行。为了 Telnet 到模拟器,需要知道模拟器的端口号。这可以通过查看 Android 模拟器窗口的标题栏获得。端口号通常以 5554 开始,每一个后续模拟器的端口号以 2 递增,例如 5556、5558 等。假定当前有一个 Android 模拟器在运行,那么可以使用以下命令 Telnet 到它上面(使用您的模拟器的实际端口号替换 5554):

```
C:\telnet localhost 5554
```

要给模拟器发送一条 SMS 消息,可使用以下命令:

```
sms send +1234567 Hello my friend!
```

sms send 命令的语法如下所示:

```
sms send <phone_number> <message>
```

图 B-14 展示了模拟器正在接收刚刚发送的 SMS 消息。

除了使用 Telnet 发送 SMS 消息之外,还可使用 Eclipse 中的 DDMS 透视图。如果 DDMS 透视图在 Eclipse 中没有显示,可以单击 Open Perspective 按钮(如图 B-15 所示)并选择 Other 显示它。

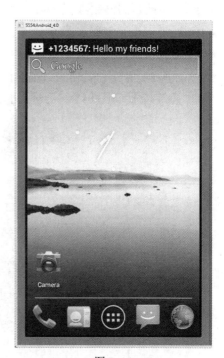

图 B-14

图 B-15

选择 DDMS 透视图(如图 B-16 所示)并单击 OK 按钮。

DDMS 透视图显示之后,就可以看到 Devices 选项卡(如图 B-17 所示),里面显示了当前正在运行的模拟器的列表。选择一个打算向其发送 SMS 消息的模拟器实例,在 Emulator Control 选项卡的下面,将看到 Telephony Actions 部分。在 Incoming number 字段中,输入一个任意电话号码并选中 SMS 单选按钮。输入一条消息并单击 Send 按钮。

现在,选中的模拟器将收到一条传入的 SMS 消息。

如果同时有多个 AVD 运行,可以使用模拟器的端口号作为电话号码在每一个 AVD 之间发送 SMS 消息。例如,如果您有分别运行在端口号 5554 和 5556 上的两个模拟器,它们的电话号码将分别是 5554 和 5556。

图 B-16

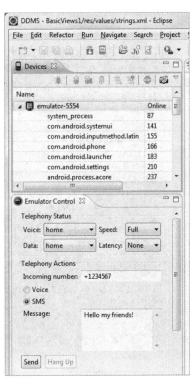

图 B-17

B.7 打电话

除了发送 SMS 消息到模拟器,还可以使用 Telnet 客户端给模拟器打电话。直接使用以下命令就可以做到这一点。

要 Telnet 到模拟器,使用下列命令(使用您的模拟器的实际端口号替换 5554):

```
C:\telnet localhost 5554
```

要给模拟器打电话,使用下列命令:

```
gsm call +1234567
```

gsm send 命令的语法如下所示:

```
gsm call <phone_number>
```

图 B-18 展示了模拟器收到一个来电。

与发送 SMS 消息类似,也可以使用 DDMS 透视图来给模拟器打电话。图 B-19 展示了如何使用 Telephony Actions 部分来打电话。

还可以使用端口号作为电话号码在 AVD 之间打电话。

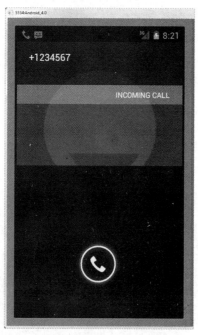

图 B-18

图 B-19

B.8 从模拟器中传入/传出文件

偶尔,您也许需要从模拟器中传入/传出文件。使用 DDMS 透视图是最容易的方法。

从 DDMS 透视图中选择模拟器(或者设备，如果您有一个连接到计算机的真实的 Android 设备的话)并单击 File Explorer 选项卡来查看它的文件系统(如图 B-20 所示)。

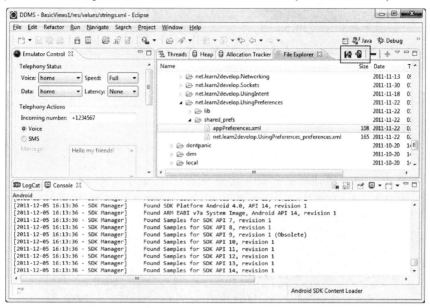

图 B-20

 注意：当使用 adb.exe 实用工具从模拟器中拖出或拖入文件时，确保只有一个 AVD 在运行。

图 B-20 中高亮显示的两个按钮可用于从模拟器中拖出或向模拟器中拖入一个文件。

或者，还可以使用 Android SDK 自带的 adb.exe 实用工具来从模拟器中拖出或拖入文件。这个实用工具位于<Android_SDK_Folder>\platform-tools\文件夹下。

要从已连接的模拟器/设备上复制文件到计算机上，可使用以下命令：

    adb.exe pull <source path on emulator>

图 B-21 展示了如何从模拟器中提取一个 XML 文件并将其保存到您的计算机上。

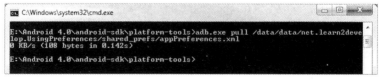

图 B-21

要将文件复制到已连接的模拟器/设备上，可使用以下命令：

adb.exe push <filename> <destination path on emulator>

图 B-22 中的命令复制了位于当前目录下的 NOTICE.txt 文件，并将其保存在模拟器的

/data/data/net.learn2develop.UsingPreferences/shared_prefs 文件夹下。

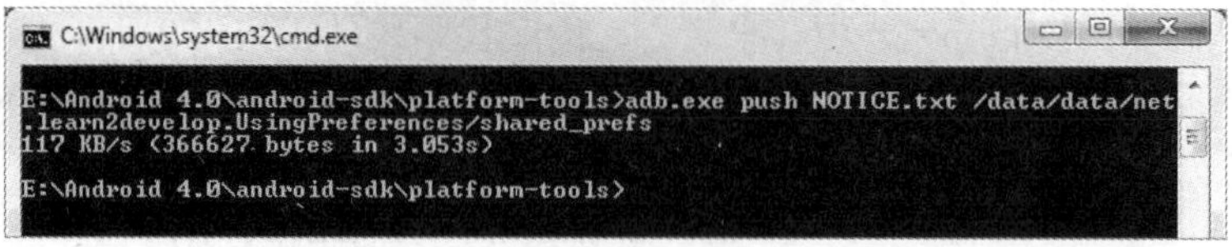

图 B-22

如果需要在模拟器中修改文件的权限，可以使用带有 shell 选项的 adb.exe 实用工具，如下所示：

adb.exe shell

图 B-23 展示了如何利用 chmod 命令修改 NOTICE.txt 文件的权限。

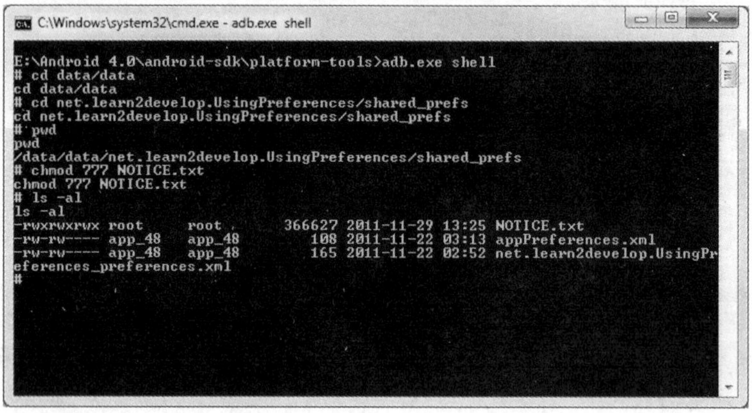

图 B-23

使用 adb.exe 实用工具，可以对 Android 模拟器发出 Unix 命令。

B.9 重置模拟器

有时候会想把应用程序安装到一个全新的 AVD 中。例如，可能以前安装的其他应用程序会影响当前应用程序(例如，一个 SMS 拦截应用程序可能会拦截发送给您的应用程序的 SMS 消息)。此时，既可以从 Settings 应用程序中卸载每个应用程序，也可以擦除模拟器的镜像，使其还原到初始状态(这是更简单的方法)。

所有部署到 Android 模拟器上的应用程序和文件都保存在一个名为 userdata-qemu.img 的文件中，此文件位于 C:\Users\<username>\.android\avd\<avd_name>.avd 文件夹下。例如，有一个名为 AndroidTabletWithMaps 的 AVD，因此 userdata-qemu.img 文件就位于 C:\Users\Wei-Meng Lee\.android\avd\AndroidTabletWithMaps.avd 文件夹下。

如果想将模拟器恢复到初始状态(也即重置)，只要删除 userdata-qemu.img 文件就行了。以前在这个 AVD 上安装的所有应用程序都将被清除。

附录 C 练习答案

本附录包括了各章最后的练习的答案。

第 1 章答案

1. AVD 指的是 Android 虚拟设备。它代表了一个 Android 模拟器，可以模拟一个实际 Android 设备的特定配置。
2. android:versionCode 属性用来以编程方式检查一个应用程序是否可以被升级。它应当包含一个顺序号(更新的应用程序的号码应该比老版本的设置得要高)。android:versionName 属性主要用来显示给用户。它是一个字符串，如 1.0.1。
3. string.xml 文件用来存储应用程序中的所有字符串常量。这使得您可以很容易地通过替换这些字符串并重新编译应用程序来本地化您的应用程序。

第 2 章答案

1. Android 操作系统将显示一个对话框，用户可以从中选择他们想使用的那一个活动。
2. 使用如下代码：

    ```
    Intent i = new
        Intent(android.content.Intent.ACTION_VIEW,
          Uri.parse("http://www.amazon.com"));
    startActivity(i);
    ```

3. 在意图筛选器中，可以指定以下内容：动作、数据、类型和类别。
4. Toast 类用来向用户显示警报，并在几秒钟后消失。NotificationManager 类用来在设备的状态栏上显示通知。由 NotificationManager 类显示的警报是持久性的，只能通过用户的选中操作来撤销。

5. 可以使用 XML 文件中的<fragment>元素或者使用 FragmentManager 类和 FragmentTransaction 类在活动中动态添加/删除碎片。
6. 活动的碎片的一个主要区别是，当活动进入后台时，会被放到 back stack 上。这样，当用户按 Back 按钮时，活动可以恢复。与之相反，碎片在进入后台时不会被自动放到 back stack 上。

第 3 章答案

1. dp 单位是与密度无关的，1dp 相当于一个 160dpi 的屏幕上的一个像素。px 单位对应于屏幕上的实际像素。一般应当使用 dp 单位，因为它可以使活动在不同屏幕大小的设备上都可以正确地缩放。
2. 随着不同屏幕大小的设备的出现，使用 AbsoluteLayout 使得您的应用程序在跨设备应用时很难保持一致的外观和体验。
3. 当一个活动被终止或转入后台时，将触发 onPause()事件。onSaveInstanceState()事件与 onPause()事件类似，除了它不总是被调用，例如当用户按下 Back 按钮来终止活动时就不会调用。
4. 3 个事件是 onPause()、onSaveInstanceState()和 onRetainNonConfigurationInstance()。一般使用 onPause()方法来保存活动的状态，因为在活动将要销毁时总是会调用这个方法。但是，对于屏幕方向的变化，使用 onSaveInstanceState()方法将活动的状态(例如用户输入的数据)保存到一个 Bundle 对象中更加方便。onRetainNonConfigurationInstance()方法对于临时保存数据(例如从 Web 服务下载的图像或文件)十分有用，这些数据可能太大，不适合放到一个 Bundle 对象中。
5. 在 Action Bar 中添加动作项类似于为选项菜单创建菜单项——只须处理 onCreateOptionsMenu()事件和 onOptionsItemSelected()事件。

第 4 章答案

1. 应该检验每一个 RadioButton 的 isChecked()方法来确定其是否被选中。
2. 可以使用 getResources()方法。
3. 以下代码片段用于获取当前日期：

    ```
    //--get the current date--
    Calendar today = Calendar.getInstance();
    yr = today.get(Calendar.YEAR);
    month = today.get(Calendar.MONTH);
    day = today.get(Calendar.DAY_OF_MONTH);
    ```

```
showDialog(DATE_DIALOG_ID);
```

4. 3个专用碎片是 ListFragment、DialogFragment 和 PreferenceFragment。ListFragment 对于显示一个项目列表很有帮助，例如 RSS 新闻列表。DialogFragment 允许以模态方式显示一个对话框窗口，这样用户必须做出回应，然后才能继续使用应用程序。PreferenceFragment 显示一个包含应用程序首选项的窗口，允许用户直接在应用程序中编辑首选项。

第 5 章答案

1. ImageSwitcher 可以使图像动画显示。您可以在图像显示时以及被另一幅图像替换时动画显示它。
2. 两个方法是 onCreateOptionsMenu()和 onOptionsItemSelected()。
3. 两个方法是 onCreateContextMenu()和 onContextItemSelected()。
4. 要防止启动设备的 Web 浏览器，需要实现 WebViewClient 类并重写 shouldOverrideUrlLoading()方法。

第 6 章答案

1. 可以使用 PreferenceActivity 类完成。
2. 方法名称是 getExternalStorageDirectory()。
3. 权限是 WRITE_EXTERNAL_STORAGE。

第 7 章答案

1. 代码如下所示：

```
Cursor c;
if (android.os.Build.VERSION.SDK_INT <11) {
    //---before Honeycomb---
    c = managedQuery(allContacts, projection,
            ContactsContract.Contacts.DISPLAY_NAME + " LIKE ?",
            new String[] {"%jack"},
            ContactsContract.Contacts.DISPLAY_NAME + " ASC");
} else {
    //---Honeycomb and later---
    CursorLoader cursorLoader = new CursorLoader(
            this,
            allContacts,
            projection,
            ContactsContract.Contacts.DISPLAY_NAME + " LIKE ?",
```

```
                new String[] {"%jack"},
                ContactsContract.Contacts.DISPLAY_NAME + " ASC");
            c = cursorLoader.loadInBackground();
        }
```

2. 方法是 getType()、onCreate()、query()、insert()、delete()和 update()。
3. 代码如下所示：
    ```
    <provider android:name="BooksProvider"
            android:authorities="net.learn2develop.provider.Books" />
    ```

第 8 章答案

1. 可以以编程方式从 Android 应用程序中发送 SMS 消息，也可以以应用程序的名义调用内置的 Messaging 应用程序来发送 SMS 信息。
2. 两个权限是 SEND_SMS 和 RECEIVE_SMS。
3. 广播接收者应该触发一个将由活动接收的新意图。活动应该实现另一个 BroadcastReceiver 来侦听这个新的意图。

第 9 章答案

1. 可能的原因如下所示：
    - 没有 Internet 连接
    - <uses-library>元素在 AndroidManifest.xml 文件中的位置错误
    - AndroidManifest.xml 文件中缺少 INTERNET 权限
2. 地理编码是将一个地址转换成其坐标(经度和纬度)。反向地理编码是将一对位置坐标转换成一个地址。
3. 两个位置服务提供商如下所示：
    - LocationManager.GPS_PROVIDER
    - LocationManager.NETWORK_PROVIDER
4. 方法是 addProximityAlert()。

第 10 章答案

1. 声明 INTERNET 许可。
2. 这些类为 JSONArray 和 JSONObject。
3. 该类为 AsyncTask。

第 11 章答案

1. 这是因为服务和主调活动运行在同一个进程上。如果服务是长时间运行的,那么需要在一个单独的线程上运行它,这样才不会阻塞活动。
2. IntentService 类与 Service 类相似,只不过前者在一个单独的线程上运行任务,并且当任务结束执行时可以自动停止服务。
3. 3 个方法是 doInBackground()、 onProgressUpdate()和 onPostExecute()。
4. 服务可以广播一个意图,而活动可以使用一个 IntentFilter 类来注册一个意图。
5. 推荐的方法是创建一个类来继承 AsyncTask 类,这可以确保 UI 以一种线程安全的方式更新。

第 12 章答案

1. 在 AndroidManifest.xml 文件中使用 minSdkVersion 属性来指定所需的 Android 最低版本。
2. 可以使用 Java SDK 的 keytool.exe 实用工具,也可以使用 Eclipse 的 Export 功能来生成证书。
3. 转到 Settings 应用程序并选择 Security 项。选中 Unknown sources 项。